Solid Mechanics and Its Applications

Volume 265

The fundamental questions arising in mechanics are: Why?, How?, and How much? The aim of this series is to provide lucid accounts written by authoritative researchers giving vision and insight in answering these questions on the subject of mechanics as it relates to solids. The scope of the series covers the entire spectrum of solid mechanics. Thus it includes the foundation of mechanics; variational formulations; computational mechanics; statics, kinematics and dynamics of rigid and elastic bodies; vibrations of solids and structures; dynamical systems and chaos; the theories of elasticity, plasticity and viscoelasticity; composite materials; rods, beams, shells and membranes; structural control and stability; soils, rocks and geomechanics; fracture; tribology; experimental mechanics; biomechanics and machine design. The median level of presentation is the first year graduate student. Some texts are monographs defining the current state of the field; others are accessible to final year undergraduates; but essentially the emphasis is on readability and clarity.

Springer and Professors Barber and Klarbring welcome book ideas from authors. Potential authors who wish to submit a book proposal should contact Dr. Mayra Castro, Senior Editor, Springer Heidelberg, Germany, email: mayra.castro@springer.com

Indexed by SCOPUS, Ei Compendex, EBSCO Discovery Service, OCLC, ProQuest Summon, Google Scholar and SpringerLink.

More information about this series at http://www.springer.com/series/6557

M. B. Rubin

Continuum Mechanics with Eulerian Formulations of Constitutive Equations

Springer

M. B. Rubin
Faculty of Mechanical Engineering
Technion – Israel Institute of Technology
Haifa, Israel

ISSN 0925-0042 ISSN 2214-7764 (electronic)
Solid Mechanics and Its Applications
ISBN 978-3-030-57775-9 ISBN 978-3-030-57776-6 (eBook)
https://doi.org/10.1007/978-3-030-57776-6

This Springer imprint is published by the registered company Springer Nature Switzerland AG
The registered company address is: Gewerbestrasse 11, 6330 Cham, Switzerland

This book is dedicated to my loving wife Laurel and my children and (grandsons): Adam & Dana (Leo & Tom); Daniel & Sefi.

Preface

My interest in mechanics was stimulated by my Scout Master Henry Layton when I was a Boy Scout. Henry was a mechanical engineer and a patent examiner who helped us build mini-bikes using bicycle parts and lawn mower engines. During my teenage years I bought a Craftsmen tool set at Sears and Roebuck, which I used to work on my cars and motorcycles. At the University of Colorado in Boulder, where I did my undergraduate degree in Mechanical Engineering, I learned that mathematics, vectors and tensors are the tools that I needed to truly understand the fundamentals of mechanics. Fortunately at Boulder, Prof. Frank Essenburg and Prof. William Wainwright helped me develop analytical skills and physical thinking needed to deepen my knowledge. They both encouraged me to continue my studies for a Ph.D. in applied mechanics after I graduated in December 1972.

I applied to the University of California at Berkeley and was accepted in the Department of Mechanical Engineering as a graduate student in applied mechanics. During my last semester at Boulder I took a course in continuum mechanics from a fluid mechanics professor who, unfortunately, really couldn't explain the deep physics of continuum mechanics. This caused me to change my major to bioengineering when I arrived at Berkeley for the fall quarter of 1973. However, I enrolled in a continuum mechanics course taught by Prof. Paul Naghdi who was clear, rigorous and explained the physical foundations. I thought then that if I studied bioengineering I would not know enough biology to formulate a problem and I would not know enough engineering to solve it. Consequently, I returned to applied mechanics and was truly fortunate to have Paul as a thesis advisor. Through my research, I have continued my interest in bioengineering. In my opinion, this interdisciplenary field requires experts from different fields to communicate and interact to make real progress.

Paul was a critical thinker who had the unique ability to read something that he had written as if he were an objective expert reading it for the first time. This talent helped him identify flaws in traditional approaches and create new ideas and formulations. My numerous discussions with Paul, both as a graduate student and as a colleague, challenged me and helped me develop as an independent researcher. I am immensely indebted to Paul for investing so much time to inspire and shape me as a researcher in continuum mechanics. Later I learned that both Frank Essenburg and William Wainwright were students of Paul so it is not surprising that I was attracted to Paul's lectures at Berkeley.

In August of 1979 I began work as a research engineer at SRI International. During my job interview I was told that as a theoretician I have to be willing to do experiments. At SRI, I was aided by a team of excellent technicians who taught me about many experimental problems as I acted as the supervisor of experiments. This exposure gave me a great appreciation for the difficulties of doing a good experiment, which has helped put a more physical perspective on my research over the years.

In October 1982 I moved to Israel with my wife Laurel to join the Faculty of Mechanical Engineering at Techion—Israel Institute of Technology, where I have spent my entire academic career, retiring as a Professor Emeritus in October 2019. I developed a friendship and working relationship with my senior colleague Prof. Sol Bodner, who was an experienced engineer with interests in both theory and experiments. My numerous discussions with Sol exposed me to the field of viscoplasticity and taught me how to think more physically about material response. I am also very much indebted to Sol for investing so much time in my development.

I have been teaching the course Introduction to Continuum Mechanics at Technion since the spring semester of 1983. The course and this book are based on the lecture notes of Paul Naghdi at Berkeley. Details of the presentation of this material have changed over the years as my understanding of continuum mechanics evolved due to my research and interactions with students, graduate students and colleagues, especially Prof. Eli Altus, with whom I had many discussions. During the first meeting of this course, I tell the students that continuum mechanics is a deep subject and that I am still learning after having been an active researcher in continuum mechanics for over 40 years. In my opinion, continuum mechanics is a theoretical umbrella for almost all of engineering because the thermomechanical theory applies to a broad range of solid materials (elastic, elastic–inelastic, elastic–viscoelastic) and fluid materials (gases, inviscid, viscous and viscoelastic liquids). Continuum mechanics provides a theoretical framework to ensure that we don't make fundamental blunders. However, the true beauty of the field is that we will always be challenged to use our theoretical expertise and physical intuition to synthesize experimental data to propose functional forms for constitutive equations that describe new important features of material response that needs to be modeled.

My experience has also been enriched by having been a regular Visiting Faculty at Lawrence Livermore National Laboratory (LLNL) since 1985. Dr. Lewis Glenn and Dr. Willy Moss were my first boss and colleague, respectively, at LLNL. Over the years I have had the opportunity to work with a number of very talented researchers at LLNL who have contributed to some of the constitutive equations presented in this book. At LLNL, I was exposed to the field of shock physics in geological materials which challenged me to develop specific functional forms for strongly coupled thermomechanical response that can be used to match experimental data. The exposure to real problems and the ability to work with excellent computational mechanics people at LLNL has enriched my ability to think physically. Often I would have a number of ideas why the simulations using the constitutive equations for a particular material do not match experimental data. In working with my colleagues at LLNL I realized that it is important to find the simplest way to "hack" the computer code to test an idea to see if it really makes a difference. Once the ideal that makes a difference has been identified, then it is necessary to develop the constitutive equations rigorously. It remains a challenge to ensure that the "hack" is removed and the rigorous equations have been programed.

In addition, at LLNL I learned the importance of numerical algorithms. This has particular relevance for the development of constitutive equations. Theoreticians can often propose different functional forms which model the same limiting cases. When working with computational mechanics it is important to choose those functional forms for modeling a specific material response which have the correct limits but also simplify the numerical algorithm.

I am also indebted to my colleague and friend Prof. Mahmood Jabareen in the Faculty of Civil and Environmental Engineering at Technion. His computational mechanics expertise was essential for the transition of the Cosserat Point Element (CPE) technology from a theoretical concept that I proposed in 1985 to algorithms that have been implemented in the commercial computer code LS-DYNA. We also collaborated on a number of papers which have shaped some of the ideas presented in this book, especially those on physical orthotropic invariants, the formulation of constitutive equations with a smooth elastic–inelastic transition and growth of biological materials. My graduate student and Post-Doctoral Fellow Dr. Mahmoud Safadi learned computational mechanics from Prof. Jabareen which was essential for his successful implementation of the constitutive equations for growth in the commercial computer code Abaqus. His expertise was used for simulations in our joint papers that highlighted the importance of the Eulerian formulation for growth. In addition, discussions with Dr. Gal Shmuel and Prof. Reuven Segev helped improve the presentation of the notions of a uniform material, a homogeneous material and a uniform material state. Prof. Roger Fosdick and Prof. Albrecht Bertram provided constructive criticism that improved the presentation of invariance under Superposed Rigid Body Motions, especially for constrained materials. Also, my wife Laurel proofread this book which helped identify and correct a number of typographical errors.

I am certain that engineers are essential to the future of Israel. Therefore, I am honored to be a Professor Emeritus from Technion, which is the best engineering university in Israel. Having taught at Technion makes me feel that I have contributed to the future of Israel through students who have been influenced by my teaching. In particular, I derive great satisfaction knowing that some of my graduate students have made significant contributions to the security and economic development of Israel. I am sure that I could not attain such personal satisfaction in my profession having been a professor anywhere else in the world.

I would also like to acknowledge the German Israel Foundation (GIF), the Israel Science Foundation (ISF) and my Gerard Swope Chair in Mechanics which provided financial support over the years.

My research on large deformation inelasticity and on growth of biological materials has caused me to develop an Eulerian formulation of constitutive equations for elastic and inelastic response. The important physical feature of the Eulerian formulation is that it removes arbitrariness of the choice of: a reference configuration, an intermediate zero-stress configuration, a total deformation measure and an inelastic deformation measure. The main new features of this book are the discussion of the importance of the Eulerian formulation and the demonstration of how it can be used to develop a broad range of specific constitutive equations in thermomechanics.

Haifa, Israel M. B. Rubin

Contents

Chapter 1
Introduction

Abstract The objective of this introductory chapter is to present an overview of the contents of this book and to discuss the importance of Eulerian formulations of constitutive equations. Specifically, simple one-dimensional examples are used to identify unphysical arbitrariness in the classical Lagrangian formulations of constitutive equations that can and should be removed.

1.1 Content of the Book

Continuum mechanics is concerned with the fundamental equations that describe the nonlinear thermomechanical response of all deformable media. Throughout this book, attention is limited to a simple material whose constitutive response does not depend on higher order gradients of deformation. Although the constitutive equations are phenomenological and are proposed to model the macroscopic response of materials, they are reasonably accurate for many studies of micro- and nano-mechanics where the typical length scales approach, but are still larger than, those of individual atoms. In this sense, the general thermomechanical theory provides a theoretical umbrella for most areas of study in mechanical engineering. In particular, continuum mechanics includes as special cases theories of: solids (elastic, inelastic, viscoelastic, etc.), fluids (compressible, incompressible, viscous) and the thermodynamics of heat conduction including dissipation due to inelastic effects.

A number of books have been written which discuss the fundamentals of continuum mechanics [5, 7, 9, 11], the theory of elasticity [1, 12], the theory of plasticity [2, 3] as well as the thermomechanical theory [18]. The new aspect of this book is its emphasis on an Eulerian formulation of constitutive equations for elastic materials, elastic–inelastic materials and growing biological tissues. The standard Lagrangian formulation of constitutive equations and the need for an Eulerian formulation will be discussed in detail from a physical point of view and specific constitutive equations will be described for different classes of materials.

Apart from this introduction, the material in this book on continuum mechanics is divided into five chapters. Chapter 2 develops a basic knowledge of tensor

M. B. Rubin, *Continuum Mechanics with Eulerian Formulations of Constitutive Equations*, Solid Mechanics and Its Applications 265,
https://doi.org/10.1007/978-3-030-57776-6_1

analysis using both indicial notation and direct notation. Although tensor operations in general curvilinear coordinates are needed to express spatial derivatives like those in the gradient and divergence operators, these special operations required to translate quantities in direct notation to component forms in special coordinate systems are merely mathematical in nature. Moreover, details of general curvilinear tensor analysis unnecessarily complicate the presentation of the fundamental physical issues in continuum mechanics. Consequently, here attention is mainly restricted to tensors expressed in terms of constant rectangular Cartesian base vectors to simplify the discussion of spatial derivatives and to concentrate on the main physical issues. However, an introduction to tensors with respect to curvilinear coordinates is presented in Appendix F.

Chapter 3 develops tools to analyze nonlinear deformation and motion of continua. Specifically, measures of deformation and their rates are introduced. Also, the group of Superposed Rigid Body Motions (SRBM) is introduced for later fundamental analysis of invariance of constitutive equations under SRBM.

Chapter 4 develops the balance laws that are applicable for simple continua, which are characterized by local measures of deformation. The notion of the stress tensor and its relationship to the traction vector is developed. Local forms of the equations of motion are derived from the global forms of the balance laws. Referential forms of the equations of motion are discussed and the relationships between different stress measures are developed for completeness, but they are not used in the Eulerian formulation of constitutive equations. Also, invariance under SRBM of the balance laws and the kinetic quantities are discussed.

Chapter 5 presents an introduction to constitutive theory. Although there is general consensus on the kinematics of continua, the notion of constitutive equations for special materials remains an active area of research in continuum mechanics. Specifically, in these sections the theoretical structure of constitutive equations for nonlinear anisotropic elastic solids, isotropic elastic solids, viscous and inviscid fluids, viscous dissipation, elastic–inelastic solids and viscoelastic solids are discussed.

Chapter 6 describes thermomechanical processes and the fundamental balance laws and restrictions of second laws of thermomechanics that control these processes. In addition, specific constitutive equations for: thermoelastic materials, thermoelastic–inelastic materials, orthotropic thermoelastic–inelastic materials, shock waves, porous materials and growing biological tissues are discussed. Also, jump conditions for the thermomechanical balance laws are developed.

1.2 Comparison of the Lagrangian and Eulerian Formulations

Unphysical arbitrariness of the choices of: the reference configuration; a zero-stress intermediate configuration; a total deformation measure and a plastic deformation measure has been discussed in a series of papers [15–17]. To simplify the discussion

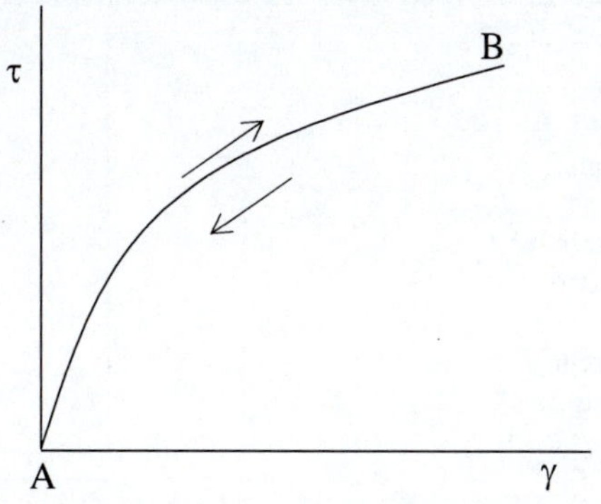

Fig. 1.1 Response of a homogeneous nonlinear elastic material to homogeneous proportional loading in shear from a uniform zero-stress material state *A* to a uniform loaded material state *B* with unloading along the same path back to the same uniform zero-stress material state *A*

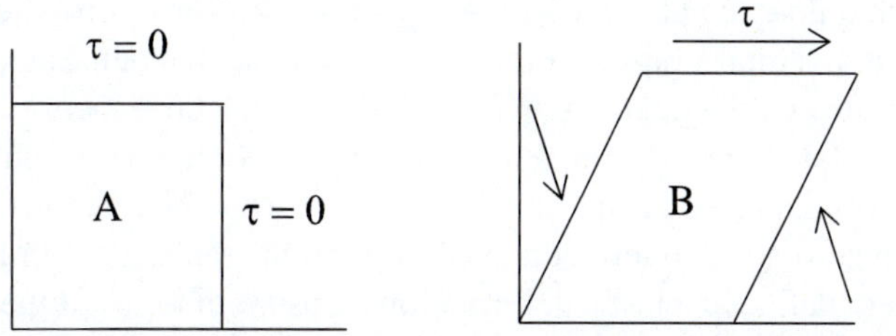

Fig. 1.2 Sketch of the deformation of a homogeneous nonlinear elastic material subjected to homogenous proportional loading in shear from a uniform zero-stress material state *A* to a uniform loaded material state *B* with unloading along the same path back to the same uniform zero-stress material state *A*

of these issues, here attention is limited to the purely mechanical theory at constant zero-stress reference temperature.

Figure 1.1 shows the shear stress τ versus the total shear strain γ for a homogeneous nonlinear elastic material subjected to homogeneous proportional loading from a uniform zero-stress material state A to a uniform loaded material state B with unloading along the same path to the same uniform zero-stress material state A. Figure 1.2 shows a sketch of the associated deformations.

These figures exhibit the property that a homogeneous nonlinear elastic material in a uniform zero-stress material state, which is loaded to a deformed state, will return to its zero-stress shape and volume when unloaded. In this sense the nonlinear elastic material remembers its zero-stress shape and density. This also suggests that the response of a homogeneous nonlinear elastic material can be characterized by a Lagrangian formulation of the constitutive equation in terms of a Lagrangian strain that measures deformations from a reference configuration with a uniform stress-free material state and vanishes in this reference configuration.

Figure 1.3 shows the shear stress τ versus the total shear strain γ for a homogeneous nonlinear elastic–plastic material subjected to homogeneous proportional

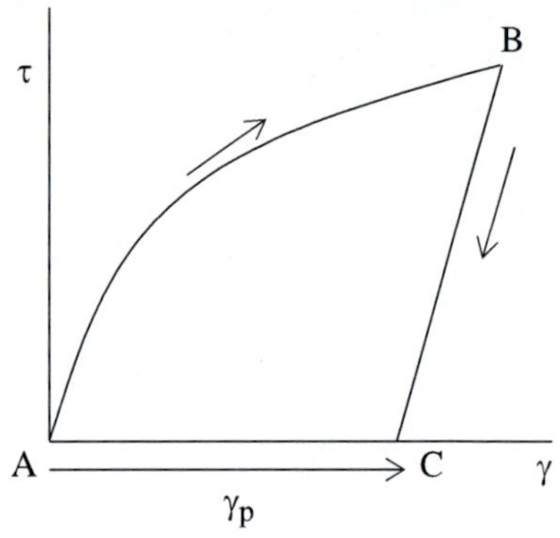

Fig. 1.3 Response of a homogeneous nonlinear elastic–plastic material to homogeneous proportional loading in shear from a uniform zero-stress material state *A* to a uniform loaded material state *B* with unloading along a different path to a uniform zero-stress material state *C* with a residual total strain γ_p

loading from a uniform zero-stress material state *A* to a uniform loaded material state *B* with unloading along a different path to a uniform zero-stress material state *C* with a residual total strain γ_p. Figure 1.4 shows a sketch of the associated deformations. Motivated by the Lagrangian formulation of elastic response, in addition to the Lagrangian total strain from the reference configuration, it is common to introduce a plastic deformation (see γ_p in Fig. 1.3) measured from the reference configuration to the uniform zero-stress intermediate configuration (see state *C* in Fig. 1.3). Also, it is common to define an elastic deformation measure in terms of the total and plastic deformation measures. In this sense, the plastic deformation measure is a history-dependent variable that is determined by integrating an evolution equation for its time rate of change.

Onat [13] discussed physical restrictions on internal state variables. This discussion proposed that internal state variables, which are determined by integrating evolution equations over time, are specified to measure properties of the material response that define the current state of the material. Moreover, since these evolution equations need initial conditions it is necessary that the values of the internal state variables be, in principle, measurable directly or indirectly by experiments on multiple identical samples of the material in its current material state. In this sense, the material state must be characterized by internal state variables whose values are measurable in the current state.

From this perspective, it is necessary to ask if the deformation measures that are used to characterize material response are acceptable internal state variables. For a homogeneous elastic material, it is common to define the deformation gradient tensor **F** from a uniform stress-free reference configuration to the current deformed configuration to characterize the constitutive equation of the elastic material. For this elastic material, it follows that since the volume and shape of the material are unique in any zero-stress material state, **F** is only known to within an arbitrary proper orthogonal rotation tensor **R** in any zero-stress material state. This means that the zero-stress value of **F** has arbitrariness due to three orientation angles associated with **R** which cannot be determined by experiments on identical samples of the same material in the current configuration. Consequently, **F** is not an acceptable internal state variable in the sense discussed by Onat [13]. For this reason, **F** should not

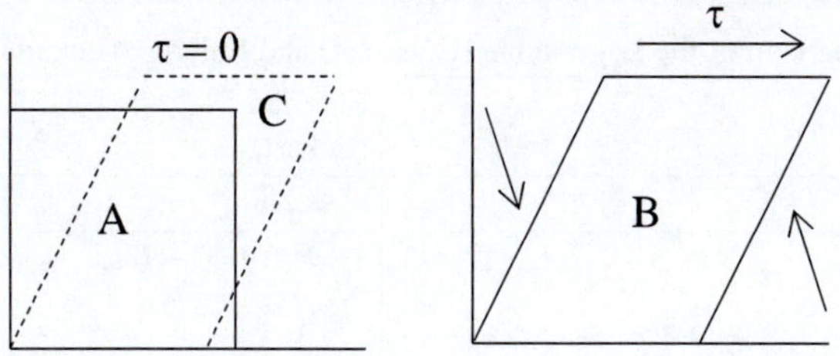

Fig. 1.4 Sketch of the deformation of a homogeneous nonlinear elastic–plastic material subjected to homogenous proportional loading in shear from a uniform zero-stress material state *A* to a uniform loaded material state *B* with unloading along a different path to a uniform zero-stress material state *C* (dashed lines) with a different shape from that in the uniform zero-stress material state *A*

appear in any constitutive equation for material response, even for nonlinear elastic materials. However, for the solution of a specific problem it is often convenient to parameterize the solution using the total deformation gradient **F** from a known specified reference configuration. In this sense, it is important to distinguish between a tensorial measure of elastic deformation from a zero-stress material state and the total deformation gradient from a specified reference configuration.

The use of **F** in constitutive equations for elastic–plastic materials is even more problematic physically. Even if plastic deformations are isochoric, a homogeneous elastic–plastic material that is loaded from a uniform zero-stress material state has no unique shape in another uniform zero-stress material state (see the initial state *A* and the intermediate state *C* in Fig. 1.4). This means that only the volumetric part of **F** can be determined in a uniform zero-stress material state so there are eight degrees of arbitrariness in **F**, three associated with orientation changes and five associated with distortional deformations. The following statement by Gilman in the discussion section in [8] refers to this physical arbitrariness.

> It seems very unfortunate to me that the theory of plasticity was ever cast into the mold of stress–strain relations because 'strain' in the plastic case has no physical meaning that is related to the material of the body in question. It is rather like trying to deduce some properties of a liquid from the shape of the container that holds it. The plastic behavior of a body depends on its structure (crystalline and defect) and on the system of stresses that is applied to it. The structure will vary with plastic strain, but not in a unique fashion. The variation will also depend on the initial structure, the values of whatever stresses are applied, and on time (some recovery occurs in almost any material at any temperature).

The Eulerian formulation of constitutive equations discussed in this book and in [15–17] is motivated by the work of Eckart [4] for elastic–inelastic solids, by Leonov [10] for polymeric liquids and is based on the work in [14]. This Eulerian formulation uses evolution equations for the material time derivative of internal state variables. More specifically, the formulation is considered to be Eulerian because the evolution equations depend only on quantities that can, in principle, be measured in the current state of the material. It will be shown that this Eulerian formulation removes arbitrariness of the choice of: the reference configuration; an intermediate configuration; a total deformation measure and an inelastic deformation measure.

Table 1.1 Comparison of the Lagrangian (Classical) and Eulerian (Eckart) formulations

Lagrangian (Classical)	Eulerian (Eckart)
$\sigma = E\varepsilon_e$	$\sigma = E\varepsilon_e$
$\dot{\varepsilon} = \partial v/\partial x$ $\dot{\varepsilon}_p = \frac{\Gamma}{E}\sigma$ $\varepsilon_e = \varepsilon - \varepsilon_p$	$\dot{\varepsilon}_e = \partial v/\partial x - \Gamma\,\varepsilon_e$
$\varepsilon(0) = ?$ $\varepsilon_p(0) = ?$	$\varepsilon_e(0) = \frac{\sigma(0)}{E}$

Table 1.1 records the basic equations needed to compare the differences between the Lagrangian (Classical) formulation and the Eulerian (Eckart) formulation for inelasticity using a simple one-dimensional model. In this model the strains are small so the notion of Lagrangian is used for quantities that are referred to a reference configuration. Specifically, the axial stress σ is determined by the axial elastic strain ε_e using Young's modulus of elasticity E in both formulations. However, in the Lagrangian formulation, it is necessary to define the total axial strain ε, the plastic or inelastic axial strain ε_p, as well as the axial elastic strain ε_e. Specifically, the total strain ε is determined by integrating an evolution equation in terms of the velocity gradient $\partial v/\partial x$. The inelastic strain ε_p is determined by integrating an evolution equation in terms of the stress σ and a non-negative function Γ that controls inelastic deformation rate, and the elastic strain ε_e is defined by the difference between the total strain and the inelastic strain. In contrast, in the Eulerian formulation the elastic strain ε_e is determined directly by integrating an evolution equation in terms of the velocity gradient $\partial v/\partial x$, the elastic strain ε_e and the function Γ.

The Eulerian evolution equation for elastic strain ε_e is consistent with the equation in the Lagrangian formulation and can be obtained by taking the time derivative $\dot{(\,)}$ of the algebraic expression for ε_e and replacing $\dot{\varepsilon}$ and $\dot{\varepsilon}_p$ with their evolution equations. However, the physics of these two formulations are different. In the Lagrangian formulation it is necessary to specify the initial values $\varepsilon(0)$ and $\varepsilon_p(0)$. But these quantities are both referred to an arbitrary choice of the reference configuration. This can be made explicit by noting that the same initial value $\varepsilon_e(0)$ of elastic strain can be obtained by changing the reference configuration using the arbitrary value A, such that

$$\varepsilon_e(0) = \varepsilon(0) - \varepsilon_p(0) = [\varepsilon(0) - A] - [\varepsilon_p(0) - A]\,, \tag{1.2.1}$$

where the scalar A in this one-dimensional model characterizes the influence of the arbitrariness of the reference configuration in a similar manner to the tensor $\mathbf{A}$ in the nonlinear three-dimensional theory discussed in (5.11.24). This arbitrariness means that the individual initial values $\varepsilon(0)$ and $\varepsilon_p(0)$ needed to integrate the evolution equations for ε and ε_p cannot be measured independently. Consequently, ε and ε_p are not internal state variables in the sense of Onat [13].

In contrast to the Lagrangian formulation, in the Eulerian formulation the elastic strain ε_e is introduced directly through an evolution equation for its rate and the initial value $\varepsilon_e(0)$ needed to integrate this equation can be determined by the measuring the initial value $\sigma(0)$ of stress. Consequently, the elastic strain ε_e is an internal state variable in the sense of Onat [13] since it is measurable. Moreover, the arbitrariness associated with the orientation of the body in a zero-stress intermediate configuration in the three-dimensional theory discussed in (5.11.25) cannot be analyzed using the simple one-dimensional model.

Following the work of Eckart [4] and Leonov [10], an Eulerian formulation for elastically isotropic inelastic materials introduces a symmetric positive-definite elastic deformation tensor $\mathbf{B}_e$ through an evolution equation for its material time derivative. Moreover, using the work of Flory [6], $\mathbf{B}_e$ is expressed in terms of the elastic dilatation J_e and the symmetric positive-definite unimodular elastic distortional deformation tensor $\mathbf{B}'_e$ defined by

$$J_e = \sqrt{\det \mathbf{B}_e}\,, \quad \mathbf{B}'_e = J_e^{-2/3}\mathbf{B}_e\,, \quad \det \mathbf{B}'_e = 1\,. \tag{1.2.2}$$

Then, for elastically isotropic thermoelastic–inelastic materials evolution equations are proposed directly for J_e and $\mathbf{B}'_e$, and the Helmholtz free energy ψ per unit mass and the Cauchy stress $\mathbf{T}$ are specified by constitutive equations which depend on $J_e, \mathbf{B}'_e$ and the absolute temperature θ in the forms

$$\psi = \psi(J_e, \mathbf{B}'_e, \theta)\,, \quad \mathbf{T} = \mathbf{T}(J_e, \mathbf{B}'_e, \theta)\,. \tag{1.2.3}$$

This constitutive equation for stress is restricted to be invertible with J_e and $\mathbf{B}'_e$ admitting the representations

$$J_e = J_e(\mathbf{T}, \theta)\,, \quad \mathbf{B}'_e = \mathbf{B}'_e(\mathbf{T}, \theta)\,. \tag{1.2.4}$$

The constitutive equation for stress is further restricted so that a zero-stress material state at zero-stress reference temperature θ_z requires

$$J_e(0, \theta_z) = 1\,, \quad \mathbf{B}'_e(0, \theta_z) = \mathbf{I}\,, \tag{1.2.5}$$

where $\mathbf{I}$ is the second-order identity tensor. These restrictions ensure that J_e and $\mathbf{B}'_e$ are internal state variables in the sense of Onat [13] since their initial values required to integrate their evolution equations can be determined by the measured values of $\mathbf{T}$ and θ in the initial state of the material.

Another specific example where it is clear that it is not sufficient to formulate constitutive equations in terms of a Lagrangian deformation measure is an anisotropic elastic material with a quadratic strain energy function. Specifically, let $\mathbf{E}$ be a Lagrangian total strain measure from a reference configuration with a uniform zero-stress material state and consider the quadratic strain energy function Σ per unit mass specified by

$$\rho_z \Sigma = \frac{1}{2} \mathbf{K} \cdot \mathbf{E} \otimes \mathbf{E}\,, \tag{1.2.6}$$

where ρ_z is a constant zero-stress mass density, $\mathbf{K}$ is a constant fourth-order stiffness tensor, $\otimes$ is the tensor product operator and $(\,\cdot\,)$ is the inner product between two tensors of any order. Referring these tensors to an arbitrary rectangular Cartesian orthonormal triad of vectors $\mathbf{e}_i$ in the reference configuration yields the expression

$$\rho_z \Sigma = \frac{1}{2} K_{ijkl} E_{ij} E_{kl}\,. \tag{1.2.7}$$

For a general anisotropic elastic material K_{ijkl} has the symmetries

$$K_{jikl} = K_{ijlk} = K_{klij} = K_{ijkl}\,, \tag{1.2.8}$$

so it is characterized by 21 independent material constants. Although this quadratic strain energy function can model general anisotropic elastic response, the representation is incomplete since it is necessary to connect the components K_{ijkl} of the stiffnesses tensor with identifiable material directions.

Following the work in [14], the elastic deformations and material orientations in the Eulerian formulation for elastically anisotropic materials discussed in this book are characterized by a right-handed triad of linearly independent microstructural vectors $\mathbf{m}_i$ with the elastic dilatation J_e defined by

$$J_e = \mathbf{m}_1 \times \mathbf{m}_2 \cdot \mathbf{m}_3 > 0\,. \tag{1.2.9}$$

Also, the elastic metric m_{ij} is defined by

$$m_{ij} = \mathbf{m}_i \cdot \mathbf{m}_j\,. \tag{1.2.10}$$

Then, for elastically anisotropic thermoelastic–inelastic materials evolution equations are proposed directly for $\mathbf{m}_i$, and the Helmholtz free energy ψ per unit mass and the Cauchy stress $\mathbf{T}$ are specified by constitutive equations which depend on $\mathbf{m}_i$ and the absolute temperature θ in the forms

$$\psi = \psi(m_{ij}, \theta)\,, \quad \mathbf{T} = \mathbf{T}(\mathbf{m}_i, \theta)\,. \tag{1.2.11}$$

This constitutive equation for stress is restricted to be invertible with $\mathbf{m}_i$ admitting the representations

$$\mathbf{m}_i = \mathbf{m}_i(\mathbf{T}, \theta)\,. \tag{1.2.12}$$

The constitutive equation for stress is further restricted so that a zero-stress material state at zero-stress reference temperature θ_z requires

$$m_{ij}(0, \theta_z) = \mathbf{m}_i(0, \theta_z) \cdot \mathbf{m}_j(0, \theta_z) = \delta_{ij}\,, \quad J_e(0, \theta_z) = 1\,, \tag{1.2.13}$$

where δ_{ij} is the Kronecker delta and $\mathbf{m}_i$ have been defined to be orthonormal in this zero-stress material state at zero-stress reference temperature. These restrictions ensure that $\mathbf{m}_i$ are internal state variables in the sense of Onat [13] since their initial values required to integrate their evolution equations can be determined by the measured values of $\mathbf{T}$ and θ in the initial state of the material. Depending on the material being modeled it may be necessary to consider the response of identical samples of the material in its current state to different loading paths to determine the values of $\mathbf{m}_i$ in the current state. Further in this regard, it is noted that symmetries of the material response characterized by the Helmholtz free energy ψ make the response of the material insensitive to any indeterminacy in the inversion (1.2.12) for $\mathbf{m}_i$.

This representation has the advantage that the indices $i = 1, 2, 3$ of these vectors characterize specific material directions. It will be shown that these microstructural vectors can be used to model elastic deformations for anisotropic elastic materials and for the rate-independent and rate-dependent response of anisotropic elastic–inelastic materials.

Details of fundamental aspects of the Eulerian formulation of constitutive equations can be found in Sects. 3.11, 3.14, 5.2, 5.3, 5.4, 5.11, 5.12 and in Chap. 6 for thermomechanical response.

References

1. Atkin RJ, Fox N (1980) An introduction to the theory of elasticity. Longman Group, Harlow
2. Bertram A (2012) Elasticity and plasticity of large deformations. Springer, Berlin
3. Besseling JF, Van Der Giessen E (1994) Mathematical modeling of inelastic deformation. CRC Press, Boca Raton
4. Eckart C (1948) The thermodynamics of irreversible processes. IV. The theory of elasticity and anelasticity. Phys Rev 73:373–382
5. Eringen AC (1967) Mechanics of continua. Wiley, Hoboken
6. Flory PJ (1961) Thermodynamic relations for high elastic materials. Trans Faraday Soc 57:829–838
7. Fung YC (1967) Continuum mechanics. Prentice-Hall, Upper Saddle River
8. Gilman JJ (1960) Physical nature of plastic flow and fracture. In: Plasticity, proceedings of 2nd symposium on naval structural mechanics, pp 43–99
9. Holzapfel GS (2000) Nonlinear solid mechanics: a continuum approach for engineering science. Wiley, Hoboken
10. Leonov AI (1976) Nonequilibrium thermodynamics and rheology of viscoelastic polymer media. Rheol Acta 15:85–98
11. Malvern LE (1969) Introduction to the mechanics of a continuous medium
12. Ogden RW (1997) Non-linear elastic deformations
13. Onat ET (1968) The notion of state and its implications in thermodynamics of inelastic solids. Irreversible aspects of continuum mechanics and transfer of physical characteristics in moving fluids, pp 292–314
14. Rubin MB (1994) Plasticity theory formulated in terms of physically based microstructural variables - Part I. Theory. Int J Solids Struct 31:2615–2634

15. Rubin MB (1996) On the treatment of elastic deformation in finite elastic-viscoplastic theory. Int J Plast 12:951–965
16. Rubin MB (2001) Physical reasons for abandoning plastic deformation measures in plasticity and viscoplasticity theory. Arch Mech 53:519–539
17. Rubin MB (2012) Removal of unphysical arbitrariness in constitutive equations for elastically anisotropic nonlinear elastic-viscoplastic solids. Int J Eng Sci 53:38–45
18. Tadmor EB, Ronald EM, Ryan SE (2012) Continuum mechanics and thermodynamics–from fundamental concepts to governing equations. Cambridge University Press

Chapter 2
Basic Tensor Analysis

Abstract Tensors are mathematical objects which ensure that mathematical equations characterizing physics are insensitive to arbitrary choices of a coordinate system. The objective of this chapter is to present a review of tensor analysis using both index and direct notations. To simplify the presentation of tensor calculus, attention is limited to tensors expressed relative to fixed rectangular Cartesian base vectors. (Some of the content in this chapter has been adapted from Rubin (Cosserat theories: shells, rods and points. Springer Science & Business Media, Berlin, 2000) with permission.)

2.1 Vector Algebra

Tensors, tensor algebra and tensor calculus are needed to formulate physical equations in continuum mechanics which are insensitive to arbitrary choices of coordinates. To understand the mathematics of tensors it is desirable to start with the use of a language called indicial notation which develops simple rules governing these tensor manipulations. For the purposes of describing this language it is convenient to introduce a fixed right-handed triad of orthonormal rectangular Cartesian base vectors denoted by $(\mathbf{e}_1, \mathbf{e}_2, \mathbf{e}_3)$. From the study of linear vector spaces, it is recalled that vectors satisfy certain laws of addition and multiplication by a scalar. Specifically, if $\mathbf{a}$ and $\mathbf{b}$ are vectors then the quantity

$$\mathbf{c} = \mathbf{a} + \mathbf{b} \tag{2.1.1}$$

is a vector defined by the parallelogram law of addition. Furthermore, recall that the operations

M. B. Rubin, *Continuum Mechanics with Eulerian Formulations of Constitutive Equations*, Solid Mechanics and Its Applications 265,
https://doi.org/10.1007/978-3-030-57776-6_2

$$
\begin{aligned}
\mathbf{a}+\mathbf{b} &= \mathbf{b}+\mathbf{a} && \text{(commutative law)}\,,\\
(\mathbf{a}+\mathbf{b})+\mathbf{c} &= \mathbf{a}+(\mathbf{b}+\mathbf{c}) && \text{(associative law)}\,,\\
\alpha\mathbf{a} &= \mathbf{a}\alpha && \text{(multiplicative law)}\,,\\
\mathbf{a}\cdot\mathbf{b} &= \mathbf{b}\cdot\mathbf{a} && \text{(commutative law)}\,,\\
\mathbf{a}\cdot(\mathbf{b}+\mathbf{c}) &= \mathbf{a}\cdot\mathbf{b}+\mathbf{a}\cdot\mathbf{c} && \text{(distributive law)}\,,\\
\alpha(\mathbf{a}\cdot\mathbf{b}) &= (\alpha\mathbf{a})\cdot\mathbf{b} && \text{(distributive law)}\,,\\
\mathbf{a}\times\mathbf{b} &= -\mathbf{b}\times\mathbf{a} && \text{(lack of commutativity)}\,,\\
\mathbf{a}\times(\mathbf{b}+\mathbf{c}) &= \mathbf{a}\times\mathbf{b}+\mathbf{a}\times\mathbf{c} && \text{(distributive law)}\,,\\
\alpha(\mathbf{a}\times\mathbf{b}) &= (\alpha\mathbf{a})\times\mathbf{b} && \text{(associative law)}
\end{aligned}
\tag{2.1.2}
$$

are satisfied for all vectors **a**, **b** and **c** and all real numbers α, where $\mathbf{a}\cdot\mathbf{b}$ denotes the scalar product (or dot product) and $\mathbf{a}\times\mathbf{b}$ denotes the vector product (or cross product) between the vectors **a** and **b**.

The Scalar Triple Product

The scalar triple product of the vectors **a**, **b** and **c** has the property that the dot and cross products can be interchanged

$$\mathbf{a}\times\mathbf{b}\cdot\mathbf{c}=\mathbf{a}\cdot\mathbf{b}\times\mathbf{c}\,. \tag{2.1.3}$$

Moreover, using the results

$$\mathbf{a}\times\mathbf{b}\cdot\mathbf{c}=-\mathbf{b}\times\mathbf{a}\cdot\mathbf{c}=-\mathbf{b}\cdot\mathbf{a}\times\mathbf{c}=\mathbf{b}\cdot\mathbf{c}\times\mathbf{a}=\mathbf{c}\times\mathbf{a}\cdot\mathbf{b}\,, \tag{2.1.4}$$

it follows that the order of the vectors in the scalar triple product can be permuted

$$\mathbf{a}\times\mathbf{b}\cdot\mathbf{c}=\mathbf{c}\times\mathbf{a}\cdot\mathbf{b}=\mathbf{b}\times\mathbf{c}\cdot\mathbf{a}\,. \tag{2.1.5}$$

The Vector Triple Product

The vector triple product of the vectors **a**, **b** and **c** can be expanded to obtain

$$\mathbf{a}\times(\mathbf{b}\times\mathbf{c})=(\mathbf{a}\cdot\mathbf{c})\mathbf{b}-(\mathbf{a}\cdot\mathbf{b})\mathbf{c}\,. \tag{2.1.6}$$

To prove this result it is noted that this vector must be perpendicular to both **a** and $\mathbf{b}\times\mathbf{c}$. But $\mathbf{b}\times\mathbf{c}$ is perpendicular to the plane containing **b** and **c** so the vector triple product must be a vector in the plane of **b** and **c**. Moreover, the vector triple product is linear in the vectors **a**, **b** and **c**. The expression (2.1.6) can be checked by considering the special case of $\mathbf{a}=\mathbf{e}_1$, $\mathbf{b}=\mathbf{e}_3$ and $\mathbf{c}=\mathbf{e}_1$.

2.2 Indicial Notation

Quantities written in indicial notation have a finite number of indices attached to them. Since the number of indices can be zero, a quantity with no index can also be considered to be written in indicial notation. The language of indicial notation is quite simple because only two types of indices can appear in any term. Either the index is a free index or it is a repeated index. Also, a simple summation convention is defined which applies only to repeated indices. These two types of indices and the summation convention are defined as follows.

Free Indices:
Indices that appear only once in a given term are known as free indices. In this regard, a term in an equation is a quantity that is separated by a plus, minus or equal sign. Here, each of these free indices will take the values $(1, 2, 3)$. For example, i is a free index in each of the following expressions

$$(x_1, x_2, x_3) = x_i \quad (i = 1, 2, 3)\,, \tag{2.2.1a}$$

$$(\mathbf{e}_1, \mathbf{e}_2, \mathbf{e}_3) = \mathbf{e}_i \quad (i = 1, 2, 3)\,. \tag{2.2.1b}$$

Repeated Index:
Indices that appear twice in a given term are known as a repeated index. For example, i and j are free indices and m and n are repeated indices in the following expressions

$$a_i b_j c_m T_{mn} d_n\,, \quad A_{immjnn}\,, \quad A_{imn} B_{jmn}\,. \tag{2.2.2}$$

It is important to emphasize that in the language of indicial notation an index *can never appear more than twice* in any term.

Einstein Summation Convention:
When an index appears as a repeated index in a term that index is understood to take on the values $(1, 2, 3)$ and the resulting terms are summed. Thus, for example,

$$x_i \mathbf{e}_i = x_1 \mathbf{e}_1 + x_2 \mathbf{e}_2 + x_3 \mathbf{e}_3\,. \tag{2.2.3}$$

Because of this summation convention, a repeated index is also known as a dummy index since its replacement by any other letter not appearing as a free index and also not appearing as another repeated index does not change the meaning of the term in which it occurs. For examples,

$$x_i \mathbf{e}_i = x_j \mathbf{e}_j\,, \quad a_i b_m c_m = a_i b_j c_j\,. \tag{2.2.4}$$

It is important to emphasize that the same free indices must appear in each term in an equation so that for example the free index i in $(2.2.4)_2$ must appear on each side of the equality.

Kronecker Delta:
The Kronecker delta δ_{ij} is defined by

$$\delta_{ij} = \mathbf{e}_i \cdot \mathbf{e}_j = \begin{bmatrix} 1 \text{ for } i = j \\ 0 \text{ for } i \neq j \end{bmatrix} . \tag{2.2.5}$$

Since the Kronecker delta δ_{ij} vanishes unless $i = j$ it exhibits the following exchange property

$$\delta_{ij} x_j = (\delta_{i1} x_1, \; \delta_{i2} x_2, \; \delta_{i3} x_3) = (x_1, x_2, x_3) = x_i . \tag{2.2.6}$$

Notice that the Kronecker delta can be removed by replacing the repeated index j in (2.2.6) by the free index i.

Recalling that an arbitrary vector $\mathbf{a}$ in Euclidean 3-Space can be expressed as a linear combination of the base vectors $\mathbf{e}_i$ it can be expressed in the form

$$\mathbf{a} = a_i \mathbf{e}_i . \tag{2.2.7}$$

Consequently, it follows that the components a_i of $\mathbf{a}$ can be calculated using the Kronecker delta

$$a_i = \mathbf{e}_i \cdot \mathbf{a} = \mathbf{e}_i \cdot (a_m \mathbf{e}_m) = (\mathbf{e}_i \cdot \mathbf{e}_m) a_m = \delta_{im} a_m = a_i . \tag{2.2.8}$$

Notice that when the expression (2.2.7) for $\mathbf{a}$ is substituted into (2.2.8) it is necessary to change the repeated index i in (2.2.7) to another letter m because the letter i already appears in (2.2.8) as a free index. It also follows that the Kronecker delta can be used to calculate the dot product between two vectors $\mathbf{a}$ and $\mathbf{b}$ with components a_i and b_i, respectively, by

$$\mathbf{a} \cdot \mathbf{b} = (a_i \mathbf{e}_i) \cdot (b_j \mathbf{e}_j) = a_i (\mathbf{e}_i \cdot \mathbf{e}_j) b_j = a_i \delta_{ij} b_j = a_i b_i . \tag{2.2.9}$$

Permutation Symbol:
The permutation symbol ε_{ijk} is defined by

$$\varepsilon_{ijk} = \mathbf{e}_i \times \mathbf{e}_j \cdot \mathbf{e}_k = \begin{bmatrix} 1 & \text{if } (i, j, k) \text{ are an even permutation of } (1, 2, 3) \\ -1 & \text{if } (i, j, k) \text{ are an odd permutation of } (1,2,3) \\ 0 & \text{if at least two of } (i, j, k) \text{ have the same value} \end{bmatrix} . \tag{2.2.10}$$

This definition suggests that the permutation symbol can be used to calculate the vector product between two vectors. To this end, it will be shown that

$$\mathbf{e}_i \times \mathbf{e}_j = \varepsilon_{ijk} \mathbf{e}_k . \tag{2.2.11}$$

Proof Since $\mathbf{e}_i \times \mathbf{e}_j$ is a vector in Euclidean 3-Space for each choice of the indices i and j, it follows that it can be represented as a linear combination of the base vectors $\mathbf{e}_k$ such that

$$\mathbf{e}_i \times \mathbf{e}_j = A_{ijk}\mathbf{e}_k \,, \tag{2.2.12}$$

where the components A_{ijk} need to be determined. In particular, taking the dot product of (2.2.12) with $\mathbf{e}_k$ and using the definition (2.2.10) yields

$$\varepsilon_{ijk} = \mathbf{e}_i \times \mathbf{e}_j \cdot \mathbf{e}_k = A_{ijm}\mathbf{e}_m \cdot \mathbf{e}_k = A_{ijm}\delta_{mk} = A_{ijk} \,, \tag{2.2.13}$$

which proves the result (2.2.11). Now using (2.2.11), it follows that the vector product between the vectors $\mathbf{a}$ and $\mathbf{b}$ can be represented in the form

$$\mathbf{a} \times \mathbf{b} = (a_i\mathbf{e}_i) \times (b_j\mathbf{e}_j) = (\mathbf{e}_i \times \mathbf{e}_j)a_i b_j = \varepsilon_{ijk} a_i b_j \mathbf{e}_k \,. \tag{2.2.14}$$

Additional Properties of the Permutation Symbol:
Using (2.1.3) and (2.1.6) it can be shown that

$$\begin{aligned} &\varepsilon_{ijk}\varepsilon_{rsk} = (\mathbf{e}_i \times \mathbf{e}_j) \cdot (\mathbf{e}_r \times \mathbf{e}_s) = \mathbf{e}_i \cdot [\mathbf{e}_j \times (\mathbf{e}_r \times \mathbf{e}_s)] = \delta_{ir}\delta_{js} - \delta_{is}\delta_{jr} \,, \\ &\varepsilon_{ijk}\varepsilon_{rjk} = 2\delta_{ir} \,, \quad \varepsilon_{ijk}\varepsilon_{ijk} = 6 \,. \end{aligned} \tag{2.2.15}$$

Also, recall that the determinant of a matrix M_{ij} can be expressed in the forms

$$\begin{aligned} \det(M_{mn}) &= \varepsilon_{ijk} M_{i1} M_{j2} M_{k3} \,, \\ \varepsilon_{rst} \det(M_{mn}) &= \varepsilon_{ijk} M_{ir} M_{js} M_{kt} \,, \\ \det(M_{mn}) &= \frac{1}{6}\varepsilon_{ijk}\varepsilon_{rst} M_{ir} M_{js} M_{kt} \,. \end{aligned} \tag{2.2.16}$$

Contraction:
Contraction is the process of replacing two free indices in a given expression with the same index together with the implied summation convention. For example, contraction on the free indices i, j in δ_{ij} yields

$$\delta_{ii} = \delta_{11} + \delta_{22} + \delta_{33} = 3 \,. \tag{2.2.17}$$

Note that contraction on the set of $9 = 3^2$ quantities T_{ij} can be performed by multiplying T_{ij} by δ_{ij} to obtain

$$T_{ij}\delta_{ij} = T_{ii} \,. \tag{2.2.18}$$

2.3 Direct Notation (Special Case)

A scalar is sometimes referred to as a zero-order tensor and a vector is sometimes referred to as a first-order tensor. Higher order tensors are defined inductively starting with the notion of a first-order tensor or vector. Specifically, since a second-order tensor is a linear operator whose domain is the space of all vectors and whose range is the space of all vectors it is possible to define the second-order tensor inductively using vector spaces.

Tensor of Order M:
The quantity $\mathbf{T}$ is called a tensor of order M ($M \geq 2$) if it is a linear operator whose domain is the space of all vectors $\mathbf{v}$ and whose range $\mathbf{Tv}$ or $\mathbf{vT}$ is a tensor of order $M-1$. Since $\mathbf{T}$ is a linear operator it satisfies the following rules

$$\mathbf{T}(\mathbf{v}+\mathbf{w}) = \mathbf{Tv} + \mathbf{Tw}\,, \tag{2.3.1a}$$

$$\alpha(\mathbf{Tv}) = (\alpha\mathbf{T})\mathbf{v} = \mathbf{T}(\alpha\mathbf{v})\,, \tag{2.3.1b}$$

$$(\mathbf{v}+\mathbf{w})\mathbf{T} = \mathbf{vT} + \mathbf{wT}\,, \tag{2.3.1c}$$

$$\alpha(\mathbf{vT}) = (\alpha\mathbf{v})\mathbf{T} = (\mathbf{vT})\alpha\,, \tag{2.3.1d}$$

where $\mathbf{v}$ and $\mathbf{w}$ are arbitrary vectors and α is an arbitrary real number. Notice that the tensor $\mathbf{T}$ can operate on its right [e.g., (2.3.1a), (2.3.1b)] or on its left [e.g., (2.3.1c), (2.3.1d)] and that, in general, operation on the right and the left is not commutative

$$\mathbf{Tv} \neq \mathbf{vT} \quad \text{Lack of commutativity}\,. \tag{2.3.2}$$

Zero Tensor of Order M:
The zero tensor of order M is denoted by $\mathbf{0}(M)$ and is a linear operator whose domain is the space of all vectors $\mathbf{v}$ and whose range $\mathbf{0}(M-1)$ is the zero tensor of order $M-1$

$$\mathbf{0}(M)\mathbf{v} = \mathbf{v}\,\mathbf{0}(M) = \mathbf{0}(M-1)\,. \tag{2.3.3}$$

Notice that these tensors are defined inductively starting with the known properties of the real number 0 which is the zero tensor $\mathbf{0}(0)$ of order 0.

Addition and Subtraction:
The usual rules of addition and subtraction of two tensors $\mathbf{A}$ and $\mathbf{B}$ apply when the two tensors have the same order. It is emphasized that tensors of different orders cannot be added or subtracted.

To define the operations of tensor product, dot product and juxtaposition for general tensors it is convenient to first consider the definitions of these properties for the special case of the tensor product of a string of M ($M \geq 2$) vectors ($\mathbf{a}_1, \mathbf{a}_2, \ldots, \mathbf{a}_M$). Also, it is necessary to define the left transpose and right transpose of the tensor product of a string of vectors.

Tensor Product (Special Case):
The tensor product operation is denoted by the symbol $\otimes$ and it is defined so that the tensor product of a string of M ($M \geq 1$) vectors ($\mathbf{a}_1, \mathbf{a}_2, \ldots, \mathbf{a}_M$) is a tensor of order M. Before presenting the general expression, consider the following simple cases

$$(\mathbf{a}_1 \otimes \mathbf{a}_2)\mathbf{v} = (\mathbf{a}_2 \cdot \mathbf{v})\mathbf{a}_1 \,, \quad \mathbf{v}(\mathbf{a}_1 \otimes \mathbf{a}_2) = (\mathbf{a}_1 \cdot \mathbf{v})\mathbf{a}_2 \,, \tag{2.3.4}$$

where $\mathbf{v}$ is an arbitrary vector and the symbol $(\cdot)$ is the usual dot product between two vectors. The more general case satisfies the properties

$$(\mathbf{a}_1 \otimes \mathbf{a}_2 \otimes \cdots \otimes \mathbf{a}_{M-1} \otimes \mathbf{a}_M)\mathbf{v} = (\mathbf{a}_M \cdot \mathbf{v})(\mathbf{a}_1 \otimes \mathbf{a}_2 \otimes \cdots \otimes \mathbf{a}_{M-1}) \,, \tag{2.3.5a}$$

$$\mathbf{v}(\mathbf{a}_1 \otimes \mathbf{a}_2 \otimes \cdots \otimes \mathbf{a}_{M-1} \otimes \mathbf{a}_M) = (\mathbf{a}_1 \cdot \mathbf{v})(\mathbf{a}_2 \otimes \cdots \otimes \mathbf{a}_M) \,, \tag{2.3.5b}$$

$$\begin{aligned} &\alpha(\mathbf{a}_1 \otimes \mathbf{a}_2 \otimes \cdots \otimes \mathbf{a}_M) = (\alpha\mathbf{a}_1 \otimes \mathbf{a}_2 \otimes \cdots \otimes \mathbf{a}_M) \\ &= \cdots = (\mathbf{a}_1 \otimes \mathbf{a}_2 \otimes \cdots \otimes \alpha\mathbf{a}_M) = (\mathbf{a}_1 \otimes \mathbf{a}_2 \otimes \cdots \otimes \mathbf{a}_M)\alpha \,, \end{aligned} \tag{2.3.5c}$$

$$\begin{aligned} &(\mathbf{a}_1 \otimes \mathbf{a}_2 \otimes \cdots \otimes \mathbf{a}_{K-1} \otimes \{\mathbf{a}_K + \mathbf{w}\} \cdots \otimes \mathbf{a}_{M-1} \otimes \mathbf{a}_M) \\ &= (\mathbf{a}_1 \otimes \mathbf{a}_2 \otimes \cdots \otimes \mathbf{a}_{K-1} \otimes \mathbf{a}_K \cdots \otimes \mathbf{a}_{M-1} \otimes \mathbf{a}_M) \\ &+ (\mathbf{a}_1 \otimes \mathbf{a}_2 \otimes \cdots \otimes \mathbf{a}_{K-1} \otimes \mathbf{w} \cdots \otimes \mathbf{a}_{M-1} \otimes \mathbf{a}_M) \\ &\qquad \text{for } 1 \leq K \leq M \,, \end{aligned} \tag{2.3.5d}$$

where $\mathbf{w}$ is another arbitrary vector and α is an arbitrary real number. It is important to note from (2.3.5a) and (2.3.5b) that, in general, the order of the operation is not commutative.

Dot Product (Special Case):
The dot product operation between two vectors can be generalized to an operation between any two tensors (including higher order tensors). Specifically, the dot product of the tensor product of a string of M vectors ($\mathbf{a}_1, \mathbf{a}_2, \ldots, \mathbf{a}_M$) with the tensor product of another string of N vectors ($\mathbf{b}_1, \mathbf{b}_2, \ldots, \mathbf{b}_N$) is a tensor of order $|M - N|$. Before presenting the general expression, consider the following simple cases

$$\begin{aligned} &(\mathbf{a}_1 \otimes \mathbf{a}_2) \cdot (\mathbf{b}_1 \otimes \mathbf{b}_2) = (\mathbf{a}_1 \cdot \mathbf{b}_1)(\mathbf{a}_2 \cdot \mathbf{b}_2) \,, \\ &(\mathbf{a}_1 \otimes \mathbf{a}_2 \otimes \mathbf{a}_3) \cdot (\mathbf{b}_1 \otimes \mathbf{b}_2) = \mathbf{a}_1(\mathbf{a}_2 \cdot \mathbf{b}_1)(\mathbf{a}_3 \cdot \mathbf{b}_2) \,, \\ &(\mathbf{a}_1 \otimes \mathbf{a}_2) \cdot (\mathbf{b}_1 \otimes \mathbf{b}_2 \otimes \mathbf{b}_3) = (\mathbf{a}_1 \cdot \mathbf{b}_1)(\mathbf{a}_2 \cdot \mathbf{b}_2)\mathbf{b}_3 \,. \end{aligned} \tag{2.3.6}$$

The more general case satisfies the properties

$$\begin{aligned} &(\mathbf{a}_1 \otimes \mathbf{a}_2 \otimes \cdots \otimes \mathbf{a}_M) \cdot (\mathbf{b}_1 \otimes \mathbf{b}_2 \otimes \cdots \otimes \mathbf{b}_N) \\ &= (\mathbf{a}_1 \otimes \mathbf{a}_2 \otimes \cdots \otimes \mathbf{a}_{M-N}) \left[\prod_{K=1}^{N} (\mathbf{a}_{M-N+K} \cdot \mathbf{b}_K) \right] \quad \text{for } M > N \,, \end{aligned} \tag{2.3.7a}$$

$$(\mathbf{a}_1 \otimes \mathbf{a}_2 \otimes \cdots \otimes \mathbf{a}_M) \cdot (\mathbf{b}_1 \otimes \mathbf{b}_2 \otimes \cdots \otimes \mathbf{b}_M) = \prod_{K=1}^{M} (\mathbf{a}_K \cdot \mathbf{b}_K) \quad \text{for } M = N\,, \tag{2.3.7b}$$

$$(\mathbf{a}_1 \otimes \mathbf{a}_2 \otimes \cdots \otimes \mathbf{a}_M) \cdot (\mathbf{b}_1 \otimes \mathbf{b}_2 \otimes \cdots \otimes \mathbf{b}_N)$$
$$= \Big[\prod_{K=1}^{M} (\mathbf{a}_K \cdot \mathbf{b}_K)\Big](\mathbf{b}_{M+1} \otimes \mathbf{b}_{M+2} \otimes \cdots \otimes \mathbf{b}_N) \quad \text{for } M < N\,, \tag{2.3.7c}$$

where $\prod$ is the usual product operator indicating the product of the series of quantities defined by the values of K

$$\prod_{K=1}^{M} (\mathbf{a}_K \cdot \mathbf{b}_K) = (\mathbf{a}_1 \cdot \mathbf{b}_1)(\mathbf{a}_2 \cdot \mathbf{b}_2) \ldots (\mathbf{a}_N \cdot \mathbf{b}_M)\,. \tag{2.3.8}$$

Note from (2.3.7a) and (2.3.7c) that if the orders of the tensors are not equal ($M \neq N$) then the order of the dot product operator is important. However, when the orders of the tensors are equal ($M = N$) then the dot product operation yields a scalar (2.3.7b) and the order of the operation is unimportant (i.e., the operation is commutative).

Cross Product (Special Case):
The cross product of the tensor product of a string of M vectors $(\mathbf{a}_1, \mathbf{a}_2, \ldots, \mathbf{a}_M)$ with the tensor product of another string of N vectors $(\mathbf{b}_1, \mathbf{b}_2, \ldots, \mathbf{b}_N)$ is a tensor of order M if $M \geq N$ and is of order N if $N > M$. Before presenting the general expression, consider the following simple cases

$$\begin{aligned}
(\mathbf{a}_1 \otimes \mathbf{a}_2) \times (\mathbf{b}_1 \otimes \mathbf{b}_2) &= (\mathbf{a}_1 \times \mathbf{b}_1) \otimes (\mathbf{a}_2 \times \mathbf{b}_2)\,,\\
(\mathbf{a}_1 \otimes \mathbf{a}_2 \otimes \mathbf{a}_3) \times (\mathbf{b}_1 \otimes \mathbf{b}_2) &= \mathbf{a}_1 \otimes (\mathbf{a}_2 \times \mathbf{b}_1) \otimes (\mathbf{a}_3 \times \mathbf{b}_2)\,,\\
(\mathbf{a}_1 \otimes \mathbf{a}_2) \times (\mathbf{b}_1 \otimes \mathbf{b}_2 \otimes \mathbf{b}_3) &= (\mathbf{a}_1 \times \mathbf{b}_1) \otimes (\mathbf{a}_2 \times \mathbf{b}_2) \otimes \mathbf{b}_3\,.
\end{aligned} \tag{2.3.9}$$

The more general case satisfies the properties

$$\begin{aligned}
&(\mathbf{a}_1 \otimes \mathbf{a}_2 \otimes \cdots \otimes \mathbf{a}_M) \times (\mathbf{b}_1 \otimes \mathbf{b}_2 \otimes \cdots \otimes \mathbf{b}_N)\\
&\qquad = (\mathbf{a}_1 \otimes \mathbf{a}_2 \otimes \cdots \otimes \mathbf{a}_{M-N}) \left[\prod_{K=1}^{N} \otimes (\mathbf{a}_{M-N+K} \times \mathbf{b}_K)\right] \quad \text{for } M > N\,,\\
&(\mathbf{a}_1 \otimes \mathbf{a}_2 \otimes \cdots \otimes \mathbf{a}_M) \times (\mathbf{b}_1 \otimes \mathbf{b}_2 \otimes \cdots \otimes \mathbf{b}_M)\\
&\qquad = (\mathbf{a}_1 \times \mathbf{b}_1) \prod_{K=2}^{M} \otimes (\mathbf{a}_K \times \mathbf{b}_K) \quad \text{for } M = N\,,\\
&(\mathbf{a}_1 \otimes \mathbf{a}_2 \otimes \cdots \otimes \mathbf{a}_M) \times (\mathbf{b}_1 \otimes \mathbf{b}_2 \otimes \cdots \otimes \mathbf{b}_N)\\
&\qquad = \left[\prod_{K=1}^{M} (\mathbf{a}_K \times \mathbf{b}_K) \otimes\right] (\mathbf{b}_{M+1} \otimes \mathbf{b}_{M+2} \otimes \cdots \otimes \mathbf{b}_N) \quad \text{for } M < N\,.
\end{aligned} \tag{2.3.10}$$

Note that the order of the cross product operation is important.

Juxtaposition (Special Case):
The operation of juxtaposition of the tensor product of a string of M ($M \geq 1$) vectors ($\mathbf{a}_1, \mathbf{a}_2, \ldots, \mathbf{a}_M$) with the tensor product of another string of N ($N \geq 1$) vectors ($\mathbf{b}_1, \mathbf{b}_2, \ldots, \mathbf{b}_N$) is a tensor of order $M + N - 2$. Before presenting the general expression, consider the following simple cases

$$\mathbf{a}_1\,\mathbf{b}_1 = \mathbf{a}_1 \cdot \mathbf{b}_1\,, \tag{2.3.11a}$$

$$(\mathbf{a}_1 \otimes \mathbf{a}_2)(\mathbf{b}_1 \otimes \mathbf{b}_2) = (\mathbf{a}_2 \cdot \mathbf{b}_1)(\mathbf{a}_1 \otimes \mathbf{b}_2)\,. \tag{2.3.11b}$$

Note from (2.3.11a) that the juxtaposition of a vector with another vector is the same as the dot product of the two vectors. In spite of this fact, the dot product between two vectors is usually expressed explicitly. The more general case satisfies the properties

$$\begin{aligned}&(\mathbf{a}_1 \otimes \mathbf{a}_2 \otimes \cdots \otimes \mathbf{a}_M)(\mathbf{b}_1 \otimes \mathbf{b}_2 \otimes \cdots \otimes \mathbf{b}_N)\\ &\quad = (\mathbf{a}_M \cdot \mathbf{b}_1)(\mathbf{a}_1 \otimes \mathbf{a}_2 \otimes \cdots \otimes \mathbf{a}_{M-1} \otimes \mathbf{b}_2 \otimes \cdots \otimes \mathbf{b}_N)\,.\end{aligned} \tag{2.3.12}$$

It is obvious that the order of the operation juxtaposition is important.

Transpose (Special Case):
The left transpose of order N of the tensor product of a string of M ($M \geq 2N$) vectors is denoted by a superscript $(\,)^{LT(N)}$ on the left-hand side of the string of vectors. Similarly, the right transpose of order N of the tensor product of a string of M ($M \geq 2N$) vectors is denoted by a superscript $(\,)^{T(N)}$ on the right-hand side of the string of vectors. When, $N = 2$ this notation is simplified by denoting $LT(2) = LT$ and $T(2) = T$. Before presenting the general expression, consider the following simples cases

$${}^{LT}(\mathbf{a}_1 \otimes \mathbf{a}_2 \otimes \mathbf{a}_3) = (\mathbf{a}_2 \otimes \mathbf{a}_1) \otimes \mathbf{a}_3\,, \tag{2.3.13a}$$

$$(\mathbf{a}_1 \otimes \mathbf{a}_2 \otimes \mathbf{a}_3)^T = \mathbf{a}_1 \otimes (\mathbf{a}_3 \otimes \mathbf{a}_2)\,, \tag{2.3.13b}$$

$${}^{LT(2)}(\mathbf{a}_1 \otimes \mathbf{a}_2 \otimes \mathbf{a}_3 \otimes \mathbf{a}_4) = (\mathbf{a}_3 \otimes \mathbf{a}_4) \otimes (\mathbf{a}_1 \otimes \mathbf{a}_2)\,, \tag{2.3.13c}$$

$$(\mathbf{a}_1 \otimes \mathbf{a}_2 \otimes \mathbf{a}_3 \otimes \mathbf{a}_4)^{T(2)} = (\mathbf{a}_3 \otimes \mathbf{a}_4) \otimes (\mathbf{a}_1 \otimes \mathbf{a}_2)\,. \tag{2.3.13d}$$

From (2.3.13c) and (2.3.13d) it can be seen that the right and left transposes of order 2 of the tensor product of a string of vectors of order $4 = 2 \times 2$ are equal. The more general case satisfies the properties

$$\begin{aligned}&{}^{LT}(\mathbf{a}_1 \otimes \mathbf{a}_2 \otimes \cdots \otimes \mathbf{a}_M) = (\mathbf{a}_2 \otimes \mathbf{a}_1) \otimes (\mathbf{a}_3 \otimes \cdots \otimes \mathbf{a}_M)\,,\\ &(\mathbf{a}_1 \otimes \mathbf{a}_2 \otimes \cdots \mathbf{a}_{M-1} \otimes \mathbf{a}_M)^T = (\mathbf{a}_1 \otimes \mathbf{a}_2 \otimes \cdots \otimes \mathbf{a}_{M-2}) \otimes (\mathbf{a}_M \otimes \mathbf{a}_{M-1})\,.\end{aligned} \tag{2.3.14}$$

Thus, in general, the right and left transposes of order N of the tensor product of a string of vectors of order $2N$ are equal.

2.4 Direct Notation (General Case)

Using the above definitions, it is possible to define the base tensors and components of tensors of any order on a Euclidean 3-space. To this end, recall that $\mathbf{e}_i$ are the orthonormal base vectors of a right-handed rectangular Cartesian coordinate system. Thus, $\mathbf{e}_i$ span the space of vectors.

Base Tensors:
It also follows inductively that the tensor product of the string of M vectors

$$(\mathbf{e}_i \otimes \mathbf{e}_j \otimes \mathbf{e}_k \otimes \cdots \otimes \mathbf{e}_r \otimes \mathbf{e}_s \otimes \mathbf{e}_t)\,, \tag{2.4.1}$$

with M free indices $(i, j, k, \ldots, r, s, t)$ are base tensors for all tensors of order M. This is because when (2.4.1) is in juxtaposition with an arbitrary vector $\mathbf{v}$ it yields a scalar times the base tensors of all tensors of order $M - 1$, such that

$$\begin{aligned}(\mathbf{e}_i \otimes \mathbf{e}_j \otimes \mathbf{e}_k \otimes \cdots \otimes \mathbf{e}_r \otimes \mathbf{e}_s \otimes \mathbf{e}_t)\mathbf{v} &= (\mathbf{e}_t \cdot \mathbf{v})(\mathbf{e}_i \otimes \mathbf{e}_j \otimes \mathbf{e}_k \otimes \cdots \otimes \mathbf{e}_r \otimes \mathbf{e}_s)\,,\\ \mathbf{v}(\mathbf{e}_i \otimes \mathbf{e}_j \otimes \mathbf{e}_k \otimes \cdots \otimes \mathbf{e}_r \otimes \mathbf{e}_s \otimes \mathbf{e}_t) &= (\mathbf{e}_i \cdot \mathbf{v})(\mathbf{e}_j \otimes \mathbf{e}_k \otimes \cdots \otimes \mathbf{e}_r \otimes \mathbf{e}_s \otimes \mathbf{e}_t)\,.\end{aligned} \tag{2.4.2}$$

Components of an Arbitrary Tensor:
By definition the base tensors (2.4.1) span the space of tensors of order M so an arbitrary tensor $\mathbf{T}$ of order M can be expressed as a linear combination of these base tensors such that

$$\mathbf{T} = T_{ijk\ldots rst}(\mathbf{e}_i \otimes \mathbf{e}_j \otimes \mathbf{e}_k \otimes \cdots \otimes \mathbf{e}_r \otimes \mathbf{e}_s \otimes \mathbf{e}_t)\,, \tag{2.4.3}$$

where the coefficients $T_{ijk\ldots rst}$ in (2.4.3) are the components of $\mathbf{T}$ relative to the coordinate system defined by the base vectors $\mathbf{e}_i$ and the summation convention is used over repeated indices in (2.4.3). The components of $\mathbf{T}$ can be calculated by

$$T_{ijk\ldots rst} = \mathbf{T} \cdot (\mathbf{e}_i \otimes \mathbf{e}_j \otimes \mathbf{e}_k \otimes \cdots \otimes \mathbf{e}_r \otimes \mathbf{e}_s \otimes \mathbf{e}_t)\,. \tag{2.4.4}$$

Notice that the components of the tensor $\mathbf{T}$ are obtained by taking the dot product of the tensor with the base tensors of the space defining the order of the tensor, just as for the case of vectors (tensors of order one). It is important to emphasize that the matrix T_{ij} of nine quantities is not a tensor. For the matrix T_{ij} to be connected to a second-order tensor $\mathbf{T}$, it is necessary to know that T_{ij} are components of $\mathbf{T}$ relative to a known set of base vectors $\mathbf{e}_i$. Only then is it possible to determine the tensor $\mathbf{T}$

$$\mathbf{T} = T_{ij}\mathbf{e}_i \otimes \mathbf{e}_j\,, \tag{2.4.5}$$

with the direct form $\mathbf{T}$ of the tensor being independent of the choice of the base vectors.

Tensor Product (General Case):
Let **A** be a tensor of order M with components $A_{ij...mn}$ and let **B** be a tensor of order N with components $B_{rs...vw}$ then the tensor product of **A** and **B**

$$\mathbf{A} \otimes \mathbf{B} = A_{ij...mn} B_{rs...vw} (\mathbf{e}_i \otimes \mathbf{e}_j \otimes \cdots \otimes \mathbf{e}_m \otimes \mathbf{e}_n) \otimes (\mathbf{e}_r \otimes \mathbf{e}_s \otimes \cdots \otimes \mathbf{e}_v \otimes \mathbf{e}_w) \tag{2.4.6}$$

is a tensor of order $(M + N)$.

Dot Product (General Case):
The dot product $\mathbf{A} \cdot \mathbf{B}$ of a tensor **A** of order M with a tensor **B** of order N is a tensor of order $|M - N|$. As examples let **A** and **B** be second-order tensors with components A_{ij} and B_{ij} and let **C** be a fourth-order tensor with components C_{ijkl}. Then

$$\begin{aligned} &\mathbf{A} \cdot \mathbf{B} = \mathbf{B} \cdot \mathbf{A} = A_{ij} B_{ij} \,, \qquad &&\mathbf{A} \cdot \mathbf{C} = A_{ij} C_{ijkl} \mathbf{e}_k \otimes \mathbf{e}_l \,, \\ &\mathbf{C} \cdot \mathbf{A} = C_{ijkl} A_{kl} \mathbf{e}_i \otimes \mathbf{e}_j \,, \qquad &&\mathbf{A} \cdot \mathbf{C} \neq \mathbf{C} \cdot \mathbf{A} \,. \end{aligned} \tag{2.4.7}$$

Cross Product (General Case):
The cross product $\mathbf{A} \times \mathbf{B}$ of a tensor **A** of order M with a tensor **B** of order N is a tensor of order M if $M \geq N$ and a tensor of order N if $N > M$. As examples, let **v** be a vector with components v_i and **A** and **B** be second-order tensors with components A_{ir} and B_{js}. Then

$$\mathbf{A} \times \mathbf{v} = A_{ir} v_s \mathbf{e}_i \otimes (\mathbf{e}_r \times \mathbf{e}_s) = \varepsilon_{rst} A_{ir} v_s (\mathbf{e}_i \otimes \mathbf{e}_t) \,, \tag{2.4.8a}$$

$$\mathbf{v} \times \mathbf{A} = v_s A_{ir} (\mathbf{e}_s \times \mathbf{e}_i) \otimes \mathbf{e}_r = \varepsilon_{sit} v_s A_{ir} (\mathbf{e}_t \otimes \mathbf{e}_r) \,, \tag{2.4.8b}$$

$$\mathbf{A} \times \mathbf{B} = A_{ir} B_{js} (\mathbf{e}_i \times \mathbf{e}_j) \otimes (\mathbf{e}_r \times \mathbf{e}_s) = \varepsilon_{ijk} \varepsilon_{rst} A_{ir} B_{js} (\mathbf{e}_k \otimes \mathbf{e}_t) \,, \tag{2.4.8c}$$

$$\mathbf{B} \times \mathbf{A} = B_{js} A_{ir} (\mathbf{e}_j \times \mathbf{e}_i) \otimes (\mathbf{e}_s \times \mathbf{e}_r) = \varepsilon_{ijk} \varepsilon_{rst} B_{js} A_{ir} (\mathbf{e}_k \otimes \mathbf{e}_t) \,, \tag{2.4.8d}$$

$$\mathbf{A} \times \mathbf{v} \neq \mathbf{v} \times \mathbf{A} \,, \tag{2.4.8e}$$

$$\mathbf{A} \times \mathbf{B} \neq \mathbf{B} \times \mathbf{A} \,. \tag{2.4.8f}$$

Note that, in general, the cross product operation is not commutative. However, from (2.4.8c) and (2.4.8d) it is observed that the cross product of two second-order tensors is commutative. Also, unlike with vectors, the cross product of two second-order tensors is not necessarily zero.

Juxtaposition (General Case):
Let **A** be a tensor of order M with components $A_{ij...mn}$ and **B** be a tensor of order N with components $B_{rs...vw}$. Then, juxtaposition of **A** with **B** is a tensor of order $(M + N - 2)$ expressed by

$$\begin{aligned}
\mathbf{A}\,\mathbf{B} &= A_{ij\ldots mn}B_{rs\ldots vw}(\mathbf{e}_i\otimes\mathbf{e}_j\otimes\cdots\otimes\mathbf{e}_m\otimes\mathbf{e}_n)\,(\mathbf{e}_r\otimes\mathbf{e}_s\otimes\cdots\otimes\mathbf{e}_v\otimes\mathbf{e}_w)\,,\\
&= A_{ij\ldots mn}B_{rs\ldots vw}(\mathbf{e}_n\cdot\mathbf{e}_r)(\mathbf{e}_i\otimes\mathbf{e}_j\otimes\cdots\otimes\mathbf{e}_m\otimes\mathbf{e}_s\otimes\cdots\otimes\mathbf{e}_v\otimes\mathbf{e}_w)\,,\\
&= A_{ij\ldots mn}B_{rs\ldots vw}\delta_{nr}(\mathbf{e}_i\otimes\mathbf{e}_j\otimes\cdots\otimes\mathbf{e}_m\otimes\mathbf{e}_s\otimes\cdots\otimes\mathbf{e}_v\otimes\mathbf{e}_w)\,,\\
&= A_{ij\ldots mn}B_{ns\ldots vw}(\mathbf{e}_i\otimes\mathbf{e}_j\otimes\cdots\otimes\mathbf{e}_m\otimes\mathbf{e}_s\otimes\cdots\otimes\mathbf{e}_v\otimes\mathbf{e}_w)\,.
\end{aligned}\tag{2.4.9}$$

Note that the juxtaposition of a tensor with a vector is the same as the dot product of the tensor with the vector.

Transpose of a Tensor:
Let $\mathbf{T}$ be a tensor of order M with components $T_{ijkl\ldots rstu}$ relative to the base vectors $\mathbf{e}_i$. Then, using the definitions of the transpose of a string of vectors, the Nth order $(2N \le M)$ left transpose tensor ${}^{LT(N)}\mathbf{T}$ and right transpose tensor $\mathbf{T}^{T(N)}$ of $\mathbf{T}$ are defined, such that

$$\begin{aligned}
\mathbf{T} &= T_{ijkl\ldots rstu}(\mathbf{e}_i\otimes\mathbf{e}_j\otimes\mathbf{e}_k\otimes\mathbf{e}_l\otimes\cdots\otimes\mathbf{e}_r\otimes\mathbf{e}_s\otimes\mathbf{e}_t\otimes\mathbf{e}_u)\,,\\
{}^{LT}\mathbf{T} &= T_{ijkl\ldots rstu}(\mathbf{e}_j\otimes\mathbf{e}_i)\otimes(\mathbf{e}_k\otimes\mathbf{e}_l\otimes\cdots\otimes\mathbf{e}_r\otimes\mathbf{e}_s\otimes\mathbf{e}_t\otimes\mathbf{e}_u)\,,\\
\mathbf{T}^T &= T_{ijkl\ldots rstu}(\mathbf{e}_i\otimes\mathbf{e}_j\otimes\mathbf{e}_k\otimes\mathbf{e}_l\otimes\cdots\otimes\mathbf{e}_r\otimes\mathbf{e}_s)\otimes(\mathbf{e}_u\otimes\mathbf{e}_t)\,,\\
{}^{LT(2)}\mathbf{T} &= T_{ijkl\ldots rstu}(\mathbf{e}_k\otimes\mathbf{e}_l)\otimes(\mathbf{e}_i\otimes\mathbf{e}_j)\otimes\cdots\otimes\mathbf{e}_r\otimes\mathbf{e}_s\otimes\mathbf{e}_t\otimes\mathbf{e}_u)\,,\\
\mathbf{T}^{T(2)} &= T_{ijkl\ldots rstu}(\mathbf{e}_i\otimes\mathbf{e}_j\otimes\mathbf{e}_k\otimes\mathbf{e}_l\otimes\cdots\otimes(\mathbf{e}_t\otimes\mathbf{e}_u)\otimes(\mathbf{e}_r\otimes\mathbf{e}_s)\,.
\end{aligned}\tag{2.4.10}$$

In particular, note that the transpose operation does not change the order of the indices of the components of the tensor but merely changes the order of the base vectors. To see this more clearly, let $\mathbf{T}$ be a second-order tensor with components T_{ij} so that

$$\mathbf{T} = T_{ij}(\mathbf{e}_i\otimes\mathbf{e}_j)\,,\quad \mathbf{T}^T = T_{ij}\mathbf{e}_j\otimes\mathbf{e}_i = {}^{LT}\mathbf{T}\,.\tag{2.4.11}$$

It follows that for a second-order tensor $\mathbf{T}$ and an arbitrary vector $\mathbf{v}$

$$\mathbf{T}\mathbf{v} = \mathbf{v}\mathbf{T}^T\,,\quad \mathbf{T}^T\mathbf{v} = \mathbf{v}\mathbf{T}\,.\tag{2.4.12}$$

Also, it is noted that the separate notation for the left transpose has been introduced to avoid confusion in interpreting an expression of the type $\mathbf{A}^T\mathbf{B}$ which is not, in general, equal to $\mathbf{A}^{LT}\mathbf{B}$.

Identity Tensor of Order 2M:
The identity tensor of order $2M$ $(M \ge 1)$ is denoted by $\mathbf{I}(2M)$ and is a tensor that has the property that the dot product of $\mathbf{I}(2M)$ with an arbitrary tensor $\mathbf{A}$ of order M yields the result $\mathbf{A}$, such that

$$\mathbf{I}(2M)\cdot\mathbf{A} = \mathbf{A}\cdot\mathbf{I}(2M) = \mathbf{A}\,.\tag{2.4.13}$$

Letting $\mathbf{e}_i \otimes \mathbf{e}_j \otimes \cdots \otimes \mathbf{e}_s \otimes \mathbf{e}_t$ be the base tensors of order M, the $2M^{th}$ order unit tensor $\mathbf{I}$ can be represented in the form

$$\mathbf{I}(2M) = (\mathbf{e}_i \otimes \mathbf{e}_j \otimes \cdots \otimes \mathbf{e}_s \otimes \mathbf{e}_t) \otimes (\mathbf{e}_i \otimes \mathbf{e}_j \otimes \cdots \otimes \mathbf{e}_s \otimes \mathbf{e}_t) \,, \tag{2.4.14}$$

where the summation convention over repeated indices is implied. Since the second-order identity tensor appears often in continuum mechanics it is convenient to denote it by $\mathbf{I}$, which can be expressed in the form

$$\mathbf{I} = \mathbf{e}_i \otimes \mathbf{e}_i \,. \tag{2.4.15}$$

It then follows that the components of the second-order identity tensor are represented by the Kronecker delta

$$\mathbf{I} \cdot (\mathbf{e}_i \otimes \mathbf{e}_j) = \delta_{ij} \,. \tag{2.4.16}$$

Zero Tensor of Order M:
Since all components of the zero tensor of order M are 0 and since the order of the tensors in a given equation will usually be obvious from the context, the symbol 0 is used to denote the zero tensor of any order.

Lack of Commutativity:
Note that, in general, the operations of tensor product, dot product, cross product and juxtaposition are not commutative so the order of these operations must be preserved. Specifically, it follows that

$$\begin{aligned} &\mathbf{A} \otimes \mathbf{B} \neq \mathbf{B} \otimes \mathbf{A} \,, \quad \mathbf{A} \cdot \mathbf{B} \neq \mathbf{B} \cdot \mathbf{A} \,, \\ &\mathbf{A} \times \mathbf{B} \neq \mathbf{B} \times \mathbf{A} \,, \quad \mathbf{A}\mathbf{B} \neq \mathbf{B}\mathbf{A} \,, \end{aligned} \tag{2.4.17}$$

for general tensors $\mathbf{A}$ and $\mathbf{B}$.

Permutation Tensor:
The permutation tensor $\boldsymbol{\varepsilon}$ is a third-order tensor that can be defined such that

$$\boldsymbol{\varepsilon} \cdot (\mathbf{e}_i \otimes \mathbf{e}_j \otimes \mathbf{e}_k) = \varepsilon_{ijk} \,, \tag{2.4.18}$$

and that for any two vectors $\mathbf{a}$ and $\mathbf{b}$

$$\boldsymbol{\varepsilon} \cdot (\mathbf{a} \otimes \mathbf{b}) = \mathbf{a} \times \mathbf{b} \,. \tag{2.4.19}$$

Hierarchy of Tensor Operations:
To simplify the notation and reduce the need for using parentheses to clarify mathematical equations, it is convenient to define the hierarchy of the tensor operations

Table 2.1 Hierarchy of tensor operations

Level	Tensor Operation
1	Left transpose $^{LT}()$ and right transpose $()^T$
2	Cross product $(\times)$
3	Juxtaposition and tensor product $(\otimes)$
4	Dot product $(\cdot)$
5	Addition and subtraction

according to Table 2.1 with level 1 operations being performed before level 2 operations and so forth. Also, as is usual, the order in which operations in the same level are performed is determined by which operation appears in the most left-hand position in the equation.

2.5 Tensor Calculus in Rectangular Cartesian Coordinates

For simplicity, attention is limited in this section to tensors that are expressed relative to fixed rectangular Cartesian base vectors $\mathbf{e}_i$ because derivatives of these base vectors with respect to $\mathbf{x}$ are zero. However, a brief introduction to tensors with respect to general curvilinear coordinates can be found in Appendix F.

Gradient:
Let x_i be the components of the position vector $\mathbf{x}$ relative to the rectangular Cartesian base vectors $\mathbf{e}_i$. The gradient of a scalar function f with respect to the position $\mathbf{x}$ is a vector denoted by grad$f$ and represented by

$$\text{grad} f = \partial f/\partial \mathbf{x} = \partial f/\partial x_m \, \mathbf{e}_m = f_{,m} \, \mathbf{e}_m \,, \tag{2.5.1}$$

where for convenience a comma is used to denote partial differentiation with respect to the indicated components of $\mathbf{x}$. Also, the gradient of a tensor function $\mathbf{T}$ of order M $(M \geq 1)$ is a tensor of order $M + 1$ denoted by grad$\mathbf{T}$ is represented by

$$\text{grad}\mathbf{T} = \partial \mathbf{T}/\partial \mathbf{x} = \mathbf{T}_{,m} \otimes \mathbf{e}_m \,. \tag{2.5.2}$$

For example, if $\mathbf{T}$ is a second-order tensor then

$$\text{grad}\mathbf{T} = \partial \mathbf{T}/\partial \mathbf{x} = \partial (T_{ij}\mathbf{e}_i \otimes \mathbf{e}_j)/\partial x_m \otimes \mathbf{e}_m = T_{ij,m} \, \mathbf{e}_i \otimes \mathbf{e}_j \otimes \mathbf{e}_m = \mathbf{T}_{,m} \otimes \mathbf{e}_m \,. \tag{2.5.3}$$

It should be noted that the gradient operator is defined so that the increment $\partial\mathbf{x}$ operates on the right-hand side of the tensor $\mathbf{T}$. In contrast, some texts define the gradient operator so that the increment $\partial\mathbf{x}$ operates on the left-hand side of the tensor $\mathbf{T}$.

Divergence:
The divergence of a tensor $\mathbf{T}$ of order M $(M \geq 1)$ is a tensor of order $M - 1$ denoted by div$\mathbf{T}$ is represented by

$$\text{div}\mathbf{T} = \mathbf{T},_m \cdot \mathbf{e}_m = (\partial\mathbf{T}/\partial\mathbf{x})\,\mathbf{e}_m \cdot \mathbf{e}_m = (\partial\mathbf{T}/\partial\mathbf{x}) \cdot (\mathbf{e}_m \otimes \mathbf{e}_m) = (\partial\mathbf{T}/\partial\mathbf{x}) \cdot \mathbf{I}\,. \tag{2.5.4}$$

For example, if $\mathbf{T}$ is a second-order tensor then

$$\text{div}\mathbf{T} = T_{ij},_j\, \mathbf{e}_i\,. \tag{2.5.5}$$

Curl:
The curl of a vector $\mathbf{v}$ with components v_i is a vector denoted by curl$\mathbf{v}$ and is represented by

$$\text{curl}\mathbf{v} = -\mathbf{v},_j \times \mathbf{e}_j = -v_i,_j\, \varepsilon_{ijk}\mathbf{e}_k = v_i,_j\, \varepsilon_{jik}\mathbf{e}_k\,. \tag{2.5.6}$$

Also, the curl of a tensor $\mathbf{T}$ of order M $(M \geq 1)$ is a tensor of order M denoted by curl$\mathbf{T}$ and is represented by

$$\text{curl}\mathbf{T} = -\mathbf{T},_k \times \mathbf{e}_k\,. \tag{2.5.7}$$

For example, if $\mathbf{T}$ is a second-order tensor with components T_{ij} then

$$\text{curl}\mathbf{T} = -T_{ij},_k\, \varepsilon_{jkm}\mathbf{e}_i \otimes \mathbf{e}_m\,. \tag{2.5.8}$$

Laplacian:
The Laplacian of a tensor $\mathbf{T}$ of order M is a tensor of order M denoted by $\nabla^2\mathbf{T}$ and is represented by

$$\nabla^2\mathbf{T} = \text{div}(\text{grad}\mathbf{T}) = (\mathbf{T},_i \otimes \mathbf{e}_i),_j \cdot \mathbf{e}_j = \mathbf{T},_{mm}\,. \tag{2.5.9}$$

Divergence Theorem:
Let $\mathbf{n}$ be the unit outward normal to a closed surface ∂P of a region P, da be the element of area of ∂P, dv be the element of volume of P, and let $\mathbf{T}$ be an arbitrary tensor of any order. Then, the divergence theorem states that

$$\int_{\partial P} \mathbf{T}\mathbf{n}da = \int_{\partial P} \text{div}\mathbf{T}dv\,. \tag{2.5.10}$$

General Curvilinear Coordinates:
Appendix F presents an introduction to tensors with respect to general curvilinear coordinates.

2.6 Additional Definitions and Results

To better understand the definition of juxtaposition and to connect this definition with the usual rules for matrix multiplication let **A**, **B** and **C** be second-order tensors with components A_{ij}, B_{ij} and C_{ij}, respectively, relative to the rectangular Cartesian base vectors $\mathbf{e}_i$, and define **C** by

$$\mathbf{C} = \mathbf{A}\mathbf{B} \,. \tag{2.6.1}$$

Expressing these tensors in terms of their components yields

$$\begin{aligned} \mathbf{C} &= A_{ij}\mathbf{e}_i \otimes \mathbf{e}_j B_{mn}\mathbf{e}_m \otimes \mathbf{e}_n = A_{ij}B_{mn}(\mathbf{e}_j \cdot \mathbf{e}_m)\mathbf{e}_i \otimes \mathbf{e}_n = A_{im}B_{mn}\mathbf{e}_i \otimes \mathbf{e}_n \,, \\ C_{ij} &= \mathbf{C} \cdot \mathbf{e}_i \otimes \mathbf{e}_j = A_{rm}B_{mn}(\mathbf{e}_r \otimes \mathbf{e}_n) \cdot (\mathbf{e}_i \otimes \mathbf{e}_j) = A_{im}B_{mj} \,. \end{aligned} \tag{2.6.2}$$

Examination of the result (2.6.2) indicates that the second index of **A** is summed with the first index of **B**, which is consistent with the usual operation of row times column in the definition of matrix multiplication.

Symmetric:
The second-order tensor **A** with the $9 = 3^2$ components A_{ij} referred to the base vectors $\mathbf{e}_i$ is said to be symmetric if

$$\mathbf{A} = \mathbf{A}^T \,, \quad A_{ij} = A_{ji} \,. \tag{2.6.3}$$

It follows from (2.4.12) that if **A** is symmetric and **v** is an arbitrary vector with components v_i then

$$\mathbf{A}\mathbf{v} = \mathbf{v}\mathbf{A} \,, \quad A_{ij}v_j = v_j A_{ji} \,. \tag{2.6.4}$$

Moreover, it is noted that since the off-diagonal components of a symmetric second-order tensor satisfy the restrictions

$$A_{21} = A_{12} \,, \quad A_{31} = A_{13} \,, \quad A_{32} = A_{23} \,, \tag{2.6.5}$$

a symmetric second-order tensor has only six independent components.

Skew-symmetric:
The second-order tensor **A** with the $9 = 3^2$ components A_{ij} referred to the base vectors $\mathbf{e}_i$ is said to be skew-symmetric if

$$\mathbf{A} = -\mathbf{A}^T \,, \quad A_{ij} = -A_{ji} \,. \tag{2.6.6}$$

It also follows from (2.4.12) that if **A** is skew-symmetric and **v** is an arbitrary vector with components v_i then

$$\mathbf{A}\mathbf{v} = -\mathbf{v}\mathbf{A} \,, \quad A_{ij}v_j = -v_j A_{ji} \,. \tag{2.6.7}$$

Moreover, it is noted that since the diagonal components of a skew-symmetric second-order tensor vanish

$$A_{11} = A_{22} = A_{33} = 0\,, \tag{2.6.8}$$

and the off-diagonal components satisfy the restrictions

$$A_{21} = -A_{12}\,, \quad A_{31} = -A_{13}\,, \quad A_{32} = -A_{23}\,, \tag{2.6.9}$$

a skew-symmetric second-order tensor has only three independent components.

Symmetric and Skew-symmetric Parts:
Using these definitions it is observed that an arbitrary second-order tensor **B**, with components B_{ij}, can be separated additively and uniquely into its symmetric part denoted by $\mathbf{B}_{\text{sym}}$, with components $B_{(ij)}$, and its skew-symmetric part denoted by $\mathbf{B}_{\text{skew}}$, with components $B_{[ij]}$, such that

$$\begin{aligned}
&\mathbf{B} = \mathbf{B}_{\text{sym}} + \mathbf{B}_{\text{skew}}\,, && B_{ij} = B_{(ij)} + B_{[ij]}\,,\\
&\mathbf{B}_{\text{sym}} = \tfrac{1}{2}(\mathbf{B} + \mathbf{B}^T) = \mathbf{B}^T_{\text{sym}}\,, && B_{(ij)} = \tfrac{1}{2}(B_{ij} + B_{ji}) = B_{(ji)}\,,\\
&\mathbf{B}_{\text{skew}} = \tfrac{1}{2}(\mathbf{B} - \mathbf{B}^T) = -\mathbf{B}^T_{\text{skew}}\,, && B_{[ij]} = \tfrac{1}{2}(B_{ij} - B_{ji}) = -B_{[ji]}\,.
\end{aligned} \tag{2.6.10}$$

Trace:
The trace operation is defined as the dot product of an arbitrary second-order tensor **T** with the second-order identity tensor **I**. Letting T_{ij} be the components of **T** yields

$$\mathbf{T} \cdot \mathbf{I} = T_{ij}(\mathbf{e}_i \otimes \mathbf{e}_j) \cdot \mathbf{I} = T_{ij}(\mathbf{e}_i \cdot \mathbf{e}_j) = T_{ij}\delta_{ij} = T_{jj}\,. \tag{2.6.11}$$

Deviatoric Tensor:
The second-order tensor **A** with the $9 = 3^2$ components A_{ij} referred to the base vectors $\mathbf{e}_i$ is said to be deviatoric if

$$\mathbf{A} \cdot \mathbf{I} = 0\,, \quad A_{mm} = 0\,. \tag{2.6.12}$$

Unless otherwise stated the deviatoric part of a second-order tensor **A** will be denoted by $\mathbf{A}''$, which is defined by

$$\mathbf{A}'' = \mathbf{A} - \frac{1}{3}(\mathbf{A} \cdot \mathbf{I})\,\mathbf{I}\,. \tag{2.6.13}$$

Here, a double prime is used to denote a deviatoric tensor instead of the more common use of a single prime in order not to misinterpret the unimodular elastic deformation tensor $\mathbf{F}'$ introduced later as a deviatoric tensor.

Spherical and Deviatoric Parts:
Using these definitions it is observed that an arbitrary second-order tensor **T**, with components T_{ij}, can be separated additively and uniquely into its spherical part

denoted by $T\,\mathbf{I}$, with components $T\delta_{ij}$, and its deviatoric part denoted by $\mathbf{T}''$, with components T''_{ij}, such that

$$\mathbf{T} = T\,\mathbf{I} + \mathbf{T}'' , \qquad T_{ij} = T\delta_{ij} + T''_{ij} , \tag{2.6.14a}$$

$$\mathbf{T}'' \cdot \mathbf{I} = 0 , \qquad T''_{mm} = 0 . \tag{2.6.14b}$$

Taking the dot product of (2.6.14a) with the second-order identity $\mathbf{I}$ it can be shown that the scalar T is the mean value of the diagonal terms of $\mathbf{T}$

$$T = \frac{1}{3}\mathbf{T} \cdot \mathbf{I} = \frac{1}{3}T_{mm} . \tag{2.6.15}$$

When $\mathbf{T}$ is the stress tensor, this trace operator can be used to define the pressure p as minus the average of the diagonal components of $\mathbf{T}$

$$p = -\frac{1}{3}\mathbf{T} \cdot \mathbf{I} , \tag{2.6.16}$$

where p is positive in compression with $\mathbf{T}$ being positive in tension.

Dilatational and Distortional Parts:
Using the work in [1], it follows that a general second-order tensor $\mathbf{F}$ with positive determinant J

$$J = \det \mathbf{F} > 0 \tag{2.6.17}$$

separates multiplicatively into its dilatational part $J^{1/3}\mathbf{I}$ and its distortional part $\mathbf{F}'$

$$\mathbf{F} = (J^{1/3}\mathbf{I})\,\mathbf{F}' , \quad \mathbf{F}' = J^{-1/3}\mathbf{F} , \quad \det \mathbf{F}' = 1 , \tag{2.6.18}$$

where $\mathbf{F}'$ is a unimodular tensor with determinant equal to unity. When, $\mathbf{F}$ is the deformation gradient, J is a pure measure of dilatational deformation and $\mathbf{F}'$ is a measure of distortional deformation which includes rotation.

Dot Product Between a String of Tensors:
For later convenience it is useful to consider properties of the dot product between strings of second-order tensors. To this end, let $\mathbf{A}, \mathbf{B}, \mathbf{C}$ and $\mathbf{D}$ be second-order tensors, with components A_{ij}, B_{ij}, C_{ij} and D_{ij}, respectively. Then, it can be shown that

$$\begin{aligned} &\mathbf{A} \cdot (\mathbf{BCD}) = A_{ij}(B_{im}C_{mn}D_{nj}) , \quad \mathbf{A} \cdot (\mathbf{BCD}) = (\mathbf{B}^T\mathbf{A}) \cdot (\mathbf{CD}) , \\ &\mathbf{A} \cdot (\mathbf{BCD}) = (\mathbf{AD}^T) \cdot (\mathbf{BC}) , \quad \mathbf{A} \cdot (\mathbf{BCD}) = (\mathbf{B}^T\mathbf{AD}^T) \cdot \mathbf{C} . \end{aligned} \tag{2.6.19}$$

2.7 Transformation Relations

Consider two right-handed orthonormal rectangular Cartesian coordinate systems with base vectors $\mathbf{e}_i$ and $\tilde{\mathbf{e}}_i$, and define the transformation tensor $\mathbf{A}$ by

$$\mathbf{A} = \mathbf{e}_m \otimes \tilde{\mathbf{e}}_m \,. \tag{2.7.1}$$

It follows from this definition that $\mathbf{A}$ is an orthogonal tensor

$$\begin{aligned} \mathbf{A}\mathbf{A}^T &= (\mathbf{e}_m \otimes \tilde{\mathbf{e}}_m)(\tilde{\mathbf{e}}_n \otimes \mathbf{e}_n) = (\tilde{\mathbf{e}}_m \cdot \tilde{\mathbf{e}}_n)(\mathbf{e}_m \otimes \mathbf{e}_n)\,, \\ &= \delta_{mn}(\mathbf{e}_m \otimes \mathbf{e}_n) = (\mathbf{e}_m \otimes \mathbf{e}_m) = \mathbf{I}\,, \end{aligned} \tag{2.7.2a}$$

$$\begin{aligned} \mathbf{A}^T\mathbf{A} &= (\tilde{\mathbf{e}}_m \otimes \mathbf{e}_m)(\mathbf{e}_n \otimes \tilde{\mathbf{e}}_n) = (\mathbf{e}_m \cdot \mathbf{e}_n)(\tilde{\mathbf{e}}_m \otimes \tilde{\mathbf{e}}_n)\,, \\ &= \delta_{mn}(\tilde{\mathbf{e}}_m \otimes \tilde{\mathbf{e}}_n) = (\tilde{\mathbf{e}}_m \otimes \tilde{\mathbf{e}}_m) = \mathbf{I}\,. \end{aligned} \tag{2.7.2b}$$

It also follows that the base vectors $\mathbf{e}_i$ and $\tilde{\mathbf{e}}_i$ are related by the expressions

$$\mathbf{e}_i = \mathbf{A}\tilde{\mathbf{e}}_i = (\mathbf{e}_m \otimes \tilde{\mathbf{e}}_m)\tilde{\mathbf{e}}_i = \mathbf{e}_m(\tilde{\mathbf{e}}_m \cdot \tilde{\mathbf{e}}_i) = \mathbf{e}_m\delta_{mi}\,, \tag{2.7.3a}$$

$$\tilde{\mathbf{e}}_i = \mathbf{A}^T\mathbf{e}_i\,, \tag{2.7.3b}$$

where in obtaining (2.7.3b) Eq. (2.7.3a) has be multiplied by $\mathbf{A}^T$ and use has been made of the fact that $\mathbf{A}$ is an orthogonal tensor.

These equations can be written in equivalent component forms by noting that the components A_{ij} of $\mathbf{A}$ referred to the base vectors $\mathbf{e}_i$ and the components $\tilde{A}_{ij}$ of $\mathbf{A}$ referred to the base vectors $\tilde{\mathbf{e}}_i$ are defined by

$$\begin{aligned} A_{ij} &= \mathbf{A} \cdot (\mathbf{e}_i \otimes \mathbf{e}_j) = (\mathbf{e}_m \otimes \tilde{\mathbf{e}}_m) \cdot (\mathbf{e}_i \otimes \mathbf{e}_j) = (\mathbf{e}_m \cdot \mathbf{e}_i)(\tilde{\mathbf{e}}_m \cdot \mathbf{e}_j)\,, \\ &= \delta_{mi}(\tilde{\mathbf{e}}_m \cdot \mathbf{e}_j) = \tilde{\mathbf{e}}_i \cdot \mathbf{e}_j\,, \end{aligned} \tag{2.7.4a}$$

$$\begin{aligned} \tilde{A}_{ij} &= \mathbf{A} \cdot (\tilde{\mathbf{e}}_i \otimes \tilde{\mathbf{e}}_j) = (\mathbf{e}_m \otimes \tilde{\mathbf{e}}_m) \cdot (\tilde{\mathbf{e}}_i \otimes \tilde{\mathbf{e}}_j) = (\mathbf{e}_m \cdot \tilde{\mathbf{e}}_i)(\tilde{\mathbf{e}}_m \cdot \tilde{\mathbf{e}}_j)\,, \\ &= (\mathbf{e}_m \cdot \tilde{\mathbf{e}}_i)\delta_{mj} = \mathbf{e}_j \cdot \tilde{\mathbf{e}}_i = \tilde{\mathbf{e}}_i \cdot \mathbf{e}_j\,. \end{aligned} \tag{2.7.4b}$$

It is important to emphasize that these results indicate that the first index of A_{ij} (or $\tilde{A}_{ij}$) is identified with the base vectors $\tilde{\mathbf{e}}_i$ and the second index is identified with the base vectors $\mathbf{e}_i$. This identification is a consequence of the definition (2.7.1) and is arbitrary in the sense that one could introduce an alternative definition where the order of the vectors in (2.7.1) is reversed. However, once the definition (2.7.1) is introduced it is essential to maintain consistency throughout the text. Also, note from (2.7.4a) and (2.7.4b) that the components of $\mathbf{A}$ referred to either of the base vectors $\mathbf{e}_i$ or $\tilde{\mathbf{e}}_i$ are equal

$$A_{ij} = \tilde{A}_{ij}\,. \tag{2.7.5}$$

This is because $\mathbf{A}$ is a two-point tensor which is defined by both of the triads $\mathbf{e}_i$ and $\tilde{\mathbf{e}}_i$.

Using these expressions, Eqs. (2.7.3a) and (2.7.3b) can be rewritten in the forms

$$\mathbf{e}_i = (A_{mn}\tilde{\mathbf{e}}_m \otimes \tilde{\mathbf{e}}_n)\tilde{\mathbf{e}}_i = A_{mi}\tilde{\mathbf{e}}_m \,, \tag{2.7.6a}$$

$$\tilde{\mathbf{e}}_i = (A_{mn}\mathbf{e}_n \otimes \mathbf{e}_m)\mathbf{e}_i = A_{in}\mathbf{e}_n \,. \tag{2.7.6b}$$

Again, it is noted that the first index of A_{ij} refers to the base vectors $\tilde{\mathbf{e}}_i$ and the second index refers to the base vectors $\mathbf{e}_i$.

One of the most fundamental properties of a tensor $\mathbf{T}$ is that the tensor is independent of the choice of the coordinate system used to express its components. Specifically, it is noted that all the tensor properties in Sect. 2.3 have been defined without regard to any particular coordinate system. Furthermore, it is emphasized that since physical laws cannot depend on an arbitrary choice of a coordinate system it is essential to express the mathematical representation of these physical laws using tensors. For this reason tensors are essential in continuum mechanics.

Although an arbitrary tensor $\mathbf{T}$ of order M is independent of the choice of a coordinate system, the components $T_{ijk...rst}$ of $\mathbf{T}$ with respect to the base vectors $\mathbf{e}_i$ are defined by (2.4.4) explicitly depend on the choice of the coordinate system defined by $\mathbf{e}_i$. It follows by analogy to (2.4.4) that the components $\tilde{T}_{ijk...rst}$ of $\mathbf{T}$ relative to the base vectors $\tilde{\mathbf{e}}_i$ are defined by

$$\tilde{T}_{ijk...rst} = \mathbf{T} \cdot (\tilde{\mathbf{e}}_i \otimes \tilde{\mathbf{e}}_j \otimes \tilde{\mathbf{e}}_k \otimes \cdots \otimes \tilde{\mathbf{e}}_r \otimes \tilde{\mathbf{e}}_s \otimes \tilde{\mathbf{e}}_t) \,, \tag{2.7.7}$$

so that $\mathbf{T}$ admits the alternative representation

$$\mathbf{T} = \tilde{T}_{ijk...rst}(\tilde{\mathbf{e}}_i \otimes \tilde{\mathbf{e}}_j \otimes \tilde{\mathbf{e}}_k \otimes \cdots \otimes \tilde{\mathbf{e}}_r \otimes \tilde{\mathbf{e}}_s \otimes \tilde{\mathbf{e}}_t) \,. \tag{2.7.8}$$

Now, since $\mathbf{T}$ admits both of the representations (2.4.3) and (2.7.8), it follows that the components $T_{ijk..rst}$ and $\tilde{T}_{ijk...rst}$ must be related to each other. To determine this relation, (2.7.8) is substituted into (2.7.7) and use is made of (2.7.6a) and (2.7.6b) to obtain

$$\begin{aligned}
T_{ijk..rst} &= \mathbf{T} \cdot (A_{li}\tilde{\mathbf{e}}_l \otimes A_{mj}\tilde{\mathbf{e}}_m \otimes A_{nk}\tilde{\mathbf{e}}_n \otimes \cdots \otimes A_{ur}\tilde{\mathbf{e}}_u \otimes A_{vs}\tilde{\mathbf{e}}_v \otimes A_{wt}\tilde{\mathbf{e}}_w) \\
&= A_{li}A_{mj}A_{nk}...A_{ur}A_{vs}A_{wt}\mathbf{T} \cdot (\tilde{\mathbf{e}}_l \otimes \tilde{\mathbf{e}}_m \otimes \tilde{\mathbf{e}}_n \otimes \cdots \otimes \tilde{\mathbf{e}}_u \otimes \tilde{\mathbf{e}}_v \otimes \tilde{\mathbf{e}}_w) \\
&= A_{li}A_{mj}A_{nk}...A_{ur}A_{vs}A_{wt}\tilde{T}_{lmn...uvw} \,, \\
\tilde{T}_{ijk..rst} &= \mathbf{T} \cdot (A_{il}\mathbf{e}_l \otimes A_{jm}\mathbf{e}_m \otimes A_{kn}\mathbf{e}_n \otimes \cdots \otimes A_{ru}\mathbf{e}_u \otimes A_{sv}\mathbf{e}_v \otimes A_{tw}\mathbf{e}_w) \\
&= A_{il}A_{jm}A_{kn}...A_{ru}A_{sv}A_{tw}\mathbf{T} \cdot (\mathbf{e}_l \otimes \mathbf{e}_m \otimes \mathbf{e}_n \otimes \cdots \otimes \mathbf{e}_u \otimes \mathbf{e}_v \otimes \mathbf{e}_w) \\
&= A_{il}A_{jm}A_{kn}...A_{ru}A_{sv}A_{tw}T_{lmn...uvw} \,.
\end{aligned} \tag{2.7.9}$$

For examples, if $\mathbf{v}$ is a vector with components v_i and $\tilde{v}_i$ then

$$\mathbf{v} = v_i\mathbf{e}_i = \tilde{v}_i\tilde{\mathbf{e}}_i \,, \quad v_i = A_{mi}\tilde{v}_m \,, \quad \tilde{v}_i = A_{im}v_m \,. \tag{2.7.10}$$

and if $\mathbf{T}$ is a second-order tensor with components T_{ij} and $\tilde{T}_{ij}$ then

$$\begin{aligned} &\mathbf{T} = T_{ij}\mathbf{e}_i \otimes \mathbf{e}_j = \tilde{T}_{ij}\tilde{\mathbf{e}}_i \otimes \tilde{\mathbf{e}}_j \, , \\ &T_{ij} = A_{mi}A_{nj}\tilde{T}_{mn} \, , \qquad T_{ij} = A^T_{im}\tilde{T}_{mn}A_{nj} \, , \\ &\tilde{T}_{ij} = A_{im}A_{jn}T_{mn} \, , \qquad \tilde{T}_{ij} = A_{im}T_{mn}A^T_{nj} \, . \end{aligned} \tag{2.7.11}$$

Again, in these transformation relations the first component of A_{ij} remains connected to $\tilde{\mathbf{e}}_i$ and its second component remains connected to $\mathbf{e}_j$. Also, use has been made of the result $A^T_{im} = A_{im}$.

References

1. Flory PJ (1961) Thermodynamic relations for high elastic materials. Trans Faraday Soc 57:829–838
2. Rubin MB (2000) Cosserat theories: shells, rods and points, vol 79. Springer Science & Business Media, Berlin

Chapter 3
Kinematics

Abstract The objective of this chapter is to discuss nonlinear kinematics of deformable continua. Bodies, configurations and motion of continua are discussed along with a definition of the material time derivative, which is used to determine the velocity and acceleration of a material point. Deformation tensors and rate of deformation tensors are defined and analyzed. The notion of Superposed Rigid Body Motions (SRBM) is presented and the associated transformation relations of specific tensors are developed. In addition, an Eulerian formulation of evolution equations for elastic deformations is proposed and strongly objective, robust numerical integration algorithms for these evolution equations are developed.

3.1 Bodies, Configurations and Motion

Bodies

In an abstract sense a body $\mathcal{B}$ is a set of material particles which are denoted by Y (see Fig. 3.1). In mechanics a body is assumed to be smooth and can be put into correspondence with a domain of Euclidean 3-Space. Bodies are often mapped to their configurations, i.e., the regions of Euclidean 3-Space they occupy at each instant of time t $(-\infty < t < \infty)$. In the following, all position vectors are referred to a fixed point inertial in space.

Current Configuration and Motion

The current configuration of the body is the region of Euclidean 3-Space occupied by the body at the current time t. Let $\mathbf{x}$ be the position vector which identifies the place occupied by the particle Y at the time t. Since it is assumed that the body can be mapped smoothly into a domain of Euclidean 3-Space, a motion of the body can be represented as

$$\mathbf{x} = \bar{\mathbf{x}}(Y, t)\,. \tag{3.1.1}$$

In this expression, Y refers to the material particle, t refers to the current time, $\mathbf{x}$ refers to the value of the function and $\bar{\mathbf{x}}$ characterizes how each particle Y moves

M. B. Rubin, *Continuum Mechanics with Eulerian Formulations of Constitutive Equations*, Solid Mechanics and Its Applications 265,
https://doi.org/10.1007/978-3-030-57776-6_3

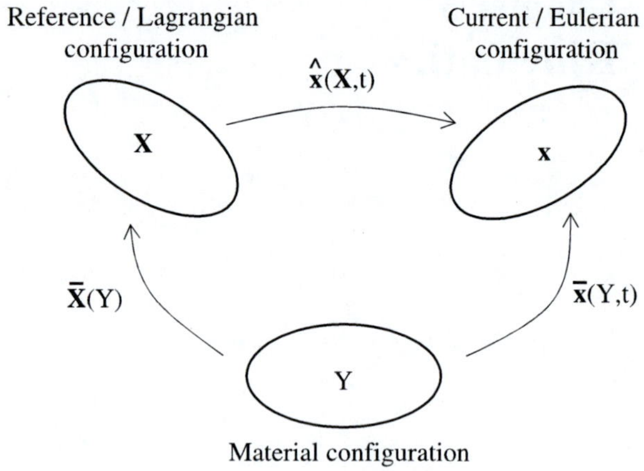

Fig. 3.1 Definition of the material Y, Referential/Lagrangian $\mathbf{X}$ and Current/Eulerian $\mathbf{x}$ configurations

through space as time progresses. It is assumed that this function is invertible so that

$$Y = \bar{\mathbf{x}}^{-1}(\mathbf{x}, t) = \tilde{Y}(\mathbf{x}, t) . \tag{3.1.2}$$

Reference Configuration
Sometimes it is convenient to select one particular configuration, called a reference configuration, and refer everything concerning the body and its motion to this configuration. The reference configuration need not necessarily be an actual configuration occupied by the body and in particular, the reference configuration need not be the *initial* configuration.

Let $\mathbf{X}$ be the position vector of the particle Y in the reference configuration. Then, the mapping from Y to the place $\mathbf{X}$ in the reference configuration can be written as

$$\mathbf{X} = \bar{\mathbf{X}}(Y) . \tag{3.1.3}$$

In this expression, $\mathbf{X}$ refers to the value of the function $\bar{\mathbf{X}}$ which characterizes the mapping. It is important to note that this mapping does not depend on time because the reference configuration is a single constant configuration. Moreover, this mapping is assumed to be invertible with its inverse given by

$$Y = \bar{\mathbf{X}}^{-1}(\mathbf{X}) = \hat{Y}(\mathbf{X}) . \tag{3.1.4}$$

Motion
It follows that the mapping from the reference configuration to the current configuration can be obtained by substituting (3.1.4) into (3.1.1) to deduce that

$$\mathbf{x} = \bar{\mathbf{x}}(\hat{Y}(\mathbf{X}), t) = \hat{\mathbf{x}}(\mathbf{X}, t) , \tag{3.1.5}$$

which characterizes the motion all material points. From this expression, it is obvious that the functional form of the mapping $\hat{\mathbf{x}}$ depends on the specific choice of the reference configuration. Further in this regard, it is emphasized that the choice of the reference configuration is similar to the choice of coordinates in that it is arbitrary to the extent that a one-to-one correspondence exists between the material particles Y and their locations $\mathbf{X}$ in the reference configuration. Also, the inverse of this mapping can be written in the form

$$\mathbf{X} = \tilde{\mathbf{X}}(\mathbf{x}, t) . \tag{3.1.6}$$

In contrast to the material configuration, which is based on the abstract notion of a material point Y, the mapping (3.1.5) expresses $\mathbf{x}$ as a vector function of $\mathbf{X}$ and t and the inverse mapping (3.1.6) expresses $\mathbf{X}$ as a vector function of $\mathbf{x}$ and t. These vector functions are mathematical functions that are assumed to be smooth functions which can be differentiated with respect to either of their arguments as many times as necessary.

3.2 Representations

Material, Lagrangian and Eulerian Representations

There are several methods of describing properties of a body. The following considers three possible representations. To this end, let f be an arbitrary scalar or tensor function characterizing a property of the body which admits the following three representations

$$f = \bar{f}(Y, t) \qquad \text{Material representation ,} \tag{3.2.1a}$$

$$f = \hat{f}(\mathbf{X}, t) \qquad \text{Lagrangian representation ,} \tag{3.2.1b}$$

$$f = \tilde{f}(\mathbf{x}, t) \qquad \text{Eulerian representation .} \tag{3.2.1c}$$

For definiteness, a symbol is used to denote different functional forms from the value of a function. Whenever this is necessary, the functions that depend on Y are denoted with an overbar $(\bar{\ })$, functions that depend on $\mathbf{X}$ are denoted with a hat $(\hat{\ })$ and functions that depend on $\mathbf{x}$ are denoted with a tilde $(\tilde{\ })$. Furthermore, the functional forms $\bar{f}$, $\hat{f}$, $\tilde{f}$ are related by the expressions

$$\hat{f}(\mathbf{X}, t) = \bar{f}(\hat{Y}(\mathbf{X}), t) , \quad \tilde{f}(\mathbf{x}, t) = \hat{f}(\tilde{\mathbf{X}}(\mathbf{x}, t), t) . \tag{3.2.2}$$

The representation (3.2.1a) is called *material* because the material point Y is used as an independent variable. The representation (3.2.1b) is called *referential* or *Lagrangian* because the position $\mathbf{X}$ of a material point in the reference configuration is an independent variable, and the representation (3.2.1c) is called *spatial* or *Eulerian* because the current position $\mathbf{x}$ in space is used as an independent variable. However, it is emphasized that in view of the invertibility of these functions, a one-to-one correspondence can be established between any two of these representations.

In this book, use is made of both the coordinate free forms of equations as well as their indicial counterparts. To this end, let $\mathbf{e}_A$ be a fixed right-handed orthonormal rectangular Cartesian basis associated with the reference configuration and let $\mathbf{e}_i$ be a fixed right-handed orthonormal rectangular Cartesian basis associated with the current configuration. Moreover, these base vectors are specified to coincide so that

$$\mathbf{e}_i \cdot \mathbf{e}_A = \delta_{iA} \,, \tag{3.2.3}$$

where δ_{iA} is the usual Kronecker delta. In the following, all tensor quantities are referred to either of these bases and for clarity use is made of upper case letters to indicate indices of quantities associated with the reference configuration and with lower case letters to indicate indices of quantities associated with the current configuration. For example,

$$\mathbf{X} = X_A \mathbf{e}_A \,, \quad \mathbf{x} = x_i \mathbf{e}_i \,, \tag{3.2.4}$$

where X_A are the rectangular Cartesian components of the position vector $\mathbf{X}$ and x_i are the rectangular Cartesian components of the position vector $\mathbf{x}$ and the usual summation convention over repeated indices is used. It follows that the motion (3.1.5) can be written in the form

$$x_i = \hat{x}_i(X_A, t) \,. \tag{3.2.5}$$

Velocity and Acceleration

The velocity $\mathbf{v}$ of a material point Y is defined as the rate of change with time t of position of the material point. Since the function $\bar{\mathbf{x}}(Y, t)$ characterizes the position of the material point Y at any time t, it follows that the velocity is defined conceptually by

$$\mathbf{v} = \dot{\mathbf{x}} = \frac{\partial \bar{\mathbf{x}}(Y, t)}{\partial t} \,, \quad v_i = \dot{x}_i = \frac{\partial \bar{x}_i(Y, t)}{\partial t} \,, \tag{3.2.6}$$

where a superposed dot $(\dot{\ })$ is used to denote partial differentiation with respect to time t holding the material particle Y fixed. Similarly, the acceleration $\mathbf{a}$ of a material point Y is defined by

$$\mathbf{a} = \dot{\mathbf{v}} = \frac{\partial \bar{\mathbf{v}}(Y, t)}{\partial t} \,, \quad a_i = \dot{v}_i = \frac{\partial \bar{v}_i(Y, t)}{\partial t} \,. \tag{3.2.7}$$

Notice that in view of the mappings (3.1.4) and (3.1.6), the velocity and acceleration can be expressed as functions of either $(\mathbf{X}, t)$ or $(\mathbf{x}, t)$.

Material Derivative

The material derivative of an arbitrary function f is defined conceptually by

$$\dot{f} \equiv \left. \frac{\partial \bar{f}(Y, t)}{\partial t} \right|_Y \,. \tag{3.2.8}$$

It is important to emphasize that the material derivative, which is denoted by a superposed dot $\dot{()}$ is defined to be the rate of change with time t of the function holding the material particle Y fixed. In this sense the velocity $\mathbf{v}$ is the material derivative of the position $\mathbf{x}$ and the acceleration $\mathbf{a}$ is the material derivative of the velocity $\mathbf{v}$. Recalling that the function f can be expressed in terms of either the Material (3.2.1a), Lagrangian (3.2.1b) or Eulerian (3.2.1c) representations, it follows from the chain rule of differentiation that $\dot{f}$ admits the additional representations

$$\dot{f} = \frac{\partial \hat{f}(\mathbf{X},t)}{\partial t}\dot{t} + [\partial \hat{f}(\mathbf{X},t)/\partial \mathbf{X}]\,\dot{\mathbf{X}} = \frac{\partial \hat{f}(\mathbf{X},t)}{\partial t}\,,$$

$$\dot{f} = \frac{\partial \hat{f}(\mathbf{X},t)}{\partial t}\dot{t} + [\partial \hat{f}(\mathbf{X},t)/\partial X_A]\,\dot{X}_A = \frac{\partial \hat{f}(\mathbf{X},t)}{\partial t}\,, \tag{3.2.9a}$$

$$\dot{f} = \frac{\partial \tilde{f}(\mathbf{x},t)}{\partial t}\dot{t} + [\partial \tilde{f}(\mathbf{x},t)/\partial \mathbf{x}]\,\dot{\mathbf{x}} = \frac{\partial \tilde{f}(\mathbf{x},t)}{\partial t} + [\partial \tilde{f}(\mathbf{x},t)/\partial \mathbf{x}]\,\mathbf{v}\,,$$

$$\dot{f} = \frac{\partial \tilde{f}(\mathbf{x},t)}{\partial t}\dot{t} + [\partial \tilde{f}(\mathbf{x},t)/\partial x_m]\,\dot{x}_m = \frac{\partial \tilde{f}(\mathbf{x},t)}{\partial t} + [\partial \tilde{f}(\mathbf{x},t)/\partial x_m]\,v_m\,, \tag{3.2.9b}$$

where in (3.2.9a) use has been made of the fact that the mapping (3.1.3) from the material point Y to its location $\mathbf{X}$ in the reference configuration is independent of time so that $\dot{\mathbf{X}}$ vanishes. It is important to emphasize that the physics of the material derivative defined by (3.2.8) remains unchanged even though its specific functional form for the different representations (3.2.9a) and (3.2.9b) changes.

3.3 Deformation Gradient and Deformation Measures

To describe the deformation of the body from the reference configuration to the current configuration, it is convenient to think of the body in its reference configuration as a finite collection of neighboring tetrahedrons. As the number of tetrahedrons increases it is possible to approximate a body having an arbitrary shape. If the deformation of each of these tetrahedrons from the reference configuration to the current configuration can be determined, then the shape (and volume) of the body in the current configuration can be determined by simply connecting the neighboring tetrahedrons. Since a tetrahedron is characterized by a triad of three vectors, the deformation of an arbitrary elemental tetrahedron (infinitesimally small) can be characterized by determining the deformation of an arbitrary material line element. This is because the material line element can be identified with each of the base vectors which represent the edges of the tetrahedron.

Deformation Gradient

For this reason it is sufficient to determine the deformation of a general material line element $d\mathbf{X}$ in the reference configuration to the material line element $d\mathbf{x}$ in the current configuration. Recalling that the mapping $\mathbf{x} = \hat{\mathbf{x}}(\mathbf{X},t)$ defines the position $\mathbf{x}$

in the current configuration of any material point $\mathbf{X}$ in the reference configuration at time t, it follows that

$$d\mathbf{x} = (\partial\hat{\mathbf{x}}/\partial\mathbf{X})d\mathbf{X} = \mathbf{F}d\mathbf{X}\,, \tag{3.3.1a}$$

$$dx_i = (\partial\hat{x}_i/\partial X_A)dX_A = x_{i,A}\,dX_A = F_{iA}dX_A\,, \tag{3.3.1b}$$

$$\mathbf{F} = (\partial\hat{\mathbf{x}}/\partial\mathbf{X}) = F_{iA}\mathbf{e}_i \otimes \mathbf{e}_A\,, \quad F_{iA} = x_{i,A}\,, \tag{3.3.1c}$$

where $\mathbf{F}$ is the deformation gradient with components F_{iA}. Unless otherwise stated, throughout the text a comma denotes partial differentiation with respect to X_A if the index is a capital letter and with respect to x_i if the index is a lower case letter. Since the mapping $\hat{\mathbf{x}}(\mathbf{X}, t)$ is invertible, $\mathbf{F}$ must satisfy the restriction

$$\det\mathbf{F} \neq 0\,, \quad \det(x_{i,A}) \neq 0\,, \tag{3.3.2}$$

for all time and all points in the spatial region occupied by the body. To ensure that the reference configuration has the possibility of coinciding with the current configuration at any time (i.e., $\mathbf{x} = \mathbf{X}$ and $\mathbf{F} = \mathbf{I}$), the deformation gradient must satisfy the restriction that

$$\det\mathbf{F} > 0\,, \quad \det(x_{i,A}) > 0\,. \tag{3.3.3}$$

Right and Left Cauchy–Green Deformation Tensors and the Cauchy Deformation Tensor
The magnitude ds of the material line element $d\mathbf{x}$ in the current configuration can be calculated using (3.3.1a), such that

$$(ds)^2 = d\mathbf{x}\cdot d\mathbf{x} = \mathbf{F}d\mathbf{X}\cdot\mathbf{F}d\mathbf{X} = d\mathbf{X}\cdot\mathbf{F}^T\mathbf{F}d\mathbf{X} = d\mathbf{X}\cdot\mathbf{C}d\mathbf{X}\,,$$
$$(ds)^2 = dx_i dx_i = F_{iA}dX_A F_{iB}dX_B = dX_A(x_{i,A}\,x_{i,B})dX_B = dX_A C_{AB} dX_B\,, \tag{3.3.4a}$$

$$\mathbf{C} = \mathbf{F}^T\mathbf{F} = C_{AB}\mathbf{e}_A \otimes \mathbf{e}_B\,, \quad C_{AB} = F_{iA}F_{iB} = x_{i,A}\,x_{i,B}\,, \tag{3.3.4b}$$

where $\mathbf{C}$ is called the *right Cauchy–Green deformation tensor*. Similarly, the magnitude dS of the material line element $d\mathbf{X}$ in the reference configuration can be calculated by inverting (3.3.1a) to obtain

$$d\mathbf{X} = \mathbf{F}^{-1}d\mathbf{x}\,, \quad dX_A = X_{A,i}\,dx_i\,, \tag{3.3.5}$$

which yields

$$(dS)^2 = d\mathbf{X} \cdot d\mathbf{X} = \mathbf{F}^{-1}d\mathbf{x} \cdot \mathbf{F}^{-1}d\mathbf{x} = d\mathbf{x} \cdot \mathbf{F}^{-T}\mathbf{F}^{-1}d\mathbf{x} = d\mathbf{x} \cdot \mathbf{c}d\mathbf{x}\,,$$
$$(dS)^2 = dX_A dX_A = X_{A,i}\, dx_i X_{A,j}\, dx_j = dx_i(X_{A,i}\, X_{A,j})dx_j = dx_i c_{ij} dx_j\,, \tag{3.3.6a}$$

$$\mathbf{c} = \mathbf{F}^{-T}\mathbf{F}^{-1} = c_{ij}\mathbf{e}_i \otimes \mathbf{e}_j\,, \quad c_{ij} = X_{A,i}\, X_{A,j}\,, \tag{3.3.6b}$$

where $\mathbf{F}^{-T}$ is the transpose of $\mathbf{F}^{-1}$ and $\mathbf{c}$ is the *Cauchy deformation tensor*. It is also convenient to define the *left Cauchy–Green deformation tensor* $\mathbf{B}$ by

$$\mathbf{B} = \mathbf{F}\mathbf{F}^T = B_{ij}\mathbf{e}_i \otimes \mathbf{e}_j\,, \quad B_{ij} = F_{iA}F_{jA} = x_{i,A}\, x_{j,A}\,, \tag{3.3.7}$$

and it is noted that

$$\mathbf{c} = \mathbf{B}^{-1}\,. \tag{3.3.8}$$

Stretch and Extension

The stretch λ of a material line element is defined in terms of the ratio of the lengths ds and dS of the line element in the present and reference configurations, respectively, such that

$$\lambda = \frac{ds}{dS}\,. \tag{3.3.9}$$

Also, the extension ε of the same material line element is defined by

$$\varepsilon = \lambda - 1 = \frac{ds - dS}{dS}\,. \tag{3.3.10}$$

It follows from these definitions that the stretch is always positive. Also, the stretch is greater than one and the extension is greater than zero when the material line element is extended relative to its reference length.

For convenience let $\mathbf{S}$ be the unit vector defining the direction of the material line element $d\mathbf{X}$ in the reference configuration and let $\mathbf{s}$ be the unit vector defining the direction of the associated material line element $d\mathbf{x}$ in the current configuration, such that

$$d\mathbf{X} = \mathbf{S}dS\,, \qquad dX_A = S_A dS\,, \qquad \mathbf{S} \cdot \mathbf{S} = S_A S_A = 1\,, \tag{3.3.11a}$$
$$d\mathbf{x} = \mathbf{s}ds\,, \qquad dx_i = s_i ds\,, \qquad \mathbf{s} \cdot \mathbf{s} = s_i s_i = 1\,. \tag{3.3.11b}$$

Thus, using (3.3.4a) and (3.3.6a) it can be shown that

$$\lambda\mathbf{s} = \mathbf{F}\mathbf{S}\,, \qquad \lambda s_i = x_{i,A}\, S_A\,, \tag{3.3.12a}$$
$$\lambda^2 = \mathbf{C} \cdot (\mathbf{S} \otimes \mathbf{S})\,, \qquad \lambda^2 = C_{AB} S_A S_B\,, \tag{3.3.12b}$$
$$\frac{1}{\lambda^2} = \mathbf{c} \cdot (\mathbf{s} \otimes \mathbf{s})\,, \qquad \frac{1}{\lambda^2} = c_{ij} s_i s_j\,. \tag{3.3.12c}$$

Since the stretch is positive, it follows from (3.3.12b) and (3.3.12c) that $\mathbf{C}$ and $\mathbf{c}$ are positive-definite tensors. Similarly, it can be shown that $\mathbf{B}$ in (3.3.7) is also a positive-definite tensor. In addition, notice from (3.3.12b) that the stretch of a material line element depends not only on the value of $\mathbf{C}$ at the material point $\mathbf{X}$ and the time t, but it depends also on the orientation $\mathbf{S}$ of the material line element in the reference configuration.

A Pure Measure of Dilatation (Volume Change)
To discuss the relative volume change of a material element, it is convenient to first prove that for any nonsingular second-order tensor $\mathbf{F}$ and any two vectors $\mathbf{a}$ and $\mathbf{b}$ that

$$\mathbf{Fa} \times \mathbf{Fb} = \det(\mathbf{F})\, \mathbf{F}^{-T} (\mathbf{a} \times \mathbf{b})\,. \tag{3.3.13}$$

To prove this result, it is noted that the quantity $\mathbf{F}^{T} (\mathbf{a} \times \mathbf{b})$ is a vector that is orthogonal to the plane formed by the vectors $\mathbf{Fa}$ and $\mathbf{Fb}$ since

$$\begin{aligned}
&\mathbf{F}^{-T} (\mathbf{a} \times \mathbf{b}) \cdot \mathbf{Fa} = (\mathbf{a} \times \mathbf{b}) \cdot (\mathbf{F}^{-T})^{T} \mathbf{Fa} = (\mathbf{a} \times \mathbf{b}) \cdot \mathbf{F}^{-1}\mathbf{Fa} = (\mathbf{a} \times \mathbf{b}) \cdot \mathbf{a} = 0\,,\\
&\mathbf{F}^{-T} (\mathbf{a} \times \mathbf{b}) \cdot \mathbf{Fb} = (\mathbf{a} \times \mathbf{b}) \cdot \mathbf{F}^{-1}\mathbf{Fb} = (\mathbf{a} \times \mathbf{b}) \cdot \mathbf{b} = 0\,.
\end{aligned} \tag{3.3.14}$$

This means that the quantity $\mathbf{Fa} \times \mathbf{Fb}$ must be a vector that is parallel to $\mathbf{F}^{-T} (\mathbf{a} \times \mathbf{b})$ so that

$$\mathbf{Fa} \times \mathbf{Fb} = \alpha\, \mathbf{F}^{-T} (\mathbf{a} \times \mathbf{b})\,. \tag{3.3.15}$$

Next, the value of the scalar α is determined by noting that both sides of Eq. (3.3.15) must be linear functions of $\mathbf{a}$ and $\mathbf{b}$. This means that α is independent of the vectors $\mathbf{a}$ and $\mathbf{b}$. Moreover, letting $\mathbf{c}$ be an arbitrary vector, it follows that

$$\mathbf{Fa} \times \mathbf{Fb} \cdot \mathbf{Fc} = \alpha\, \mathbf{F}^{-T} (\mathbf{a} \times \mathbf{b}) \cdot \mathbf{Fc} = \alpha (\mathbf{a} \times \mathbf{b}) \cdot \mathbf{c}\,. \tag{3.3.16}$$

The proof is finished by recognizing that one definition of the determinant of $\mathbf{F}$ is

$$\alpha = \det \mathbf{F} = \frac{(\mathbf{Fa} \times \mathbf{Fb}) \cdot \mathbf{Fc}}{(\mathbf{a} \times \mathbf{b}) \cdot \mathbf{c}}\,, \tag{3.3.17}$$

for any set of linearly independent vectors $\mathbf{a}$, $\mathbf{b}$ and $\mathbf{c}$. Specifically, using the rectangular Cartesian base vectors $\mathbf{e}_i$ and taking $\mathbf{a} = \mathbf{e}_1$, $\mathbf{b} = \mathbf{e}_2$ and $\mathbf{c} = \mathbf{e}_3$, it follows that

$$\det \mathbf{F} = (\mathbf{Fe}_1 \times \mathbf{Fe}_2) \cdot \mathbf{Fe}_3\,, \tag{3.3.18}$$

which can be recognized as the scalar triple product of the columns of $\mathbf{F}$.

Now, it will be shown that the determinant J of the deformation gradient $\mathbf{F}$

$$J = \det \mathbf{F}\,, \tag{3.3.19}$$

is a pure measure of dilatation. To this end, consider an elemental material volume defined by the linearly independent material line elements $d\mathbf{X}^1$, $d\mathbf{X}^2$ and $d\mathbf{X}^3$ in the reference configuration and defined by the associated linearly independent material line elements $d\mathbf{x}^1$, $d\mathbf{x}^2$ and $d\mathbf{x}^3$ in the current configuration. Thus, the elemental volumes dV in the reference configuration and dv in the current configuration are given by

$$dV = d\mathbf{X}^1 \times d\mathbf{X}^2 \cdot d\mathbf{X}^3\,, \tag{3.3.20a}$$

$$dv = d\mathbf{x}^1 \times d\mathbf{x}^2 \cdot d\mathbf{x}^3\,. \tag{3.3.20b}$$

Since (3.3.1a) defines the mapping of each material line element from the reference configuration to the current configuration, it follows that

$$\begin{aligned} dv &= \mathbf{F}d\mathbf{X}^1 \times \mathbf{F}d\mathbf{X}^2 \cdot \mathbf{F}d\mathbf{X}^3 = J\mathbf{F}^{-T}(d\mathbf{X}^1 \times d\mathbf{X}^2) \cdot \mathbf{F}d\mathbf{X}^3\,, \\ dv &= J(d\mathbf{X}^1 \times d\mathbf{X}^2) \cdot \mathbf{F}^{-1}\mathbf{F}d\mathbf{X}^3 = J d\mathbf{X}^1 \times d\mathbf{X}^2 \cdot d\mathbf{X}^3\,, \\ dv &= J dV\,. \end{aligned} \tag{3.3.21}$$

This means that J is a pure measure of dilatation. It also follows from (3.3.4b) and (3.3.19) that

$$J^2 = \det \mathbf{C}\,. \tag{3.3.22}$$

Pure Measures of Distortion (Shape Change)

In general, the deformation gradient $\mathbf{F}$ characterizes the dilatation (volume change), distortion (shape change) and the orientation of a material region. Therefore, whenever $\mathbf{F}$ is a unimodular tensor (i.e., its determinant J equals unity), $\mathbf{F}$ is a measure of distortion and orientation. Using this idea, which originated with Flory [1], it is possible to separate $\mathbf{F}$ into its dilatational part $J^{1/3}\mathbf{I}$ and its distortional part $\mathbf{F}'$ such that

$$\mathbf{F} = (J^{1/3}\mathbf{I})\mathbf{F}' = J^{1/3}\mathbf{F}'\,, \quad \mathbf{F}' = J^{-1/3}\mathbf{F}\,, \quad \det \mathbf{F}' = 1\,. \tag{3.3.23}$$

Note that since $\mathbf{F}'$ is unimodular (3.3.23), it is a pure measure of distortion and orientation. Similarly, the deformation tensor $\mathbf{C}$ can be separated into its dilatational part $J^{2/3}\mathbf{I}$ and its distortional part $\mathbf{C}'$ such that

$$\mathbf{C} = (J^{2/3}\mathbf{I})\mathbf{C}' = J^{2/3}\mathbf{C}'\,, \quad \mathbf{C}' = J^{-2/3}\mathbf{C}\,, \quad \det \mathbf{C}' = 1\,, \tag{3.3.24}$$

in contrast to $\mathbf{F}'$, $\mathbf{C}'$ is a pure measure of distortional deformation only.

Strain Measures

Using (3.3.4a) and (3.3.6a), it follows that the change in length of a material line element can be expressed in the following forms

$$ds^2 - dS^2 = d\mathbf{X} \cdot (\mathbf{C} - \mathbf{I})d\mathbf{X} = d\mathbf{X} \cdot (2\mathbf{E})d\mathbf{X} = 2\mathbf{E} \cdot (d\mathbf{X} \otimes d\mathbf{X}) ,$$
$$ds^2 - dS^2 = dX_A(C_{AB} - \delta_{AB})dX_B = dX_A(2E_{AB})dX_B , \tag{3.3.25a}$$
$$ds^2 - dS^2 = d\mathbf{x} \cdot (\mathbf{I} - \mathbf{c})d\mathbf{x} = d\mathbf{x} \cdot (2\mathbf{e})d\mathbf{x} = 2\mathbf{e} \cdot (d\mathbf{x} \otimes d\mathbf{x}) ,$$
$$ds^2 - dS^2 = dx_i(\delta_{ij} - c_{ij})dx_j = dx_i(2e_{ij})dx_j , \tag{3.3.25b}$$

where the Lagrangian strain $\mathbf{E}$ and the Almansi strain $\mathbf{e}$ are defined by

$$2\mathbf{E} = \mathbf{C} - \mathbf{I} , \tag{3.3.26a}$$
$$2\mathbf{e} = \mathbf{I} - \mathbf{c} . \tag{3.3.26b}$$

Furthermore, in view of the separation (3.3.24) it is sometimes convenient to define a scalar measure of dilatational strain E_v and a tensorial measure of distortional strain $\mathbf{E}'$ by

$$2E_v = J^2 - 1 , \quad 2\mathbf{E}' = \mathbf{C}' - \mathbf{I} . \tag{3.3.27}$$

Eigenvalues of **C** *and* **B**

The notions of eigenvalues, eigenvectors and the principal invariants of a tensor are briefly reviewed in Appendix A. Using the definitions (3.3.4b), (3.3.7) and (A.1.3) it is first shown that the principal invariants of **C** and **B** are equal. To this end, use is made of the properties of the dot product given by (2.6.19) to deduce that

$$\mathbf{C} \cdot \mathbf{I} = \mathbf{F}^T\mathbf{F} \cdot \mathbf{I} = \mathbf{F} \cdot \mathbf{F} = \mathbf{F}\mathbf{F}^T \cdot \mathbf{I} = \mathbf{B} \cdot \mathbf{I} ,$$
$$\mathbf{C} \cdot \mathbf{C} = \mathbf{F}^T\mathbf{F} \cdot \mathbf{F}^T\mathbf{F} = \mathbf{F} \cdot \mathbf{F}\mathbf{F}^T\mathbf{F} = \mathbf{F}\mathbf{F}^T \cdot \mathbf{F}\mathbf{F}^T = \mathbf{B} \cdot \mathbf{B} , \tag{3.3.28}$$
$$\det \mathbf{C} = \det(\mathbf{F}^T\mathbf{F}) = \det \mathbf{F}^T \det \mathbf{F} = (\det \mathbf{F})^2 = \det(\mathbf{F}\mathbf{F}^T) = \det \mathbf{B} .$$

It follows from (A.1.3) that the principal invariants of **C** and **B** are equal

$$I_1(\mathbf{C}) = I_1(\mathbf{B}) , \quad I_2(\mathbf{C}) = I_2(\mathbf{B}) , \quad I_3(\mathbf{C}) = I_3(\mathbf{B}) . \tag{3.3.29}$$

Furthermore, using (3.3.12b) it can be seen that the eigenvalues of **C** are also the squares of the principal values of stretch λ, which are determined by the characteristic equation

$$\det(\mathbf{C} - \lambda^2 \mathbf{I}) = -\lambda^6 + \lambda^4 I_1(\mathbf{C}) - \lambda^2 I_2(\mathbf{C}) + I_3(\mathbf{C}) = \det(\mathbf{B} - \lambda^2 \mathbf{I}) = 0 . \tag{3.3.30}$$

Displacement Vector

The displacement vector **u** is the vector that connects the position **X** of a material point in the reference configuration to its position **x** in the current configuration so that

$$\mathbf{u} = \mathbf{x} - \mathbf{X} , \quad \mathbf{x} = \mathbf{X} + \mathbf{u} ,$$
$$\mathbf{X} = \mathbf{x} - \mathbf{u} , \quad \mathbf{u} = u_A\mathbf{e}_A = u_i\mathbf{e}_i . \tag{3.3.31}$$

It then follows from the definition (3.3.1c) of the deformation gradient **F** that

$$\mathbf{F} = \partial\mathbf{x}/\partial\mathbf{X} = \partial(\mathbf{X}+\mathbf{u})/\partial\mathbf{X} = \mathbf{I} + \partial\hat{\mathbf{u}}/\partial\mathbf{X}\,, \tag{3.3.32a}$$

$$\mathbf{F}^{-1} = \partial\mathbf{X}/\partial\mathbf{x} = \partial(\mathbf{x}-\mathbf{u})/\partial\mathbf{x} = \mathbf{I} - \partial\tilde{\mathbf{u}}/\partial\mathbf{x}\,, \tag{3.3.32b}$$

$$\begin{aligned} \mathbf{C} &= \mathbf{F}^T\mathbf{F} = (\mathbf{I} + \partial\hat{\mathbf{u}}/\partial\mathbf{X})^T(\mathbf{I} + \partial\hat{\mathbf{u}}/\partial\mathbf{X})\,,\\ &= \mathbf{I} + \partial\hat{\mathbf{u}}/\partial\mathbf{X} + (\partial\hat{\mathbf{u}}/\partial\mathbf{X})^T + (\partial\hat{\mathbf{u}}/\partial\mathbf{X})^T(\partial\hat{\mathbf{u}}/\partial\mathbf{X})\,,\\ C_{AB} &= \delta_{AB} + \hat{u}_{A},_{B} + \hat{u}_{B},_{A} + \hat{u}_{M},_{A}\,\hat{u}_{M},_{B}\,, \end{aligned} \tag{3.3.32c}$$

$$\begin{aligned} \mathbf{c} &= \mathbf{B}^{-1} = \mathbf{F}^{-T}\mathbf{F}^{-1} = (\mathbf{I} - \partial\tilde{\mathbf{u}}/\partial\mathbf{x})^T(\mathbf{I} - \partial\tilde{\mathbf{u}}/\partial\mathbf{x})\,,\\ &= \mathbf{I} - \partial\tilde{\mathbf{u}}/\partial\mathbf{x} - (\partial\tilde{\mathbf{u}}/\partial\mathbf{x})^T + (\partial\tilde{\mathbf{u}}/\partial\mathbf{x})^T(\partial\tilde{\mathbf{u}}/\partial\mathbf{x})\,,\\ c_{ij} &= \delta_{ij} - \tilde{u}_{i},_{j} - \tilde{u}_{j},_{i} + \tilde{u}_{m},_{i}\,\tilde{u}_{m},_{j}\,. \end{aligned} \tag{3.3.32d}$$

Then, with the help of the definitions (3.3.26a) and (3.3.26b), the strains **E** and **e** can be expressed in terms of the displacement gradients $\partial\hat{\mathbf{u}}/\partial\mathbf{X}$ and $\partial\tilde{\mathbf{u}}/\partial\mathbf{x}$ by

$$\begin{aligned} \mathbf{E} &= \frac{1}{2}[\partial\hat{\mathbf{u}}/\partial\mathbf{X} + (\partial\hat{\mathbf{u}}/\partial\mathbf{X})^T + (\partial\hat{\mathbf{u}}/\partial\mathbf{X})^T(\partial\hat{\mathbf{u}}/\partial\mathbf{X})] = E_{AB}\mathbf{e}_A \otimes \mathbf{e}_B\,,\\ E_{AB} &= \frac{1}{2}(\hat{u}_{A},_{B} + \hat{u}_{B},_{A} + \hat{u}_{M},_{A}\,\hat{u}_{M},_{B})\,, \end{aligned} \tag{3.3.33a}$$

$$\begin{aligned} \mathbf{e} &= \frac{1}{2}[\partial\tilde{\mathbf{u}}/\partial\mathbf{x} + (\partial\tilde{\mathbf{u}}/\partial\mathbf{x})^T - (\partial\tilde{\mathbf{u}}/\partial\mathbf{x})^T(\partial\tilde{\mathbf{u}}/\partial\mathbf{x})] = e_{ij}\mathbf{e}_i \otimes \mathbf{e}_j\,,\\ e_{ij} &= \frac{1}{2}(\tilde{u}_{i},_{j} + \tilde{u}_{j},_{i} - \tilde{u}_{m},_{i}\,\tilde{u}_{m},_{j})\,. \end{aligned} \tag{3.3.33b}$$

Since these expressions have been obtained without any approximation they are exact and are sometimes referred to as finite strain measures. Notice the different signs in front of the quadratic terms in the displacement gradients appearing in the expressions (3.3.33a) and (3.3.33b).

Material Area Element

The material area element dA formed by the elemental parallelogram associated with the linearly independent material line elements $d\mathbf{X}^1$ and $d\mathbf{X}^2$ in the reference configuration, and the material area element da formed by the corresponding linearly independent material line elements $d\mathbf{x}^1$ and $d\mathbf{x}^2$ in the current configuration are given by

$$\mathbf{N}dA = d\mathbf{X}^1 \otimes d\mathbf{X}^2\,, \quad \mathbf{n}da = \mathbf{F}d\mathbf{X}^1 \otimes \mathbf{F}d\mathbf{X}^2\,, \tag{3.3.34}$$

where **N** and **n** are the unit vectors normal to the material surfaces defined by $d\mathbf{X}^1$, $d\mathbf{X}^2$ and $d\mathbf{x}^1$, $d\mathbf{x}^2$, respectively. It follows from (3.3.1a) and (3.3.13) that

$$\mathbf{n}da = \mathbf{F}d\mathbf{X}^1 \otimes \mathbf{F}d\mathbf{X}^2 = J\mathbf{F}^{-T}(d\mathbf{X}^1 \otimes d\mathbf{X}^2) = J\mathbf{F}^{-T}\mathbf{N}dA\,, \tag{3.3.35}$$

which is called Nanson's formula. It is important to emphasize that a material line element that was normal to the material surface in the reference configuration does not necessarily remain normal to the same material surface in the current configuration.

3.4 Polar Decomposition Theorem

The polar decomposition theorem states that any invertible second-order tensor $\mathbf{F}$ can be uniquely decomposed into its polar form

$$\mathbf{F} = \mathbf{RU} = \mathbf{VR}\,, \quad F_{iA} = R_{iM}U_{MA} = V_{im}R_{mA}\,, \tag{3.4.1}$$

where $\mathbf{R}$ is an orthogonal tensor

$$\begin{aligned} &\mathbf{R} = R_{iA}\mathbf{e}_i \otimes \mathbf{e}_A\,, \\ &\mathbf{R}^T\mathbf{R} = \mathbf{I}\,, \qquad R_{mA}R_{mB} = \delta_{AB}\,, \\ &\mathbf{R}\mathbf{R}^T = \mathbf{I}\,, \qquad R_{iA}R_{jA} = \delta_{ij}\,, \end{aligned} \tag{3.4.2}$$

$\mathbf{U}$ is the right stretch tensor and $\mathbf{V}$ is the left stretch tensor. These stretch tensors are symmetric, positive-definite tensors so that for an arbitrary vector $\mathbf{v}$, it follows that

$$\begin{aligned} &\mathbf{U}^T = \mathbf{U} = U_{AB}\mathbf{e}_A \otimes \mathbf{e}_B\,, \quad U_{BA} = U_{AB}\,, \\ &\mathbf{v}\cdot\mathbf{U}\mathbf{v} = \mathbf{U}\cdot\mathbf{v}\otimes\mathbf{v} > 0\,, \quad v_A U_{AB} v_B = U_{AB}v_A v_B > 0 \text{ for } \mathbf{v} \neq 0\,, \\ &\mathbf{V}^T = \mathbf{V} = V_{ij}\mathbf{e}_i \otimes \mathbf{e}_j\,, \quad V_{ji} = V_{ij}\,, \\ &\mathbf{v}\cdot\mathbf{V}\mathbf{v} = \mathbf{V}\cdot\mathbf{v}\otimes\mathbf{v} > 0\,, \quad v_i V_{ij} v_j = V_{ij}v_1 v_j > 0 \text{ for } \mathbf{v} \neq 0\,. \end{aligned} \tag{3.4.3}$$

From these expressions and the definitions (3.3.4b) and (3.3.7) it can be deduced that

$$\mathbf{C} = \mathbf{U}^2\,, \quad \mathbf{B} = \mathbf{V}^2\,, \tag{3.4.4}$$

which explains why $\mathbf{C}$ is called the right Cauchy–Green deformation tensor and $\mathbf{B}$ is called the left Cauchy–Green deformation tensor.

To prove this theorem it is convenient to first consider the following Lemma.

Lemma *If* $\mathbf{S}$ *is an invertible second-order tensor then* $\mathbf{S}^T\mathbf{S}$ *and* $\mathbf{S}\mathbf{S}^T$ *are positive-definite symmetric tensors.*

Proof (i) Consider two vectors $\mathbf{v}$ and $\mathbf{w}$ defined by

$$\mathbf{w} = \mathbf{S}\mathbf{v}\,, \quad w_i = S_{ij}v_j\,. \tag{3.4.5}$$

Since $\mathbf{S}$ is invertible, it follows that

$$\begin{aligned} &\mathbf{w} = 0 \quad \text{if and only if} \quad \mathbf{v} = 0\,, \\ &\mathbf{w} \neq 0 \quad \text{if and only if} \quad \mathbf{v} \neq 0\,. \end{aligned} \tag{3.4.6}$$

Now, consider

$$\mathbf{w} \cdot \mathbf{w} = \mathbf{S}\mathbf{v} \cdot \mathbf{S}\mathbf{v} = \mathbf{v} \cdot \mathbf{S}^T\mathbf{S}\mathbf{v} = \mathbf{S}^T\mathbf{S} \cdot (\mathbf{v} \otimes \mathbf{v}) \,,$$

$$w_m w_m = S_{mi} v_i S_{mj} v_j = v_i (S^T_{im} S_{mj}) v_j \,. \tag{3.4.7}$$

Since $\mathbf{w} \cdot \mathbf{w} > 0$ whenever $\mathbf{v} \neq 0$, it follows that $\mathbf{S}^T\mathbf{S}$ is positive-definite.

(ii) Alternatively, define the two vectors $\mathbf{v}$ and $\mathbf{w}$ by

$$\mathbf{w} = \mathbf{S}^T\mathbf{v} \,, \quad w_i = S^T_{ij} v_j = S_{ji} v_j \,. \tag{3.4.8}$$

Since $\mathbf{S}$ is invertible, it follows that

$$\mathbf{w} \cdot \mathbf{w} = \mathbf{S}^T\mathbf{v} \cdot \mathbf{S}^T\mathbf{v} = \mathbf{v} \cdot \mathbf{S}\mathbf{S}^T\mathbf{v} = \mathbf{S}\mathbf{S}^T \cdot (\mathbf{v} \otimes \mathbf{v}) \,,$$

$$w_m w_m = S_{im} v_i S_{jm} v_j = v_i (S_{im} S^T_{mj}) v_j \,. \tag{3.4.9}$$

Moreover, since $\mathbf{w} \cdot \mathbf{w} > 0$ whenever $\mathbf{v} \neq 0$ the tensor $\mathbf{S}^T\mathbf{S}$ is positive-definite.

To prove the polar decomposition theorem it is convenient to first prove existence of the forms $\mathbf{F} = \mathbf{R}\mathbf{U}$ and $\mathbf{F} = \mathbf{V}\mathbf{R}$ and then prove uniqueness of the quantities $\mathbf{R}$, $\mathbf{U}$ and $\mathbf{V}$.

Existence

(i) Since $\mathbf{F}$ is invertible the tensor $\mathbf{F}^T\mathbf{F}$ is symmetric and positive-definite so there exists a unique symmetric positive-definite square root $\mathbf{U}$ defined by

$$\mathbf{U} = (\mathbf{F}^T\mathbf{F})^{1/2} \,, \quad \mathbf{U}^2 = \mathbf{F}^T\mathbf{F} \,, \quad U_{AM} U_{MB} = F_{mA} F_{mB} \,. \tag{3.4.10}$$

Now, let $\mathbf{R}_1$ be defined by

$$\mathbf{R}_1 = \mathbf{F}\mathbf{U}^{-1} \,, \quad \mathbf{F} = \mathbf{R}_1\mathbf{U} \,. \tag{3.4.11}$$

To prove that $\mathbf{R}_1$ is an orthogonal tensor consider

$$\mathbf{R}_1\mathbf{R}_1^T = \mathbf{F}\mathbf{U}^{-1}(\mathbf{F}\mathbf{U}^{-1})^T = \mathbf{F}\mathbf{U}^{-1}\mathbf{U}^{-T}\mathbf{F}^{-T} = \mathbf{F}(\mathbf{U}^2)^{-1}\mathbf{F}^T \,,$$

$$= \mathbf{F}(\mathbf{F}^T\mathbf{F})^{-1}\mathbf{F}^T = \mathbf{F}(\mathbf{F}^{-1}\mathbf{F}^{-T})\mathbf{F}^T = \mathbf{I} \,, \tag{3.4.12a}$$

$$\mathbf{R}_1^T\mathbf{R}_1 = \mathbf{U}^{-T}\mathbf{F}^T\mathbf{F}\mathbf{U}^{-1} = \mathbf{U}^{-1}\mathbf{U}^2\mathbf{U}^{-1} = \mathbf{I} \,. \tag{3.4.12b}$$

(ii) Similarly, since $\mathbf{F}$ is invertible the tensor $\mathbf{F}\mathbf{F}^T$ is symmetric and positive-definite there exists a unique symmetric, positive-definite square root $\mathbf{V}$

$$\mathbf{V} = (\mathbf{F}\mathbf{F}^T)^{1/2} \,, \quad \mathbf{V}^2 = \mathbf{F}\mathbf{F}^T \,, \quad V_{im} V_{mj} = F_{iM} F_{jM} \,. \tag{3.4.13}$$

Now, let $\mathbf{R}_2$ be defined by

$$\mathbf{R}_2 = \mathbf{V}^{-1}\mathbf{F}\,, \quad \mathbf{F} = \mathbf{V}\mathbf{R}_2\,. \tag{3.4.14}$$

To prove that $\mathbf{R}_2$ is an orthogonal tensor consider

$$\mathbf{R}_2\mathbf{R}_2^T = \mathbf{V}^{-1}\mathbf{F}(\mathbf{V}^{-1}\mathbf{F})^T = \mathbf{V}^{-1}\mathbf{F}\mathbf{F}^T\mathbf{V}^{-1} = \mathbf{V}^{-1}\mathbf{V}^2\mathbf{V}^{-1} = \mathbf{I}\,,$$
$$\mathbf{R}_2^T\mathbf{R}_2 = \mathbf{F}^T\mathbf{V}^{-T}\mathbf{V}^{-1}\mathbf{F} = \mathbf{F}^T\mathbf{V}^{-2}\mathbf{F} = \mathbf{F}^T(\mathbf{F}\mathbf{F}^T)^{-1}\mathbf{F} = \mathbf{I}\,. \tag{3.4.15a}$$

Uniqueness

(i) Assume that $\mathbf{R}_1$ and $\mathbf{U}$ are not unique so that

$$\mathbf{F} = \mathbf{R}_1\mathbf{U} = \mathbf{R}_1^*\mathbf{U}^*\,. \tag{3.4.16}$$

Then consider

$$\mathbf{F}^T\mathbf{F} = \mathbf{U}^2 = (\mathbf{R}_1^*\mathbf{U}^*)^T(\mathbf{R}_1^*\mathbf{U}^*) = \mathbf{U}^{*T}\mathbf{R}_1^{*T}\mathbf{R}_1^*\mathbf{U}^* = \mathbf{U}^{*2}\,. \tag{3.4.17}$$

However, since $\mathbf{U}$ and $\mathbf{U}^*$ are both symmetric and positive-definite it can be deduced that $\mathbf{U}$ is unique

$$\mathbf{U}^* = \mathbf{U}\,. \tag{3.4.18}$$

Next, substituting (3.4.18) into (3.4.16) yields

$$\mathbf{R}_1\mathbf{U} = \mathbf{R}_1^*\mathbf{U}\,. \tag{3.4.19}$$

Then, multiplication of (3.4.19) on the right by $\mathbf{U}^{-1}$ proves that $\mathbf{R}_1$ is unique

$$\mathbf{R}_1 = \mathbf{R}_1^*\,. \tag{3.4.20}$$

(ii) Similarly, assume that $\mathbf{R}_2$ and $\mathbf{V}$ are not unique so that

$$\mathbf{F} = \mathbf{V}\mathbf{R}_2 = \mathbf{V}^*\mathbf{R}_2^*\,. \tag{3.4.21}$$

Then, consider

$$\mathbf{F}\mathbf{F}^T = \mathbf{V}^2 = (\mathbf{V}^*\mathbf{R}_2^*)(\mathbf{V}^*\mathbf{R}_2^*)^T = \mathbf{V}^*\mathbf{R}_2^*\mathbf{R}_2^{*T}\mathbf{V}^{*T} = \mathbf{V}^{*2}\,. \tag{3.4.22}$$

However, since $\mathbf{V}$ and $\mathbf{V}^*$ are both symmetric and positive-definite it can be deduced that $\mathbf{V}$ is unique

$$\mathbf{V}^* = \mathbf{V}\,. \tag{3.4.23}$$

Next, substituting (3.4.23) into (3.4.21) yields

$$\mathbf{VR}_2 = \mathbf{VR}_2^* . \tag{3.4.24}$$

Then, multiplication of (3.4.24) on the left by $\mathbf{V}^{-1}$ proves that $\mathbf{R}_2$ is unique

$$\mathbf{R}_2 = \mathbf{R}_2^* . \tag{3.4.25}$$

(iii) Finally, it is necessary to prove that $\mathbf{R}_1 = \mathbf{R}_2 = \mathbf{R}$. To this end, define the auxiliary tensor $\mathbf{A}$ by

$$\mathbf{A} = \mathbf{R}_1\mathbf{U}\mathbf{R}_1^T = \mathbf{F}\mathbf{R}_1^T . \tag{3.4.26}$$

Clearly, $\mathbf{A}$ is symmetric so that

$$\mathbf{A}^2 = \mathbf{A}\mathbf{A}^T = \mathbf{F}\mathbf{R}_1^T(\mathbf{F}\mathbf{R}_1^T)^T = \mathbf{F}\mathbf{R}_1^T\mathbf{R}_1\mathbf{F}^T = \mathbf{F}\mathbf{F}^T = \mathbf{V}^2 . \tag{3.4.27}$$

Since $\mathbf{A}$ and $\mathbf{V}$ are symmetric and nonsingular, it follows with the help of (3.4.14) and (3.4.26) that

$$\mathbf{V} = \mathbf{A} = \mathbf{F}\mathbf{R}_1^T = \mathbf{V}\mathbf{R}_2\mathbf{R}_1^T . \tag{3.4.28}$$

Now, multiplying (3.4.28) on the left by $\mathbf{V}^{-1}$ and on the right by $\mathbf{R}_1$, it follows that

$$\mathbf{R}_1 = \mathbf{R}_2 = \mathbf{R} , \tag{3.4.29}$$

which completes the proof.

Example As an example, consider the simple deformation field for which $\mathbf{F}$ is given by

$$\mathbf{F} = F_{11}\mathbf{e}_1 \otimes \mathbf{e}_1 + F_{12}\mathbf{e}_1 \otimes \mathbf{e}_2 + F_{21}\mathbf{e}_2 \otimes \mathbf{e}_1 + F_{22}\mathbf{e}_2 \otimes \mathbf{e}_2 + F_{33}\mathbf{e}_3 \otimes \mathbf{e}_3 . \tag{3.4.30}$$

For this deformation field the rotation tensor $\mathbf{R}$ can be written in the form

$$\mathbf{R} = \cos\gamma(\mathbf{e}_1 \otimes \mathbf{e}_1 + \mathbf{e}_2 \otimes \mathbf{e}_2) + \sin\gamma(\mathbf{e}_1 \otimes \mathbf{e}_2 - \mathbf{e}_2 \otimes \mathbf{e}_1) + \mathbf{e}_3 \otimes \mathbf{e}_3 , \tag{3.4.31}$$

where the angle γ is determined by requiring $\mathbf{U} = \mathbf{R}^T\mathbf{F}$ to be a symmetric tensor

$$\gamma = \tan^{-1}\left(\frac{F_{12} - F_{21}}{F_{11} + F_{22}}\right) . \tag{3.4.32}$$

It then follows that $\mathbf{R}$, $\mathbf{U}$ and $\mathbf{V}$ for this deformation can be expressed in the forms

$$\mathbf{R} = \frac{1}{\sqrt{(F_{11}+F_{22})^2+(F_{12}-F_{21})^2}}\Big[(F_{11}+F_{22})(\mathbf{e}_1\otimes\mathbf{e}_1+\mathbf{e}_2\otimes\mathbf{e}_2)$$
$$+(F_{12}-F_{21})(\mathbf{e}_1\otimes\mathbf{e}_2-\mathbf{e}_2\otimes\mathbf{e}_1)\Big]+\mathbf{e}_3\otimes\mathbf{e}_3\,,$$
$$\mathbf{U} = \frac{1}{\sqrt{(F_{11}+F_{22})^2+(F_{12}-F_{21})^2}}\Big[\big\{F_{11}(F_{11}+F_{22})-F_{21}(F_{12}-F_{21})\big\}\mathbf{e}_1\otimes\mathbf{e}_1$$
$$+\big\{F_{22}(F_{11}+F_{22})+F_{12}(F_{12}-F_{21})\big\}\mathbf{e}_2\otimes\mathbf{e}_2$$
$$+(F_{11}F_{12}+F_{22}F_{21})(\mathbf{e}_1\otimes\mathbf{e}_2+\mathbf{e}_2\otimes\mathbf{e}_1)\Big]+F_{33}\mathbf{e}_3\otimes\mathbf{e}_3\,,$$
$$\mathbf{V} = \frac{1}{\sqrt{(F_{11}+F_{22})^2+(F_{12}-F_{21})^2}}\Big[\big\{F_{11}(F_{11}+F_{22})+F_{12}(F_{12}-F_{21})\big\}\mathbf{e}_1\otimes\mathbf{e}_1$$
$$+\big\{F_{22}(F_{11}+F_{22})-F_{21}(F_{12}-F_{21})\big\}\mathbf{e}_2\otimes\mathbf{e}_2$$
$$+(F_{11}F_{21}+F_{22}F_{12})(\mathbf{e}_1\otimes\mathbf{e}_2+\mathbf{e}_2\otimes\mathbf{e}_1)\Big]+F_{33}\mathbf{e}_3\otimes\mathbf{e}_3\,. \tag{3.4.33}$$

Physical Interpetation

To explain the physical interpretation of the polar decomposition theorem recall from (3.3.1a) that a material line element $d\mathbf{X}$ in the reference configuration is transformed by $\mathbf{F}$ into the material line element $d\mathbf{x}$ in the current configuration and define the elemental vectors $d\mathbf{X}'$ and $d\mathbf{x}'$ such that

$$d\mathbf{x} = \mathbf{RU}d\mathbf{X} \quad \Rightarrow \quad d\mathbf{X}' = \mathbf{U}d\mathbf{X}\,, \quad d\mathbf{x} = \mathbf{R}d\mathbf{X}'\,,$$
$$dx_i = R_{iA}U_{AB}dX_B \quad \Rightarrow \quad dX'_A = U_{AB}dX_B\,, \quad dx_i = R_{iA}dX'_A\,, \tag{3.4.34a}$$
$$d\mathbf{x} = \mathbf{VR}d\mathbf{X} \quad \Rightarrow \quad d\mathbf{x}' = \mathbf{R}d\mathbf{X}\,, \quad d\mathbf{x} = \mathbf{V}d\mathbf{x}'\,,$$
$$dx_i = V_{ij}R_{jB}dX_B \quad \Rightarrow \quad dx'_j = R_{jB}dX_B\,, \quad dx_i = V_{ij}dx'_j\,. \tag{3.4.34b}$$

In general, a material line element experiences both stretch and rotation as it deforms from $d\mathbf{X}$ to $d\mathbf{x}$. However, the polar decomposition theorem indicates that part of the deformation can be described as a pure rotation. To see this, use (3.3.4a) together with (3.4.34a) and (3.4.34b) and consider

$$ds^2 = d\mathbf{x}\cdot d\mathbf{x} = \mathbf{R}d\mathbf{X}'\cdot\mathbf{R}d\mathbf{X}' = d\mathbf{X}'\cdot\mathbf{R}^T\mathbf{R}d\mathbf{X}' = d\mathbf{X}'\cdot d\mathbf{X}'\,. \tag{3.4.35}$$

It follows that the magnitude of $d\mathbf{X}'$ is the same as that of $d\mathbf{x}$ so that all the stretching occurs during the transformation from $d\mathbf{X}$ to $d\mathbf{X}'$ and that the transformation from $d\mathbf{X}'$ to $d\mathbf{x}$ is a pure rotation. Similarly, with the help of (3.3.6a) and (3.4.34b) it can be shown that

$$d\mathbf{x}'\cdot d\mathbf{x}' = \mathbf{R}d\mathbf{X}\cdot\mathbf{R}d\mathbf{X} = d\mathbf{X}\cdot\mathbf{R}^T\mathbf{R}d\mathbf{X} = d\mathbf{X}\cdot d\mathbf{X} = dS^2\,. \tag{3.4.36}$$

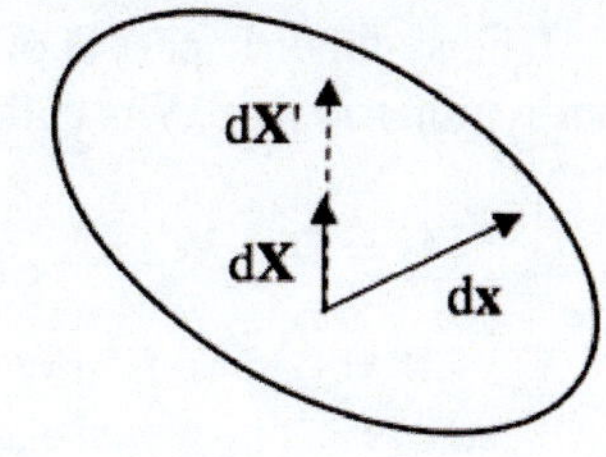

Fig. 3.2 Pure stretching followed by pure rotation: $\mathbf{F} = \mathbf{RU}$; $d\mathbf{X}' = \mathbf{U}d\mathbf{X} = \lambda d\mathbf{X}$; $d\mathbf{x} = \mathbf{R}d\mathbf{X}'$

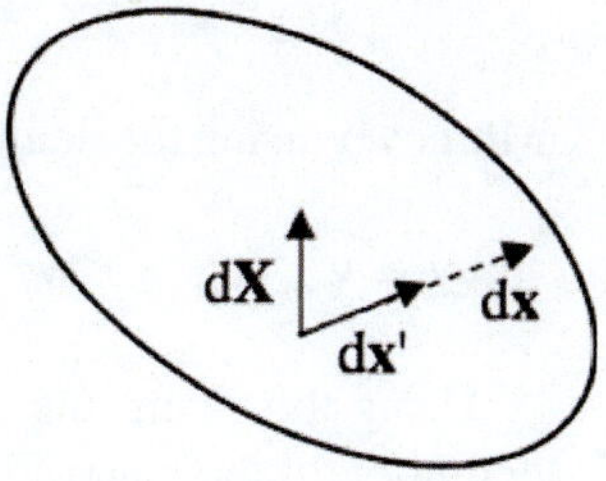

Fig. 3.3 Pure rotation followed by pure stretching: $\mathbf{F} = \mathbf{VR}$; $d\mathbf{x}' = \mathbf{R}d\mathbf{X}$; $d\mathbf{x} = \mathbf{V}d\mathbf{x}' = \lambda d\mathbf{x}'$

Consequently, it follows that the magnitude of $d\mathbf{x}'$ is the same as that of $d\mathbf{X}$ so that all the stretching occurs during the transformation from $d\mathbf{x}'$ to $d\mathbf{x}$ and that the transformation from $d\mathbf{X}$ to $d\mathbf{x}'$ is a pure rotation.

Although the transformations from $d\mathbf{X}$ to $d\mathbf{X}'$ and from $d\mathbf{x}'$ to $d\mathbf{x}$ contain all of the stretching, they also tend to rotate a general line element. However, the special line element $d\mathbf{X}$ which is parallel to any of the three principal directions of $\mathbf{U}$ transforms $d\mathbf{X}$ to $d\mathbf{X}'$ as a pure stretch without rotation (see Fig. 3.2) because

$$d\mathbf{X}' = \mathbf{U}d\mathbf{X} = \lambda d\mathbf{X}\,, \tag{3.4.37}$$

where λ is the stretch defined by (3.3.9). It then follows that for this line element

$$\begin{aligned} d\mathbf{x} &= \mathbf{F}d\mathbf{X} = \mathbf{RU}d\mathbf{X} = \mathbf{R}\lambda d\mathbf{X} = \lambda d\mathbf{x}'\,, \\ d\mathbf{x} &= \mathbf{F}d\mathbf{X} = \mathbf{VR}d\mathbf{X} = \mathbf{V}d\mathbf{x}' = \lambda d\mathbf{x}'\,, \end{aligned} \tag{3.4.38}$$

so that $d\mathbf{x}'$ is also parallel to a principal direction of $\mathbf{V}$, which means that the transformation from $d\mathbf{x}'$ to $d\mathbf{x}$ is a pure stretch without rotation (see Fig. 3.3). This also means that the rotation tensor $\mathbf{R}$ describes the complete rotation of material line elements which are either parallel to principal directions of $\mathbf{U}$ in the reference configuration or parallel to principal directions of $\mathbf{V}$ in the current configuration.

3.5 Velocity Gradient and Rate of Deformation Tensors

The gradient of the velocity $\mathbf{v}$ with respect to the present position $\mathbf{x}$ is denoted by $\mathbf{L}$ and is defined by

$$\mathbf{L} = \partial\mathbf{v}/\partial\mathbf{x}\,, \quad L_{ij} = \frac{\partial v_i}{\partial x_j} = v_{i,j}\,. \tag{3.5.1}$$

The symmetric part **D** of **L** is called the rate of deformation tensor, while its skew-symmetric part **W** is called the spin tensor, which are defined by

$$\mathbf{L} = \mathbf{D} + \mathbf{W}\,, \qquad L_{ij} = v_i,_j = D_{ij} + W_{ij}\,, \tag{3.5.2a}$$

$$\mathbf{D} = \frac{1}{2}(\mathbf{L} + \mathbf{L}^T) = \mathbf{D}^T\,, \qquad D_{ij} = \frac{1}{2}(v_i,_j + v_j,_i) = D_{ji}\,, \tag{3.5.2b}$$

$$\mathbf{W} = \frac{1}{2}(\mathbf{L} - \mathbf{L}^T) - \mathbf{W}^T\,, \qquad W_{ij} = \frac{1}{2}(v_i,_j - v_j,_i) = -W_{ji}\,. \tag{3.5.2c}$$

Moreover, using the definition (2.5.4) of the divergence operator it can be shown that

$$\mathrm{div}\mathbf{v} = \mathbf{v},_m \cdot \mathbf{e}_m = (\partial \mathbf{v}/\partial \mathbf{x})\,\mathbf{e}_m \cdot \mathbf{e}_m = \mathbf{L}\mathbf{e}_m \cdot \mathbf{e}_m = \mathbf{L} \cdot (\mathbf{e}_m \otimes \mathbf{e}_m) = \mathbf{L} \cdot \mathbf{I} = \mathbf{D} \cdot \mathbf{I}\,. \tag{3.5.3}$$

Using the chain rule of differentiation, the continuity of the derivatives and the definition of the material derivative yields the expressions

$$\dot{\mathbf{F}} = \frac{\partial}{\partial t}(\partial \hat{\mathbf{x}}/\partial \mathbf{X}) = \partial^2 \hat{\mathbf{x}}/\partial \mathbf{X}\partial t = \partial(\partial \hat{\mathbf{x}}/\partial t)/\partial \mathbf{X} = \partial \hat{\mathbf{v}}/\partial \mathbf{X} = (\partial \tilde{\mathbf{v}}/\partial \mathbf{x})(\partial \hat{\mathbf{x}}/\partial \mathbf{X}) = \mathbf{L}\mathbf{F}\,,$$
$$\overline{x_i,_A}^{\,\cdot} = \frac{\partial}{\partial t}(\hat{x}_i,_A) = \frac{\partial^2 \hat{x}_i}{\partial X_A \partial t} = \frac{\partial}{\partial X_A}\left(\frac{\partial \hat{x}_i}{\partial t}\right) = \hat{v}_i,_A = \tilde{v}_i,_m \hat{x}_m,_A\,. \tag{3.5.4}$$

It then follows that the material derivative of **C** can be expressed in the form

$$\dot{\mathbf{C}} = \overline{\mathbf{F}^T\mathbf{F}}^{\,\cdot} = \dot{\mathbf{F}}^T\mathbf{F} + \mathbf{F}^T\dot{\mathbf{F}} = (\mathbf{L}\mathbf{F})^T\mathbf{F} + \mathbf{F}^T(\mathbf{L}\mathbf{F}) = \mathbf{F}^T(\mathbf{L}^T + \mathbf{L})\mathbf{F} = 2\mathbf{F}^T\mathbf{D}\mathbf{F}\,,$$
$$\dot{C}_{AB} = \overline{x_i,_A}^{\,\cdot} x_i,_B + x_i,_A \overline{x_i,_B}^{\,\cdot} = v_i,_m x_m,_A x_i,_B + x_i,_A v_i,_m x_m,_B\,,$$
$$= x_m,_A (v_i,_m + v_m,_i) x_i,_B = 2 x_m,_A D_{mi} x_i,_B\,. \tag{3.5.5}$$

Notice that the direct notation avoids the complications of changing repeated indices.

Furthermore, since the spin tensor **W** is skew-symmetric there exits a unique vector $\boldsymbol{\omega}$, called the axial vector of **W**, such that for any vector **a**

$$\mathbf{W}\mathbf{a} = \boldsymbol{\omega} \times \mathbf{a}\,, \quad W_{ij}a_j = \varepsilon_{ikj}\omega_k a_j = -\varepsilon_{ijk}\omega_j a_k\,. \tag{3.5.6}$$

Since this equation must be true for any vector **a**, and **W** and $\boldsymbol{\omega}$ are independent of **a**, it follows that

$$\mathbf{W} = \boldsymbol{\varepsilon}^T\boldsymbol{\omega} = -\boldsymbol{\varepsilon}\boldsymbol{\omega}\,, \quad W_{ij} = \varepsilon_{ikj}\omega_k = -\varepsilon_{ijk}\omega_k\,. \tag{3.5.7}$$

Multiplying (3.5.7) by ε_{ijm} and using the identity

$$\varepsilon_{ijk}\varepsilon_{ijm} = 2\delta_{km}\,, \tag{3.5.8}$$

it is possible to solve for ω_m in terms of W_{ij} to obtain

$$\boldsymbol{\omega} = -\frac{1}{2}\boldsymbol{\varepsilon} \cdot \mathbf{W}\,, \quad \omega_m = -\frac{1}{2}\varepsilon_{ijm} W_{ij} = -\frac{1}{2}\varepsilon_{mij} W_{ij}\,. \tag{3.5.9}$$

Next, substituting (3.5.2c) into this equation and using (2.5.6) yields

$$\omega_m = -\frac{1}{4}\varepsilon_{mij}(v_{i,j} - v_{j,i}) = -\frac{1}{4}(-\varepsilon_{mij}v_{j,i} - \varepsilon_{mij}v_{j,i}) = \frac{1}{2}\varepsilon_{mij}v_{j,i} = \frac{1}{2}\varepsilon_{mji}v_{i,j}\,,$$
$$\boldsymbol{\omega} = \frac{1}{2}\mathrm{curl}\mathbf{v} = \frac{1}{2}\,\nabla \times \mathbf{v}\,, \tag{3.5.10}$$

where the symbol ∇ denotes the gradient operator

$$\nabla\,\phi = \partial\phi/\partial\mathbf{x} = \phi_{,i}\,\mathbf{e}_i\,. \tag{3.5.11}$$

For later reference, use is made of (3.3.13), (3.3.18) and (3.3.19) to deduce that

$$\dot{J} = \mathbf{F}\mathbf{e}_2 \times \mathbf{F}\mathbf{e}_3 \cdot \dot{\mathbf{F}}\mathbf{e}_1 + \mathbf{F}\mathbf{e}_3 \times \mathbf{F}\mathbf{e}_1 \cdot \dot{\mathbf{F}}\mathbf{e}_2 + \mathbf{F}\mathbf{e}_1 \times \mathbf{F}\mathbf{e}_2 \cdot \dot{\mathbf{F}}\mathbf{e}_3 = J\mathbf{F}^{-T} \cdot \dot{\mathbf{F}}\,. \tag{3.5.12}$$

Next, thinking of J as a function of $\mathbf{F}$ and using the chain rule of differentiation it can be shown that

$$\dot{J} = \frac{\partial J}{\partial \mathbf{F}} \cdot \dot{\mathbf{F}}\,, \tag{3.5.13}$$

so that

$$\left(\frac{\partial J}{\partial \mathbf{F}} - J\mathbf{F}^{-T}\right) \cdot \dot{\mathbf{F}} = 0\,. \tag{3.5.14}$$

Since this equation must be valid for all values of $\mathbf{F}$ and $\dot{\mathbf{F}}$, and the coefficient of $\dot{\mathbf{F}}$ is independent of the rate $\dot{\mathbf{F}}$, it follows that

$$\frac{\partial J}{\partial \mathbf{F}} = J\mathbf{F}^{-T}\,. \tag{3.5.15}$$

This procedure of using the material derivative of a scalar function to determine its derivative respect to its tensorial argument is often easier than differentiating the scalar function directly with respect to its argument. Now, with the help of (3.5.4) it can be shown that

$$\dot{J} = J\,\mathbf{D} \cdot \mathbf{I}\,. \tag{3.5.16}$$

Derivative of a Unimodular Tensor

With the help of (3.3.23) and (3.5.16), it follows that the unimodular tensor $\mathbf{F}'$ satisfies the evolution equation

$$\dot{\mathbf{F}}' = \mathbf{L}''\,\mathbf{F}'\,, \quad \mathbf{L}'' = \mathbf{L} - \frac{1}{3}(\mathbf{L} \cdot \mathbf{I})\,\mathbf{I}\,, \tag{3.5.17}$$

so that $\dot{\mathbf{F}}'$ is orthogonal to $\mathbf{F}'^{-T}$

$$\dot{\mathbf{F}}' \cdot \mathbf{F}'^{-T} = 0\,. \tag{3.5.18}$$

Rates of Stretch and Rotation of a Material Line Element
Using the expression (3.3.1a) and the result (3.5.4) it can be shown that the material derivative of a material line element $d\mathbf{x}$ is given by

$$\dot{d\mathbf{x}} = \mathbf{L}\, d\mathbf{x}\,. \tag{3.5.19}$$

Next, consider a material line elment which in the current configuration has stretch λ and unit direction $\mathbf{s}$. Taking the material derivative of (3.3.12a) and using (3.5.4), it follows that

$$\dot{\lambda}\mathbf{s} + \lambda\dot{\mathbf{s}} = \lambda\mathbf{L}\mathbf{s}\,. \tag{3.5.20}$$

Now, taking the dot product of this equation with $\mathbf{s}$ and using the fact that $\mathbf{s}$ is a unit vector so that $\dot{\mathbf{s}}$ is orthogonal to $\mathbf{s}$ yields an expression for the rate of stretch

$$\frac{\dot{\lambda}}{\lambda} = \mathbf{D} \cdot \mathbf{s} \otimes \mathbf{s}\,. \tag{3.5.21}$$

Then, substituting this result into (3.5.20) yields an equation for the rate of rotation of a material line element

$$\dot{\mathbf{s}} = [\mathbf{L} - (\mathbf{D} \cdot \mathbf{s} \otimes \mathbf{s})\mathbf{I}]\,\mathbf{s}\,. \tag{3.5.22}$$

Rates of Material Area Stretch and Rotation of the Normal to a Material Surface
Consider a material surface with unit normal $\mathbf{n}$ and element of area da in the present configuration. Taking the material derivative of Nanson's formula (3.3.35) and using (3.5.4) and (3.5.16) and the result

$$\dot{\overline{\mathbf{F}^{-1}}} = -\mathbf{F}^{-1}\mathbf{L}\,, \tag{3.5.23}$$

it follows that

$$\dot{\mathbf{n}}da + \mathbf{n}\dot{da} = -\mathbf{L}^T\mathbf{n}da + (\mathbf{D} \cdot \mathbf{I})\,\mathbf{n}da\,. \tag{3.5.24}$$

Next, taking the dot product of this equation with $\mathbf{n}$ and using the fact that $\mathbf{n}$ is a unit vector so that $\dot{\mathbf{n}}$ is orthogonal to $\mathbf{n}$ yields an expression for the rate of material area stretch

$$\frac{\dot{da}}{da} = (\mathbf{I} - \mathbf{n} \otimes \mathbf{n}) \cdot \mathbf{D}\,. \tag{3.5.25}$$

Then, substituting this result into (3.5.24) yields an equation for the rate of rotation of the normal $\mathbf{n}$ to the material surface

$$\dot{\mathbf{n}} = -[\mathbf{L}^T - (\mathbf{D} \cdot \mathbf{n} \otimes \mathbf{n})\,\mathbf{I}]\,\mathbf{n}\,. \tag{3.5.26}$$

3.6 Deformation: Interpretations and Examples

To interpret the various deformation measures, it is recalled from (3.3.9), (3.3.11a) and (3.3.11b) that

$$\begin{aligned} &\lambda \mathbf{s} = \mathbf{F}\,, \qquad \lambda s_i = x_{i,A} S_A\,, \qquad \lambda = \frac{ds}{dS}\,, \\ &\mathbf{s} = \frac{d\mathbf{x}}{ds}\,, \qquad \mathbf{s}\cdot\mathbf{s} = 1\,, \qquad \mathbf{S} = \frac{d\mathbf{X}}{dS}\,, \qquad \mathbf{S}\cdot\mathbf{S} = 1\,, \end{aligned} \tag{3.6.1}$$

where $\mathbf{S}$ is the unit vector in the direction of the material line element $d\mathbf{X}$ of length dS in the reference configuration, $\mathbf{s}$ is the unit vector in the direction of the same material line element $d\mathbf{x}$ of length ds in the current configuration, and λ is the stretch of the material line element. Now, from (3.3.12b) and the definition (3.3.26a) of Lagrangian strain $\mathbf{E}$, it follows that

$$\lambda^2 = \mathbf{C}\cdot(\mathbf{S}\otimes\mathbf{S}) = 1 + 2\mathbf{E}\cdot(\mathbf{S}\otimes\mathbf{S}) = 1 + 2E_{AB}S_A S_B\,. \tag{3.6.2}$$

Also, the extension ε defined in (3.3.10) becomes

$$\varepsilon = \frac{ds - dS}{dS} = \lambda - 1 = \sqrt{1 + 2\mathbf{E}\cdot\mathbf{S}\otimes\mathbf{S}} - 1 = \sqrt{1 + 2E_{AB}S_A S_B} - 1\,. \tag{3.6.3}$$

For the purpose of interpreting the diagonal components of the strain tensor E_{AB}, it is convenient to calculate the extensions ε_1, ε_2 and ε_3 of the material line elements which were parallel to the coordinate axes with base vectors $\mathbf{e}_A$ in the reference configuration

$$\begin{aligned} \varepsilon &= \varepsilon_1 = \sqrt{1 + 2E_{11}} - 1 \quad \text{for} \quad \mathbf{S} = \mathbf{e}_1\,, \\ \varepsilon &= \varepsilon_2 = \sqrt{1 + 2E_{22}} - 1 \quad \text{for} \quad \mathbf{S} = \mathbf{e}_2\,, \\ \varepsilon &= \varepsilon_3 = \sqrt{1 + 2E_{33}} - 1 \quad \text{for} \quad \mathbf{S} = \mathbf{e}_3\,. \end{aligned} \tag{3.6.4}$$

This clearly shows that the diagonal components of the strain tensor are measures of the extensions of material line elements which were parallel to the coordinate directions in the reference configuration.

To interpret the off-diagonal components of the strain tensor E_{AB} as measures of shear, consider two material line elements $d\mathbf{X}$ and $d\bar{\mathbf{X}}$ in the reference configuration which are deformed, respectively, into $d\mathbf{x}$ and $d\bar{\mathbf{x}}$ in the present configuration. Letting $\bar{\mathbf{S}}$ and $d\bar{S}$ and $\bar{\mathbf{s}}$ and $d\bar{s}$ be the directions and lengths of the material line elements $d\bar{\mathbf{X}}$ and $d\bar{\mathbf{x}}$, respectively, it follows from (3.6.1) that

$$\bar{\lambda}\,\bar{\mathbf{s}} = \mathbf{F}\bar{\mathbf{S}}\,, \quad \bar{\lambda} = \frac{d\bar{s}}{d\bar{S}}\,. \tag{3.6.5}$$

Notice that there is no over bar on $\mathbf{F}$ in this equation because (3.6.1) is valid for any material line element, including the particular material line element $d\bar{\mathbf{X}}$. Moreover, it follows that the angle Θ between the undeformed material line elements $d\mathbf{X}$ and $d\bar{\mathbf{X}}$ and the angle θ between the deformed material line elements $d\mathbf{x}$ and $d\bar{\mathbf{x}}$ can be calculated by (see Fig. 3.4)

$$\cos\Theta = \frac{d\mathbf{X}}{dS} \cdot \frac{d\bar{\mathbf{X}}}{d\bar{S}} = \mathbf{S} \cdot \bar{\mathbf{S}}\,, \quad \cos\theta = \frac{d\mathbf{x}}{ds} \cdot \frac{d\bar{\mathbf{x}}}{d\bar{s}} = \mathbf{s} \cdot \bar{\mathbf{s}}\,. \tag{3.6.6}$$

Furthermore, using (3.6.1), it follows that

$$\cos\theta = \frac{\mathbf{C} \cdot \mathbf{S} \otimes \bar{\mathbf{S}}}{\sqrt{\mathbf{C} \cdot \mathbf{S} \otimes \mathbf{S}}\sqrt{\mathbf{C} \cdot \bar{\mathbf{S}} \otimes \bar{\mathbf{S}}}} = \frac{\cos\Theta + 2\mathbf{E} \cdot \mathbf{S} \otimes \bar{\mathbf{S}}}{\sqrt{1 + 2\mathbf{E} \cdot \mathbf{S} \otimes \mathbf{S}}\sqrt{1 + 2\mathbf{E} \cdot \bar{\mathbf{S}} \otimes \bar{\mathbf{S}}}}\,. \tag{3.6.7}$$

Defining the reduction angle ψ between the two material line elements, this equation can be rewritten in the form

$$\begin{gathered} \theta = \Theta - \psi\,, \\ \cos\Theta\cos\psi + \sin\Theta\sin\psi = \frac{\cos\Theta + 2E_{AB}S_A\bar{S}_B}{\sqrt{1 + 2E_{MN}S_MS_N}\sqrt{1 + 2E_{RS}\bar{S}_R\bar{S}_S}}\,. \end{gathered} \tag{3.6.8}$$

Notice that, in general, the reduction angle ψ depends on the reference angle Θ and on all of the components of strain.

As a specific example, consider two material line elements which in the reference configuration were orthogonal and aligned along the coordinate axes such that (see Fig. 3.4)

$$\mathbf{S} = \mathbf{e}_1\,, \quad \bar{\mathbf{S}} = \mathbf{e}_2\,, \quad \Theta = \frac{\pi}{2}\,. \tag{3.6.9}$$

Then, (3.6.8) yields

$$\sin\psi = \frac{2E_{12}}{\sqrt{1 + 2E_{11}}\sqrt{1 + 2E_{22}}}\,. \tag{3.6.10}$$

This shows that the shear depends on the off-diagonal components of strain as well as on the normal components of strain. However, if the strain is small (i.e., $E_{AB} << 1$) then (3.6.10) can be approximated by

$$\psi \approx 2E_{12}\,, \tag{3.6.11}$$

which shows that the off-diagonal components of strain are related to shear deformations.

Using the work in [1], it follows that in the absence of distortional deformation the unimodular part $\mathbf{C}'$ of the deformation tensor $\mathbf{C}$ is the identity

$$\mathbf{C}' = J^{-2/3}\mathbf{C} = \mathbf{I}\,, \quad \mathbf{C} = J^{2/3}\mathbf{I}\,, \tag{3.6.12}$$

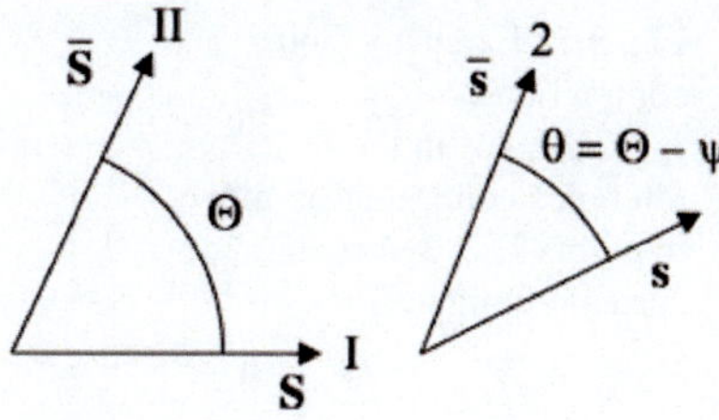

Fig. 3.4 Shear angle: Points I, II in the reference configuration move to points $1, 2$ in the current configuration. Notice that the plane of $\mathbf{s}$ and $\bar{\mathbf{s}}$ is not necessarily parallel to the plane of $\mathbf{S}$ and $\bar{\mathbf{S}}$

so the associated deformation gradient $\mathbf{F}$ is determined by the total dilatation J and an arbitrary proper orthogonal rotation tensor $\mathbf{R}$, such that

$$\mathbf{F} = J^{1/3}\mathbf{R}\,, \quad \mathbf{R}^T\mathbf{R} = \mathbf{I}\,, \quad \det \mathbf{R} = +1\,. \tag{3.6.13}$$

Using this expression for $\mathbf{C}$ in (3.6.7) yields

$$\cos\theta = \cos\Theta\,, \tag{3.6.14}$$

which means that the angle between any two material line elements remains the same unless there is some distortional deformation ($\mathbf{C}' \neq \mathbf{I}$).

3.7 Rate of Deformation: Interpretations and Examples

Recall the expressions (3.5.21) for the rate of stretch $\dot{\lambda}$ and (3.5.22) for the rate of rotation $\dot{\mathbf{s}}$ of a material line element

$$\frac{\dot{\lambda}}{\lambda} = \mathbf{D} \cdot \mathbf{s} \otimes \mathbf{s}\,, \tag{3.7.1a}$$

$$\dot{\mathbf{s}} = [\mathbf{L} - (\mathbf{D} \cdot \mathbf{s} \otimes \mathbf{s})\mathbf{I}]\,\mathbf{s}\,. \tag{3.7.1b}$$

It follows from (3.7.1a) that the logarithmic derivative of the stretch is determined by the rate of deformation tensor $\mathbf{D}$ for the material line element that is currently in the $\mathbf{s}$ direction. Moreover, substituting (3.5.2a) into (3.7.1b) yields

$$\dot{\mathbf{s}} = \mathbf{W}\mathbf{s} + [\mathbf{D} - (\mathbf{D} \cdot \mathbf{s} \otimes \mathbf{s})\mathbf{I}]\,\mathbf{s}\,, \tag{3.7.2}$$

which shows that, in general, the rate of rotation of the material line element which is currently in the direction $\mathbf{s}$ is dependent on both the rate of deformation tensor $\mathbf{D}$ and the spin tensor $\mathbf{W}$. However, if $\mathbf{s}$ is parallel to a principal direction of $\mathbf{D}$ then

$$\mathbf{D}\mathbf{s} = (\mathbf{D} \cdot \mathbf{s} \otimes \mathbf{s})\mathbf{s}\,, \quad \dot{\mathbf{s}} = \mathbf{W}\mathbf{s}\,. \tag{3.7.3}$$

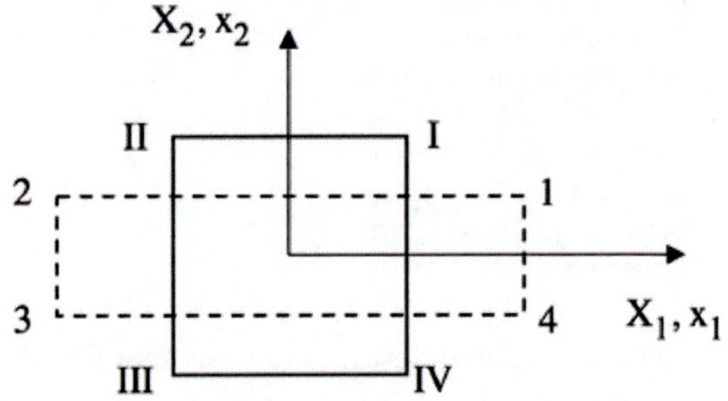

Fig. 3.5 Extension and contraction: Points I, II, II, IV in the reference configuration move to points 1, 2, 3, 4 in the current configuration

This shows that the spin tensor **W** controls the rate of rotation of the material line element $d\mathbf{x}$ which in the current configuration is parallel to a principal direction of **D**. Furthermore, using (3.5.6) it can be seen that for this case the axial vector $\boldsymbol{\omega}$ of the **W** determines the rate of rotation of **s**

$$\dot{\mathbf{s}} = \boldsymbol{\omega} \times \mathbf{s} \quad \text{for} \quad \mathbf{D}\mathbf{s} = (\mathbf{D} \cdot \mathbf{s} \otimes \mathbf{s})\mathbf{s} \,. \tag{3.7.4}$$

Example: Extension and Contraction (Fig. 3.5)
By way of example, let X_A be the Cartesian components of **X**, x_i be the Cartesian components of **x** and the Cartesian base vectors $\mathbf{e}_A$ and $\mathbf{e}_i$ coincide ($\mathbf{e}_i = \delta_{iA}\mathbf{e}_A$). Also, consider the motion defined by

$$x_1 = e^{at} X_1 \,, \quad x_2 = e^{-bt} X_2 \,, \quad x_3 = X_3 \,, \tag{3.7.5}$$

where a, b are positive numbers. The inverse mapping is given by

$$X_1 = e^{-at} x_1 \,, \quad X_2 = e^{bt} x_2 \,, \quad X_3 = x_3 \,. \tag{3.7.6}$$

It then follows that

$$F_{iA} = \begin{pmatrix} e^{at} & 0 & 0 \\ 0 & e^{-bt} & 0 \\ 0 & 0 & 1 \end{pmatrix} , \quad C_{AB} = \begin{pmatrix} e^{2at} & 0 & 0 \\ 0 & e^{-2bt} & 0 \\ 0 & 0 & 1 \end{pmatrix} , \tag{3.7.7}$$

$$E_{AB} = \frac{1}{2} \begin{pmatrix} e^{2at} - 1 & 0 & 0 \\ 0 & e^{-2bt} - 1 & 0 \\ 0 & 0 & 0 \end{pmatrix} . \tag{3.7.8}$$

To better understand this deformation, it is convenient to calculate the stretch λ and the extension ε of line elements which were parallel to the coordinate directions in the reference configuration

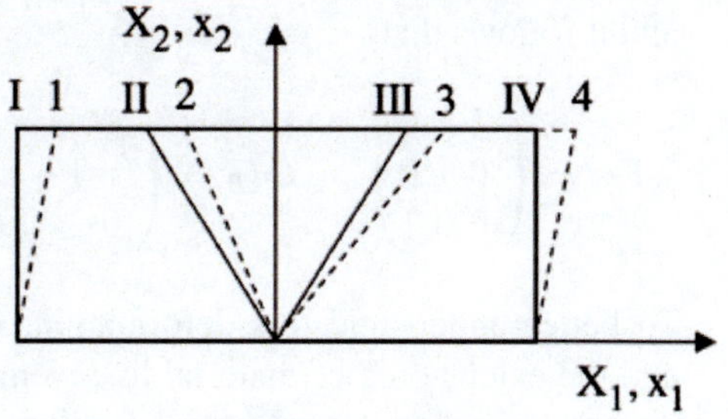

Fig. 3.6 Simple shear: Points I, II, II, IV in the reference configuration move to points 1, 2, 3, 4 in the current configuration

$$\begin{array}{llll} \text{For } \mathbf{S}=\mathbf{e}_1\,, & \lambda = e^{at} \geq 1\,, & \varepsilon = e^{at}-1 \geq 0\,, & \text{(extension)}\,, \\ \text{For } \mathbf{S}=\mathbf{e}_2\,, & \lambda = e^{-bt} \leq 1\,, & \varepsilon = e^{-bt}-1 \leq 0\,, & \text{(contraction)}\,, \\ \text{For } \mathbf{S}=\mathbf{e}_3\,, & \lambda = 1\,, & \varepsilon = 0\,, & \text{(no deformation)}\,. \end{array} \tag{3.7.9}$$

Next, consider the rate of deformation to deduce that

$$\begin{gathered} v_1 = ax_1\,, \qquad v_2 = -bx_2\,,\quad v_3 = 0\,, \\ L_{ij} = D_{ij} = \begin{pmatrix} a & 0 & 0 \\ 0 & -b & 0 \\ 0 & 0 & 0 \end{pmatrix}\,,\quad \mathbf{W} = 0\,, \qquad \boldsymbol{\omega} = 0\,. \end{gathered} \tag{3.7.10}$$

The principal directions of **D** are $\mathbf{e}_1$, $\mathbf{e}_2$ and $\mathbf{e}_3$ so since $\mathbf{W} = 0$, it follows that the material line elements that are parallel to these principal directions in the current configuration experience pure stretching without rotation

$$\begin{array}{llll} \text{For } \mathbf{s}=\mathbf{e}_1\,, & \frac{\dot{\lambda}}{\lambda} = a > 0\,, & \dot{\mathbf{s}} = 0\,, & \text{(rate of extension)}\,, \\ \text{For } \mathbf{s}=\mathbf{e}_2\,, & \frac{\dot{\lambda}}{\lambda} = -b > 0\,, & \dot{\mathbf{s}} = 0\,, & \text{(rate of contraction)}\,, \\ \text{For } \mathbf{s}=\mathbf{e}_3\,, & \frac{\dot{\lambda}}{\lambda} = 0\,, & \dot{\mathbf{s}} = 0\,, & \text{(no deformation)}\,. \end{array} \tag{3.7.11}$$

It is emphasized that although **W** vanishes, other material line elements can rotate during this motion.

Example: Simple Shear (Fig. 3.6)
To clarify the meaning of the spin tensor **W** consider a simple shearing deformation which is defined by

$$x_1 = X_1 + \kappa(t)X_2\,,\quad x_2 = X_2\,,\quad x_3 = X_3\,, \tag{3.7.12}$$

where $\kappa(t)$ is a monotonically increasing nonnegative function of time

$$\kappa \geq 0\,,\quad \dot{\kappa} > 0\,. \tag{3.7.13}$$

The inverse mapping is given by

$$X_1 = x_1 - \kappa x_2\,,\quad X_2 = x_2\,,\quad X_3 = x_3\,, \tag{3.7.14}$$

and it follows that

$$F_{iA} = \begin{pmatrix} 1 & \kappa & 0 \\ 0 & 1 & 0 \\ 0 & 0 & 1 \end{pmatrix}, \quad C_{AB} = \begin{pmatrix} 1 & \kappa & 0 \\ \kappa & 1+\kappa^2 & 0 \\ 0 & 0 & 1 \end{pmatrix}, \quad E_{AB} = \frac{1}{2}\begin{pmatrix} 0 & \kappa & 0 \\ \kappa & \kappa^2 & 0 \\ 0 & 0 & 0 \end{pmatrix}. \tag{3.7.15}$$

To better understand this deformation, it is convenient to calculate the stretch λ and the extension ε of material line elements which were parallel to the coordinate directions in the reference configuration

$$\begin{array}{llll} \text{For } \mathbf{S} = \mathbf{e}_1\,, & \lambda = 1\,, & \varepsilon = 0\,, & \text{(no deformation)}\,, \\ \text{For } \mathbf{S} = \mathbf{e}_2\,, & \lambda = \sqrt{1+\lambda^2}\,, & \varepsilon = \sqrt{1+\lambda^2} - 1 \geq 0\,, & \text{(extension)}\,, \\ \text{For } \mathbf{S} = \mathbf{e}_3\,, & \lambda = 1\,, & \varepsilon = 0\,, & \text{(no deformation)}\,. \end{array} \tag{3.7.16}$$

Notice that the result for $\mathbf{S} = \mathbf{e}_2$ could be obtained by direct calculation using elementary geometry. Next, consider the rate of deformation to deduce that

$$\begin{aligned} &v_1 = \dot{\kappa} x_2\,, \quad v_2 = 0\,, \quad v_3 = 0\,, \\ &L_{ij} = \begin{pmatrix} 0 & \dot{\kappa} & 0 \\ 0 & 0 & 0 \\ 0 & 0 & 0 \end{pmatrix}, \quad D_{ij} = \frac{1}{2}\begin{pmatrix} 0 & \dot{\kappa} & 0 \\ \dot{\kappa} & 0 & 0 \\ 0 & 0 & 0 \end{pmatrix}, \\ &W_{ij} = \frac{1}{2}\begin{pmatrix} 0 & \dot{\kappa} & 0 \\ -\dot{\kappa} & 0 & 0 \\ 0 & 0 & 0 \end{pmatrix}, \quad \boldsymbol{\omega} = -\frac{1}{2}\dot{\kappa}\mathbf{e}_3\,. \end{aligned} \tag{3.7.17}$$

Since the principal directions of $\mathbf{D}$ are $\frac{1}{\sqrt{2}}(\mathbf{e}_1 + \mathbf{e}_2)$, $\frac{1}{\sqrt{2}}(-\mathbf{e}_1 + \mathbf{e}_2)$ and $\mathbf{e}_3$, with the help of (3.7.1a), it follows that

$$\begin{array}{lll} \text{For} & \mathbf{s} = \frac{1}{\sqrt{2}}(\mathbf{e}_1 + \mathbf{e}_2)\,, & \frac{\dot{\lambda}}{\lambda} = \frac{1}{2}\dot{\kappa} > 0\,, \\ & \dot{\mathbf{s}} = (\frac{1}{2}\dot{\kappa})\frac{1}{\sqrt{2}}(\mathbf{e}_1 - \mathbf{e}_2)\,, & \text{(rate of extension)}\,, \\ \text{For} & \mathbf{s} = \frac{1}{\sqrt{2}}(-\mathbf{e}_1 + \mathbf{e}_2)\,, & \frac{\dot{\lambda}}{\lambda} = -\frac{1}{2}\dot{\kappa} < 0\,, \\ & \dot{\mathbf{s}} = (\frac{1}{2}\dot{\kappa})\frac{1}{\sqrt{2}}(\mathbf{e}_1 + \mathbf{e}_2)\,, & \text{(rate of contraction)}\,, \\ \text{For} & \mathbf{s} = \mathbf{e}_3\,, & \frac{\dot{\lambda}}{\lambda} = 0\,, \\ & \dot{\mathbf{s}} = 0\,, & \text{(no deformation)}\,. \end{array} \tag{3.7.18}$$

Thus, from (3.7.17), it follows that these special material line elements in (3.7.18) are rotating in the clockwise direction about the $\mathbf{e}_3$ axis with angular speed $\frac{1}{2}\dot{\kappa}$. In addition, it noted that this motion is isochoric (3.5.16) (no change in volume) with

$$J = \det \mathbf{F} = 1\,, \quad \mathbf{D} \cdot \mathbf{I} = 0\,. \tag{3.7.19}$$

3.8 Superposed Rigid Body Motions (SRBM)

This section develops the kinematics of Superposed Rigid Body Motions (SRBM) which will be used later to place restrictions on constitutive equations for material response. Consider a group of motions associated with configurations P^+ which differ from an arbitrary prescribed motion such as (3.1.5)

$$\mathbf{x} = \hat{\mathbf{x}}(\mathbf{X}, t)\,, \tag{3.8.1}$$

by SRBM of the entire body, (i.e., motions which in addition to the prescribed motion include purely rigid motions of the body).

To this end, consider a material point $\mathbf{X}$ of the body, which in $\mathcal{P}$ at time t occupies the location $\mathbf{x}$ as specified by (3.8.1). Suppose that under a SRBM the material point, which occupies the location $\mathbf{x}$ at time t in the configuration $\mathcal{P}$, moves to the location $\mathbf{x}^+$ at time t^+

$$t^+ = t + c\,, \tag{3.8.2}$$

in the superposed configuration $\mathcal{P}^+$, where c is a constant time shift. Throughout the text, quantities associated with the superposed configuration $\mathcal{P}^+$ are denoted using the same symbol as associated with the configuration P but with a superscript $(\,)^+$. In particular, the position $\mathbf{x}^+$ of the same material point in the superposed configuration is written in the form

$$\mathbf{x}^+ = \hat{\hat{\mathbf{x}}}^+(\mathbf{X}, t^+) = \hat{\mathbf{x}}^+(\mathbf{X}, t)\,, \tag{3.8.3}$$

where the notation $\hat{\hat{\mathbf{x}}}^+$ and $\hat{\mathbf{x}}^+$ has been used to distinguish between the function $\hat{\hat{\mathbf{x}}}^+$, which depends on t^+, and the function $\hat{\mathbf{x}}^+$, which depends on t and includes the influence of c .

Similarly, consider another material point $\mathbf{Y}$ of the body, which in the current configuration P at time t occupies the location $\mathbf{y}$ specified by

$$\mathbf{y} = \hat{\mathbf{x}}(\mathbf{Y}, t)\,. \tag{3.8.4}$$

It is important to emphasize that the function $\hat{\mathbf{x}}$ in (3.8.4) is the same function as that in (3.8.1). Furthermore, suppose that under the same SRBM the material point which occupies the location $\mathbf{y}$ at time t in the configuration $\mathcal{P}$ moves to the location $\mathbf{y}^+$ at time t^+. Then, with the help of (3.8.3), it follows that

$$\mathbf{y}^+ = \hat{\hat{\mathbf{x}}}^+(\mathbf{Y}, t^+) = \hat{\mathbf{x}}^+(\mathbf{Y}, t)\,. \tag{3.8.5}$$

Recalling the inverse relationships

$$\mathbf{X} = \tilde{\mathbf{X}}(\mathbf{x}, t)\,, \quad \mathbf{Y} = \tilde{\mathbf{X}}(\mathbf{y}, t)\,, \tag{3.8.6}$$

the function $\hat{\mathbf{x}}^+$ on the right-hand sides of (3.8.3) and (3.8.6) can be expressed as different functions of $\mathbf{x}$ and t and $\mathbf{y}$ and t, respectively, such that

$$\mathbf{x}^+ = \hat{\mathbf{x}}^+(\tilde{\mathbf{X}}(\mathbf{x}, t), t) = \tilde{\mathbf{x}}^+(\mathbf{x}, t), \quad \mathbf{y}^+ = \hat{\mathbf{x}}^+(\tilde{\mathbf{X}}(\mathbf{y}, t), t) = \tilde{\mathbf{x}}^+(\mathbf{y}, t). \tag{3.8.7}$$

Since the superposed motion of the body is restricted to be rigid, the magnitude of the relative displacement $\mathbf{y}^+ - \mathbf{x}^+$ must remain equal to the magnitude of the relative displacement $\mathbf{y} - \mathbf{x}$ for all pairs of material points $\mathbf{X}$ and $\mathbf{Y}$, and for all time. Thus,

$$[\tilde{\mathbf{x}}^+(\mathbf{y}, t) - \tilde{\mathbf{x}}^+(\mathbf{x}, t)] \cdot [\tilde{\mathbf{x}}^+(\mathbf{y}, t) - \tilde{\mathbf{x}}^+(\mathbf{x}, t)] = (\mathbf{y} - \mathbf{x}) \cdot (\mathbf{y} - \mathbf{x}). \tag{3.8.8}$$

Recognizing that $\mathbf{x}$ and $\mathbf{y}$ are independent, (3.8.8) can be differentiated first with respect to $\mathbf{x}$ and then with respect to $\mathbf{y}$ to obtain

$$\begin{gathered} -2[\partial\tilde{\mathbf{x}}^+(\mathbf{x}, t)/\partial\mathbf{x}]^T [\tilde{\mathbf{x}}^+(\mathbf{y}, t) - \tilde{\mathbf{x}}^+(\mathbf{x}, t)] = -2(\mathbf{y} - \mathbf{x}), \\ [\partial\tilde{\mathbf{x}}^+(\mathbf{x}, t)/\partial\mathbf{x}]^T [\partial\tilde{\mathbf{x}}^+(\mathbf{y}, t)/\partial\mathbf{y}] = \mathbf{I}. \end{gathered} \tag{3.8.9}$$

In this equation the transpose has been used to retain the inner product of $\tilde{\mathbf{x}}^+(\mathbf{x}, t)$ with $\tilde{\mathbf{x}}^+(\mathbf{y}, t)$. Moreover, it follows that the determinant of the tensor $\partial\tilde{\mathbf{x}}^+(\mathbf{x}, t)/\partial\mathbf{x}$ does not vanish so that this tensor is invertible and (3.8.9) can be rewritten in the alternative form

$$[\partial\tilde{\mathbf{x}}^+(\mathbf{x}, t)/\partial\mathbf{x}]^T = [\partial\tilde{\mathbf{x}}^+(\mathbf{y}, t)/\partial\mathbf{y}]^{-1}, \tag{3.8.10}$$

for all $\mathbf{x}$ and $\mathbf{y}$ in the region and all t. Thus, each side of this equation must be a tensor function of time only, say $\mathbf{Q}^T(t)$, so that

$$\partial\tilde{\mathbf{x}}^+(\mathbf{x}, t)/\partial\mathbf{x} = \mathbf{Q}(t), \tag{3.8.11}$$

for all $\mathbf{x}$ in the region and all time t. Using the fact that $\mathbf{Q}$ in (3.8.11) is independent of $\mathbf{x}$, it also follows that

$$\partial\tilde{\mathbf{x}}^+(\mathbf{y}, t)/\partial\mathbf{y} = \mathbf{Q}(t), \tag{3.8.12}$$

so that (3.8.9) restricts $\mathbf{Q}$ to be an orthogonal tensor

$$\mathbf{Q}^T(t)\mathbf{Q}(t) = \mathbf{I}, \quad \det\mathbf{Q} = \pm 1. \tag{3.8.13}$$

Since (3.8.7) represents a SRBM it must include the trivial motion

$$\tilde{\mathbf{x}}^+(\mathbf{x}, t) = \mathbf{x}, \quad \mathbf{Q} = \mathbf{I}, \quad \det\mathbf{Q} = +1. \tag{3.8.14}$$

Furthermore, since the motions are assumed to be continuous and $\det\mathbf{Q}$ cannot vanish, $\mathbf{Q}$ must remain a proper orthogonal tensor function of time only

$$\mathbf{Q}^T(t)\mathbf{Q}(t) = \mathbf{Q}(t)\mathbf{Q}^T(t) = \mathbf{I}, \quad \det\mathbf{Q} = +1. \tag{3.8.15}$$

Next, integrating (3.8.11) yields the general solution for SRBM

$$\mathbf{x}^+ = \tilde{\mathbf{x}}^+(\mathbf{x}, t) = \mathbf{c}(t) + \mathbf{Q}(t)\mathbf{x}\,, \tag{3.8.16}$$

where $\mathbf{c}(t)$ is an arbitrary vector function of time only representing an arbitrary translation of the body and $\mathbf{Q}(t)$ represents an arbitrary rotation of the body.

By definition, the superposed part of the motion defined by (3.8.16) is a rigid body motion. This means that the lengths of line elements are preserved

$$\begin{aligned}|\mathbf{x}^+ - \mathbf{y}^+|^2 &= (\mathbf{x}^+ - \mathbf{y}^+)\cdot(\mathbf{x}^+ - \mathbf{y}^+) = \mathbf{Q}(\mathbf{x}-\mathbf{y})\cdot\mathbf{Q}(\mathbf{x}-\mathbf{y}) = (\mathbf{x}-\mathbf{y})\cdot\mathbf{Q}^T\mathbf{Q}(\mathbf{x}-\mathbf{y})\,,\\ &= (\mathbf{x}-\mathbf{y})\cdot\mathbf{I}(\mathbf{x}-\mathbf{y}) = (\mathbf{x}-\mathbf{y})\cdot(\mathbf{x}-\mathbf{y}) = |\mathbf{x}-\mathbf{y}|^2\,,\end{aligned} \tag{3.8.17}$$

and the angles between two material line elements are also preserved so that

$$\begin{aligned}\cos\theta^+ &= \frac{(\mathbf{x}^+ - \mathbf{y}^+)}{|\mathbf{x}^+ - \mathbf{y}^+|}\cdot\frac{(\mathbf{x}^+ - \mathbf{z}^+)}{|\mathbf{x}^+ - \mathbf{z}^+|} = \frac{\mathbf{Q}(\mathbf{x}-\mathbf{y})}{|(\mathbf{x}-\mathbf{y})|}\cdot\frac{\mathbf{Q}(\mathbf{x}-\mathbf{z})}{|(\mathbf{x}-\mathbf{z})|}\,,\\ &= \frac{(\mathbf{x}-\mathbf{y})}{|(\mathbf{x}-\mathbf{y})|}\cdot\frac{\mathbf{Q}^T\mathbf{Q}(\mathbf{x}-\mathbf{z})}{|(\mathbf{x}-\mathbf{z})|} = \frac{(\mathbf{x}-\mathbf{y})}{|(\mathbf{x}-\mathbf{y})|}\cdot\frac{(\mathbf{x}-\mathbf{z})}{|(\mathbf{x}-\mathbf{z})|} = \cos\theta\,,\end{aligned} \tag{3.8.18}$$

where $\mathbf{x}$, $\mathbf{y}$ and $\mathbf{z}$ are material points in the body which move to $\mathbf{x}^+$, $\mathbf{y}^+$ and $\mathbf{z}^+$ under SRBM. Furthermore, this means that material areas, and volumes are preserved under SRBM. To show this use is made of (3.8.16) with $\mathbf{x} = \hat{\mathbf{x}}(\mathbf{X}, t)$ to calculate the deformation gradient $\mathbf{F}^+$ from the reference configuration to the superposed configuration

$$\mathbf{F}^+ = \partial\hat{\mathbf{x}}^+(\mathbf{X}, t)/\partial\mathbf{X} = \mathbf{Q}(\partial\mathbf{x}/\partial\mathbf{X}) = \mathbf{Q}\mathbf{F}\,, \tag{3.8.19}$$

so that from (3.3.21), (3.3.35) and (3.8.19), it follows that

$$\begin{aligned}&J^+ = \frac{dv^+}{dV} = \det\mathbf{F}^+ = \det(\mathbf{Q}\mathbf{F}) = \det\mathbf{Q}\det\mathbf{F} = J\,,\\ &\mathbf{n}^+ da^+ = d\mathbf{x}^{1+}\times d\mathbf{x}^{2+} = J^+(\mathbf{F}^+)^{-T}\mathbf{N}dA = J\mathbf{Q}\mathbf{F}^{-T}\mathbf{N}dA = \mathbf{Q}\mathbf{n}da\,,\\ &(da^+)^2 = \mathbf{n}^+da^+\cdot\mathbf{n}^+da^+ = \mathbf{Q}\mathbf{n}da\cdot\mathbf{Q}\mathbf{n}da = \mathbf{n}\cdot\mathbf{Q}^T\mathbf{Q}(da)^2 = (da)^2\,,\\ &\mathbf{n}^+ = \mathbf{Q}\mathbf{n}\,.\end{aligned} \tag{3.8.20}$$

For later convenience it is desirable to calculate expressions for the velocity and rate of deformation tensors associated with the superposed configuration. To this end, take the material derivative of (3.8.13) to deduce that

$$\dot{\mathbf{Q}}^T\mathbf{Q} + \mathbf{Q}^T\dot{\mathbf{Q}} = 0 \quad\Rightarrow\quad \dot{\mathbf{Q}} = \boldsymbol{\Omega}\mathbf{Q} \quad\Rightarrow\quad \boldsymbol{\Omega}^T = -\boldsymbol{\Omega}\,, \tag{3.8.21}$$

where $\boldsymbol{\Omega}(t)$ is a skew-symmetric tensor function of time only. Letting $\boldsymbol{\omega}$ be the axial vector of $\boldsymbol{\Omega}$ it is recalled from (3.5.6) that for an arbitrary vector $\mathbf{a}$

$$\boldsymbol{\Omega}\mathbf{a} = \boldsymbol{\omega}\times\mathbf{a}\,. \tag{3.8.22}$$

Thus, by taking the material derivative of (3.8.16) the velocity $\mathbf{v}^+$ of the material point in the superposed configuration can be expressed in the forms

$$\begin{aligned}\mathbf{v}^+ &= \dot{\mathbf{x}}^+ = \dot{\mathbf{c}} + \dot{\mathbf{Q}}\mathbf{x} + \mathbf{Q}\dot{\mathbf{x}} = \dot{\mathbf{c}} + \boldsymbol{\Omega}\mathbf{Q}\mathbf{x} + \mathbf{Q}\mathbf{v}\,,\\ \mathbf{v}^+ &= \dot{\mathbf{c}} + \boldsymbol{\Omega}(\mathbf{x}^+ - \mathbf{c}) + \mathbf{Q}\mathbf{v} = \dot{\mathbf{c}} + \boldsymbol{\omega} \times (\mathbf{x}^+ - \mathbf{c}) + \mathbf{Q}\mathbf{v}\,.\end{aligned} \tag{3.8.23}$$

It follows that the velocity gradient $\mathbf{L}^+$ and rate of deformation $\mathbf{D}^+$ and spin $\mathbf{W}^+$ tensors associated with the superposed configuration are given by

$$\begin{aligned}&\mathbf{L}^+ = \partial\mathbf{v}^+/\partial\mathbf{x}^+ = \mathbf{Q}(\partial\mathbf{v}/\partial\mathbf{x})(\partial\mathbf{x}/\partial\mathbf{x}^+) + \boldsymbol{\Omega} = \mathbf{Q}\mathbf{L}\mathbf{Q}^T + \boldsymbol{\Omega}\,,\\ &\mathbf{D}^+ = \mathbf{Q}\mathbf{D}\mathbf{Q}^T\,, \quad \mathbf{W}^+ = \mathbf{Q}\mathbf{W}\mathbf{Q}^T + \boldsymbol{\Omega}\,,\end{aligned} \tag{3.8.24}$$

where use has been made of the condition (3.8.21) and (3.8.16) has been differentiated to obtain

$$\partial\mathbf{x}^+/\partial\mathbf{x} = \mathbf{Q}\,, \quad \partial\mathbf{x}/\partial\mathbf{x}^+ = \mathbf{Q}^T\,. \tag{3.8.25}$$

In general, SRBM are in addition to the general motion $\mathbf{x}(\mathbf{X}, t)$ of a deformable body. However, the kinematics of rigid body motions can be obtained as a special case by identifying $\mathbf{x}$ with its value $\mathbf{X}$ in the fixed reference configuration so that distortion and dilatation of the body are eliminated and (3.8.23) yields

$$\mathbf{x} = \mathbf{X} \quad \Rightarrow \quad \dot{\mathbf{x}}^+ = \dot{\mathbf{c}} + \boldsymbol{\omega} \times (\mathbf{x}^+ - \mathbf{c})\,. \tag{3.8.26}$$

In this form, it is easy to recognize that $\mathbf{c}(t)$ represents the translation of a point moving with the rigid body and $\boldsymbol{\omega}$ is the absolute angular velocity of the rigid body.

In summary, the most general SRBM is characterized by Eqs. (3.8.16), (3.8.13) and (3.8.21)

$$\mathbf{x}^+ = \mathbf{c}(t) + \mathbf{Q}(t)\mathbf{x}\,, \quad \mathbf{Q}^T\mathbf{Q} = \mathbf{I}\,, \quad \det\mathbf{Q} = +1\,, \quad \dot{\mathbf{Q}} = \boldsymbol{\Omega}\mathbf{Q}\,, \quad \boldsymbol{\Omega}^T = -\boldsymbol{\Omega}\,. \tag{3.8.27}$$

3.9 Material Line, Material Surface and Material Volume

Recall that a material point Y is mapped into its location $\mathbf{X}$ in the reference configuration and that this mapping is independent of time. Consequently, lines, surfaces and volumes, which remain constant in the reference configuration, always contain the same material points and therefore are called material.

Material Line

A material line is a fixed curve in the reference configuration that can be parameterized by its archlength S, which is independent of time. It follows that the Lagrangian representation of a material line becomes

$$\mathbf{X} = \mathbf{X}(S) \,. \tag{3.9.1}$$

Alternatively, using the mapping (3.1.5) the current positions of material points on the same material line are determined by

$$\mathbf{x} = \mathbf{x}(S, t) = \hat{\mathbf{x}}(\mathbf{X}(S), t) \,. \tag{3.9.2}$$

Material Surface
A material surface is a fixed surface in the reference configuration that can be parameterized by two coordinates S_1 and S_2 that are independent of time. It follows that the Lagrangian representation of a material surface becomes

$$\mathbf{X} = \mathbf{X}(S_1, S_2) \quad \text{or} \quad \hat{f}(\mathbf{X}) = 0 \,, \tag{3.9.3}$$

where $\hat{f}(\mathbf{X}) = 0$ is a constraint on the three components of $\mathbf{X}$ which ensures that $\mathbf{X}$ identifies points in the space of the reference configuration on the material surface. Alternatively, using the mapping (3.1.5) and its inverse (3.1.6), the current positions of material points on this surface and the Eulerian representation of the same material surface can be characterized by the expressions

$$\mathbf{x} = \mathbf{x}(S_1, S_2, t) = \hat{\mathbf{x}}(\mathbf{X}(S_1, S_2), t) \quad \text{or} \quad \tilde{f}(\mathbf{x}, t) = \hat{f}(\tilde{\mathbf{X}}(\mathbf{x}, t)) = 0 \,, \tag{3.9.4}$$

where $\tilde{f}(\mathbf{x}, t) = 0$ is a constraint on the three components of $\mathbf{x}$ which ensures that $\mathbf{x}$ identifies points in the space of the current configuration on the material surface.

Lagrange's Criterion for a Material Surface
The surface defined by the constraint $\tilde{f}(\mathbf{x}, t) = 0$ is material if and only if

$$\dot{\tilde{f}} = \frac{\partial \tilde{f}}{\partial t} + \partial \tilde{f} / \partial \mathbf{x} \cdot \mathbf{v} = 0 \,. \tag{3.9.5}$$

Proof In general, the mapping (3.1.5) can be used to deduce that

$$\hat{f}(\mathbf{X}, t) = \tilde{f}(\hat{\mathbf{x}}(\mathbf{X}, t), t) \,, \tag{3.9.6}$$

which can be used to rewrite (3.9.5) in the form

$$\dot{\hat{f}}(\mathbf{X}, t) = \frac{\partial \hat{f}}{\partial t} = \dot{\tilde{f}} = 0 \,, \tag{3.9.7}$$

so that $\hat{f}$ is independent of time and the surface $\hat{f} = 0$ is fixed in the reference configuration, which means that $\hat{f} = \tilde{f} = 0$ characterizes a material surface. Alternatively, if $\hat{f}$ is independent of time, then $\dot{\hat{f}} = 0$ and $\dot{\tilde{f}} = 0$.

Material Region

A material region is a region of space bounded by a closed material surface. For example, if ∂P_0 is a closed material surface in the reference configuration then the region of space P_0 enclosed by ∂P_0 is a material region that contains the same material points for all time if P_0 and ∂P_0 are fixed in the reference configuration. Alternatively, using the mapping (3.1.5) each point of the material surface ∂P_0 maps into a point on the closed material surface ∂P in the current configuration so the region P enclosed by ∂P is the associated material region in the current configuration.

3.10 Reynolds Transport Theorem

Reynolds transport theorem is used to calculate the time derivative of an integral over a material region P in the current configuration whose closed boundary ∂P is changing with time.

Leibniz's Rule
By way of introduction, consider the simpler one-dimensional case of Leibniz's rule and recall that

$$\frac{d}{dt}\int_{\alpha(t)}^{\beta(t)} \phi(x,t)dx = \int_{\alpha(t)}^{\beta(t)} \frac{\partial \phi(x,t)}{\partial t}dx + \phi(\beta(t),t)\dot{\beta} - \phi(\alpha(t),t)\dot{\alpha}\,, \quad (3.10.1)$$

where $\phi(x,t)$ is an arbitrary function of position x and time t, and $\alpha(t)$ and $\beta(t)$ define the changing boundaries of integration. It is important to notice that the rates of change of the boundaries enter the expression in (3.10.1).
Reynolds Transport Theorem for a Material Region
To develop the generalization of (3.10.1) for a three-dimensional material region, it is convenient to consider an arbitrary scalar or tensor valued function $\boldsymbol{\phi}$ which admits the representations

$$\boldsymbol{\phi} = \tilde{\boldsymbol{\phi}}(\mathbf{x},t) = \hat{\boldsymbol{\phi}}(\mathbf{X},t)\,. \quad (3.10.2)$$

By mapping the material region P from the current configuration back to the reference configuration P_0, it is possible to calculate the derivative of the integral of $\boldsymbol{\phi}$ over the changing region P as follows

$$\begin{aligned}\frac{d}{dt}\int_P \tilde{\boldsymbol{\phi}}(\mathbf{x},t)dv &= \frac{d}{dt}\int_{P_0} \hat{\boldsymbol{\phi}}(\mathbf{X},t)JdV\,,\\ &= \int_{P_0} \frac{\partial\{\hat{\boldsymbol{\phi}}(\mathbf{X},t)J\}}{\partial t}\Big|_{\mathbf{X}}dV = \int_{P_0}(\dot{\hat{\boldsymbol{\phi}}} + \hat{\boldsymbol{\phi}}\mathrm{div}\mathbf{v})JdV\,,\end{aligned} \quad (3.10.3)$$

which can be transformed back to an integral over the present region P to obtain

$$\frac{d}{dt}\int_P \boldsymbol{\phi}(\mathbf{x},t)dv = \int_P (\dot{\boldsymbol{\phi}} + \boldsymbol{\phi}\,\mathrm{div}\mathbf{v})dv\,, \quad (3.10.4)$$

where $\dot{\boldsymbol{\phi}}$ is the usual material derivative of $\boldsymbol{\phi}$

$$\dot{\boldsymbol{\phi}} = \frac{\partial \hat{\boldsymbol{\phi}}(\mathbf{X}, t)}{\partial t} = \frac{\partial \tilde{\boldsymbol{\phi}}(\mathbf{x}, t)}{\partial t} + (\partial \tilde{\boldsymbol{\phi}}(\mathbf{x}, t)/\partial \mathbf{x}) \cdot \mathbf{v} \,. \tag{3.10.5}$$

Next, substituting (3.10.5) into (3.10.4) yields

$$\begin{aligned} \frac{d}{dt}\int_P \boldsymbol{\phi}(\mathbf{x}, t)dv &= \int_P [\frac{\partial \tilde{\boldsymbol{\phi}}(\mathbf{x}, t)}{\partial t} + (\partial \tilde{\boldsymbol{\phi}}(\mathbf{x}, t)/\partial \mathbf{x}) \cdot \mathbf{v} + \boldsymbol{\phi}\, \mathrm{div}\mathbf{v}]dv \\ &= \int_P [\frac{\partial \tilde{\boldsymbol{\phi}}(\mathbf{x}, t)}{\partial t} + \mathrm{div}(\tilde{\boldsymbol{\phi}} \otimes \mathbf{v})]dv \,, \end{aligned} \tag{3.10.6}$$

which with the help of the divergence theorem (2.5.10) can be written in the form

$$\frac{d}{dt}\int_P \boldsymbol{\phi}(\mathbf{x}, t)dv = \int_P \frac{\partial \tilde{\boldsymbol{\phi}}(\mathbf{x}, t)}{\partial t}dv + \int_{\partial P} \tilde{\boldsymbol{\phi}}\, (\mathbf{v} \cdot \mathbf{n})da \,, \tag{3.10.7}$$

where $\mathbf{n}$ is the unit outward normal to the material surface ∂P. It should be emphasized that the time differentiation and the integration over space operations commute in (3.10.3) because the region P_0 is independent of time. In contrast, the time differentiation and the integration over space operations in (3.10.7) do not commute because the region P depends on time. However, sometimes in fluid mechanics the region P in space at time t is considered to be a control volume and is identified as a fixed region $\bar{P}$ with fixed boundary $\partial \bar{P}$ which instantaneously coincide with the material region $\mathcal{P}$ and the material boundary $\partial \mathcal{P}$. Then, the time differentiation is interchanged with the integration over space operation to obtain

$$\frac{d}{dt}\int_P \boldsymbol{\phi}(\mathbf{x}, t)dv = \frac{\partial}{\partial t}\int_{\bar{P}} \tilde{\boldsymbol{\phi}}(\mathbf{x}, t)dv + \int_{\partial \bar{P}} \tilde{\boldsymbol{\phi}}\, (\mathbf{v} \cdot \mathbf{n})da \,, \tag{3.10.8}$$

where P on the left-hand side of this equation represents a material region that changes with time. In this regard, it is *essential* to interpret the partial differentiation operation in (3.10.8) as differentiation with respect to time holding $\mathbf{x}$ fixed. To avoid possible confusion, it is preferable to use the form (3.10.7) instead of (3.10.8).

Transport Theorem for a Non-material Region
To develop a generalized version of Leibnitz's rule (3.10.1) consider a general non-material region $\mathcal{V}(t)$ with general non-material closed boundary $\partial \mathcal{V}(t)$ for which

$$\frac{d}{dt}\int_{\mathcal{V}(t)} \tilde{\boldsymbol{\phi}}(\mathbf{x}, t)dv = \int_{\mathcal{V}(t)} \frac{\partial \tilde{\boldsymbol{\phi}}}{\partial t}dv + \int_{\partial \mathcal{V}(t)} \tilde{\boldsymbol{\phi}}\, (\mathbf{w} \cdot \mathbf{n})da \,, \tag{3.10.9}$$

where $\tilde{\boldsymbol{\phi}}(\mathbf{x}, t)$ is a general tensor field and $\mathbf{w}$ is the velocity of points on the moving boundary $\partial \mathcal{V}(t)$. Next, using the divergence theorem (2.5.10), it follows that

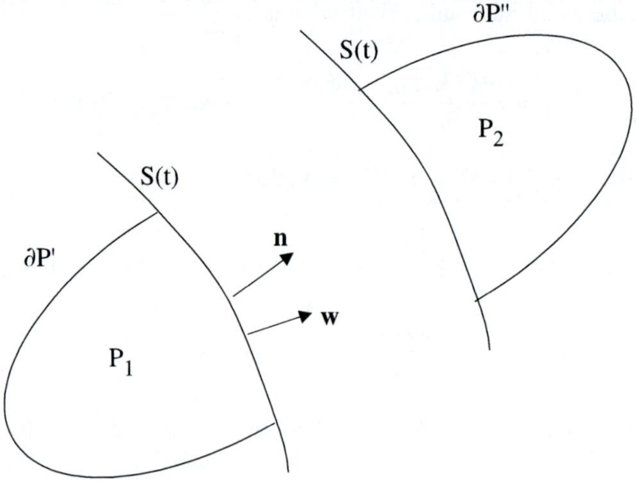

Fig. 3.7 A material region with a singular moving surface $S(t)$

$$\int_{\mathcal{V}(t)} \operatorname{div}(\boldsymbol{\phi} \otimes \mathbf{v})dv = \int_{\partial\mathcal{V}(t)} \tilde{\boldsymbol{\phi}}\ (\mathbf{v} \cdot \mathbf{n})da\,, \tag{3.10.10}$$

where $\mathbf{v}$ is the velocity of material points $\mathbf{x}$ in the region $\mathcal{V}(t)$ or the velocity of material points which instantaneously lie on the moving surface $\mathcal{V}(t)$. Thus, (3.10.9) can be rewritten in the form

$$\frac{d}{dt}\int_{\mathcal{V}(t)} \tilde{\boldsymbol{\phi}}(\mathbf{x},t)dv = \int_{\mathcal{V}(t)} \big[\frac{\partial\tilde{\boldsymbol{\phi}}}{\partial t} + \operatorname{div}(\boldsymbol{\phi} \otimes \mathbf{v})\big]dv + \int_{\partial\mathcal{V}(t)} \tilde{\boldsymbol{\phi}}\ [(\mathbf{w}-\mathbf{v})\cdot\mathbf{n}]da\,. \tag{3.10.11}$$

Moreover, using (2.5.4) and (3.10.5) it can be shown that

$$\operatorname{div}(\tilde{\boldsymbol{\phi}} \otimes \mathbf{v}) = (\partial\tilde{\boldsymbol{\phi}}/\partial\mathbf{x})\cdot\mathbf{v} + \tilde{\boldsymbol{\phi}}\operatorname{div}\mathbf{v} = \dot{\boldsymbol{\phi}} + \tilde{\boldsymbol{\phi}}\operatorname{div}\mathbf{v} - \frac{\partial\tilde{\boldsymbol{\phi}}}{\partial t}\,. \tag{3.10.12}$$

Then, using this expression the generalized transport theorem for a non-material region becomes

$$\frac{d}{dt}\int_{\mathcal{V}(t)} \boldsymbol{\phi}(\mathbf{x},t)dv = \int_{\mathcal{V}(t)} (\dot{\boldsymbol{\phi}} + \boldsymbol{\phi}\operatorname{div}\mathbf{v})dv + \int_{\partial\mathcal{V}(t)} \boldsymbol{\phi}\ [(\mathbf{w}-\mathbf{v})\cdot\mathbf{n}]da\,, \tag{3.10.13}$$

where the last term in this equation represents the flux of $\boldsymbol{\phi}$ entering $\mathcal{V}(t)$ through the moving boundary $\partial\mathcal{V}(t)$. When $\mathcal{V}$ is a material region P and $\partial\mathcal{V}$ is a material boundary ∂P then $(\mathbf{w}-\mathbf{v})\cdot\mathbf{n} = 0$ and (3.10.13) reduces to the simple form (3.10.4).

Transport Theorem for a Material Region with a Singular Moving Surface
Impulsive loading of materials cause shock waves that travel through the material region. At the front of a shock wave the state of the material can change rapidly.

Mathematically, it is convenient to approximate the front of the shock wave as a singular surface $S(t)$ moving through the material at which quantities other than the positions of material particles can be discontinuous across the surface $S(t)$. Figure 3.7 shows a material region P with closed material boundary ∂P that is divided by a singular moving surface $S(t)$ into two parts P_1 and P_2 with closed boundaries ∂P_1 and ∂P_2, respectively. Furthermore, let the intersection of ∂P_1 with ∂P be denoted by $\partial P'$ and the intersection of ∂P_2 with ∂P be denoted by $\partial P''$. Mathematically, this separation is summarized by

$$\begin{aligned} P &= P_1 \cup P_2\,, & \partial P' &= \partial P_1 \cap \partial P\,, & \partial P'' &= \partial P_2 \cap \partial P\,, \\ \partial P &= \partial P' \cup P''\,, & \partial P_1 &= \partial P' \cup S\,, & \partial P_2 &= \partial P'' \cup S\,. \end{aligned} \tag{3.10.14}$$

Points on this singular surface move with velocity $\mathbf{w}$ and the unit normal to $S(t)$ outward from the part P_1 is denoted by $\mathbf{n}$.

Application of the generalized transport theorem (3.10.13) to each of the parts P_1 and P_2 yields

$$\begin{aligned} \frac{d}{dt}\int_{P_1} \boldsymbol{\phi}(\mathbf{x},t)dv &= \int_{P_1} (\dot{\boldsymbol{\phi}} + \boldsymbol{\phi}\,\mathrm{div}\mathbf{v})dv + \int_{S(t)} \boldsymbol{\phi}_1\,\{(\mathbf{w}-\mathbf{v}_1)\cdot\mathbf{n}\}da\,, \\ \frac{d}{dt}\int_{P_2} \boldsymbol{\phi}(\mathbf{x},t)dv &= \int_{P_2} (\dot{\boldsymbol{\phi}} + \boldsymbol{\phi}\,\mathrm{div}\mathbf{v})dv - \int_{S(t)} \boldsymbol{\phi}_2\,\{(\mathbf{w}-\mathbf{v}_2)\cdot\mathbf{n}\}da\,, \end{aligned} \tag{3.10.15}$$

where $\boldsymbol{\phi}_1$ and $\mathbf{v}_1$ are the values of $\boldsymbol{\phi}$ and $\mathbf{v}$ in part P_1 and $\boldsymbol{\phi}_2$ and $\mathbf{v}_2$ are the values of $\boldsymbol{\phi}$ and $\mathbf{v}$ in part P_2, all on the singular surface $S(t)$. Next, adding these expressions yields

$$\begin{aligned} \frac{d}{dt}\int_{P} \boldsymbol{\phi}(\mathbf{x},t)dv &= \int_{P_1} (\dot{\boldsymbol{\phi}} + \boldsymbol{\phi}\,\mathrm{div}\mathbf{v})dv + \int_{P_2} (\dot{\boldsymbol{\phi}} + \boldsymbol{\phi}\,\mathrm{div}\mathbf{v})dv \\ &\quad - \int_{S(t)} [[\boldsymbol{\phi}\,\{(\mathbf{w}-\mathbf{v})\cdot\mathbf{n}\}]]da\,, \end{aligned} \tag{3.10.16}$$

where the jump operator $[[\boldsymbol{\phi}]]$ is defined by

$$[[\boldsymbol{\phi}]] = \boldsymbol{\phi}_2 - \boldsymbol{\phi}_1\,. \tag{3.10.17}$$

In addition, $\mathbf{w}$ and $\mathbf{n}$ are the same on both sides of $S(t)$

$$\mathbf{w}_1 = \mathbf{w}_2 = \mathbf{w}\,, \quad \mathbf{n}_1 = \mathbf{n}_2 = \mathbf{n}\,. \tag{3.10.18}$$

3.11 An Eulerian Formulation of Evolution Equations for Elastic Deformations

Recall from (3.5.4) that the deformation gradient $\mathbf{F}$ from the reference configuration satisfies the evolution equation

$$\dot{\mathbf{F}} = \mathbf{L}\mathbf{F}\,, \tag{3.11.1}$$

where $\mathbf{L}$ is the velocity gradient. Also, recall that the total dilatation J and the unimodular part $\mathbf{F}'$ of $\mathbf{F}$, both from the reference configuration, satisfy the evolution equations (3.5.16) and (3.5.17), respectively

$$\dot{J} = J\,\mathbf{D}\cdot\mathbf{I}\,, \quad \dot{\mathbf{F}}' = \mathbf{L}''\mathbf{F}'\,, \tag{3.11.2}$$

where $\mathbf{L}''$ is the deviatoric part of $\mathbf{L}$. To integrate these equations from an arbitrary time $t = t_1$ it is necessary to know the initial values

$$\mathbf{F}(t_1)\,, \quad J(t_1) = \det\mathbf{F}(t_1) > 0\,, \quad \mathbf{F}'(t_1)\,, \tag{3.11.3}$$

where the dependence on space has been suppressed for notational convenience. These initial values depend on an arbitrary choice of the reference configuration, with $\mathbf{F}(t_1)$ depending explicitly on the choice of the orientation of the body in the reference configuration.

Onat [4] discussed physical restrictions on internal state variables. This discussion proposed that internal state variables, which are determined by integrating time evolution equations, are specified to measure properties of the material response that define the current state of the material. Moreover, since these evolution equations need initial conditions, it is necessary that the values of the internal state variables be, in principle, measurable directly or indirectly by experiments on multiple identical samples of the material in its current state. Thus, all variables that define the current material state must be characterized by internal state variables whose values in the current state are measurable.

In this regard, it is noted that the reference configuration can be chosen to be an arbitrary configuration which admits a one-to-one mapping between material points in the reference configuration and the same material points in the current configuration. This requires $\mathbf{F}$ to be nonsingular with $\det\mathbf{F} > 0$. For example, let $\mathbf{A}$ be an arbitrary second-order tensor function of $\mathbf{X}$ only with positive determinant $\det\mathbf{A} > 0$. It then follows that $\mathbf{FA}$ satisfies the evolution equation (3.11.1)

$$\dot{\overline{\mathbf{FA}}} = \mathbf{L}(\mathbf{FA})\,. \tag{3.11.4}$$

However, since the choice of the reference configuration is arbitrary, it is not possible to determine the value of $\mathbf{FA}$ in the current state from experiments on identical samples of the material in its current state. This is true even if it is known that the the material in the reference configuration is in a uniform stress-free material state, since **FA** in the reference state could have an arbitrary orientation described by three arbitrary orientation angles of a proper orthogonal rotation tensor. This means that **F** is not an internal state variable in the sense of Onat [4] and therefore should not be used in constitutive equations, even for an elastic material. Similarly, the total dilatation $J = \det \mathbf{F}$ and the unimodular tensor $\mathbf{F}'$ from the reference configuration are also not internal state variables. However, $\mathbf{F}$, J and $\mathbf{F}'$ can be used to *parameterize* the solution of a particular problem for which the initial value of **F** is specified.

The Eulerian formulation for the purely mechanical theory of a compressible elastic material proposes an evolution equation for the elastic dilatation J_e in the form

$$\dot{J}_e = J_e \mathbf{D} \cdot \mathbf{I} \,. \tag{3.11.5}$$

Since the constitutive equation for stress is restricted to be invertible (1.2.12), it follows from (1.2.9) that J_e is an internal state variable in the sense of Onat [4] since it can be measured by experiments on identical samples of the material in the current configuration. Moreover, the evolution equation (3.11.5) is considered to be an Eulerian formulation of an evolution equation for the elastic dilatation J_e since it depends only on the current state of the material characterized by the values of J_e and **D**, which are measurable in the current state.

Anisotropic Elastic Solids

Following the work in [6] for elastically anisotropic materials, consider a triad of linearly independent microstructural vectors $\mathbf{m}_i$ $(i = 1, 2, 3)$ defined by the evolution equations

$$\dot{\mathbf{m}}_i = \mathbf{L}\, \mathbf{m}_i \,. \tag{3.11.6}$$

From (3.5.19) it is clear that $\mathbf{m}_i$ deform like material line elements. Moreover, since $\mathbf{m}_i$ are linearly independent they can be defined so that they form a right-handed triad with the elastic dilatation defined by (1.2.9)

$$J_e = \mathbf{m}_1 \times \mathbf{m}_2 \cdot \mathbf{m}_3 > 0 \,. \tag{3.11.7}$$

These vectors characterize both elastic deformations and rotations of material line elements. In particular, the elastic deformations can be defined by the elastic metric

$$m_{ij} = \mathbf{m}_i \cdot \mathbf{m}_j = m_{ji} \,, \tag{3.11.8}$$

and the vectors $\mathbf{m}_i$ can be specified so that they form an orthonormal triad in any zero-stress material state with

$$m_{ij} = \delta_{ij} \quad \text{for any zero-stress material state} \,. \tag{3.11.9}$$

Moreover, using (3.11.6) it can be shown that the elastic metric satisfies the evolution equation

$$\dot{m}_{ij} = 2(\mathbf{m}_i \otimes \mathbf{m}_j) \cdot \mathbf{D} \,. \tag{3.11.10}$$

In contrast to the total deformation gradient $\mathbf{F}$, which is not an internal state variable, the microstructural vectors $\mathbf{m}_i$ are internal state variables in the sense of Onat [4]. Specifically, since the constitutive equation for stress is restricted to be invertible (1.2.12), it follows that the values of $\mathbf{m}_i$ are measurable by experiments on identical samples of a material in its current state (see the more detailed discussion in Sect. 5.3).

As the material deforms, $\mathbf{m}_i$ do not remain orthonormal. However, since $\mathbf{m}_i$ are linearly independent, their reciprocal vectors $\mathbf{m}^i$ can be defined by

$$\begin{aligned} &\mathbf{m}^1 = J_e^{-1}\mathbf{m}_2 \times \mathbf{m}_3 , \quad \mathbf{m}^2 = J_e^{-1}\mathbf{m}_3 \times \mathbf{m}_1 , \quad \mathbf{m}^3 = J_e^{-1}\mathbf{m}_1 \times \mathbf{m}_2 , \\ &J_e^{-1} = \mathbf{m}^1 \times \mathbf{m}^2 \cdot \mathbf{m}^3 , \end{aligned} \tag{3.11.11}$$

which have the properties that

$$\mathbf{m}_i \otimes \mathbf{m}^i = \mathbf{I} . \tag{3.11.12}$$

Then, taking the material derivative of J_e in (3.11.7) and using the evolution equation (3.11.6), the definitions (3.11.11) and the result (3.11.12), it follows that

$$\begin{aligned} \dot{J}_e &= \dot{\mathbf{m}}_1 \cdot \mathbf{m}_2 \times \mathbf{m}_3 + \dot{\mathbf{m}}_2 \cdot \mathbf{m}_3 \times \mathbf{m}_1 + \dot{\mathbf{m}}_3 \cdot \mathbf{m}_1 \times \mathbf{m}_2 , \\ &= J_e \dot{\mathbf{m}}_i \cdot \mathbf{m}^i = J_e \mathbf{L} \cdot \mathbf{m}^i \otimes \mathbf{m}_i = J_e \mathbf{L} \cdot \mathbf{I} = J_e \mathbf{D} \cdot \mathbf{I} , \end{aligned} \tag{3.11.13}$$

which proves that the specification (3.11.7) satisfies the evolution equation (3.11.5) for elastic dilatation and the condition that $J_e = 1$ in any zero-stress material state.

Next, using the work of Flory [1] it is possible to develop pure measures of elastic distortional deformation. Specifically, the elastic distortional vectors $\mathbf{m}'_i$ are defined by

$$\mathbf{m}'_i = J_e^{-1/3} \mathbf{m}_i , \quad \mathbf{m}'_1 \times \mathbf{m}'_2 \cdot \mathbf{m}'_3 = 1 , \tag{3.11.14}$$

which satisfy the evolution equations

$$\dot{\mathbf{m}}'_i = \mathbf{L}'' \mathbf{m}'_i , \tag{3.11.15}$$

where $\mathbf{L}''$ is the deviatoric part of $\mathbf{L}$. Also, the elastic distortional deformation metric m'_{ij} is defined by

$$m'_{ij} = \mathbf{m}'_i \cdot \mathbf{m}'_j = m'_{ji} . \tag{3.11.16}$$

This metric satisfies the evolution equation

$$\dot{m}'_{ij} = 2\mathbf{m}'_i \otimes \mathbf{m}'_j \cdot \mathbf{D}'' = 2(\mathbf{m}'_i \otimes \mathbf{m}'_j - \frac{1}{3} m'_{ij} \mathbf{I}) \cdot \mathbf{D} , \tag{3.11.17}$$

where $\mathbf{D}''$ is the deviatoric part of $\mathbf{D}$. In addition, the associated reciprocal vectors $\mathbf{m}^{i\prime}$ satisfy equations

$$\mathbf{m}^{i\prime} = J_e^{1/3}\mathbf{m}^i \, , \quad \mathbf{m}^{1\prime} \times \mathbf{m}^{2\prime} \cdot \mathbf{m}^{3\prime} = 1 \, . \tag{3.11.18}$$

Isotropic Elastic Solids
For isotropic elastic solids, it is not possible to distinguish between the three microstructural vectors $\mathbf{m}_i$ and it is convenient to introduce a symmetric, positive deformation elastic deformation tensor $\mathbf{B}_e$ defined by

$$\mathbf{B}_e = \mathbf{m}_i \otimes \mathbf{m}_i \, , \tag{3.11.19}$$

which with the help of (3.11.6) can be shown to satisfy the evolution equation

$$\dot{\mathbf{B}}_e = \mathbf{L}\mathbf{B}_e + \mathbf{B}_e\mathbf{L}^T \, . \tag{3.11.20}$$

Since the constitutive equation for stress is restricted to be invertible (1.2.4), it follows that the value of $\mathbf{B}_e$ is measurable by experiments on identical samples of a material in its current state (see the more detailed discussion in Sect. 5.8). Consequently, $\mathbf{B}_e$ is an internal state variable in the sense of Onat [4].

Next, using the fact that (3.5.12) is valid for any nonsingular tensor, it follows that

$$\dot{\overline{\det\mathbf{B}_e}} = (\det\mathbf{B}_e)\,\mathbf{B}_e^{-1} \cdot \dot{\mathbf{B}}_e = 2(\det\mathbf{B}_e)(\mathbf{D} \cdot \mathbf{I}) \, , \tag{3.11.21}$$

so that J_e in (3.11.5) can be identified as

$$J_e = (\det\mathbf{B}_e)^{1/2} \, . \tag{3.11.22}$$

In addition, using the work of Flory [1] it is convenient to define the symmetric, positive-definite, elastic distortional deformation tensor $\mathbf{B}_e'$ by

$$\mathbf{B}_e' = J_e^{-2/3}\mathbf{B}_e = \mathbf{m}_i' \otimes \mathbf{m}_i' \, , \tag{3.11.23}$$

which can be seen to be a unimodular tensor

$$\begin{aligned} \det\mathbf{B}_e' &= \det(J_e^{-2/3}\mathbf{B}_e) = (J_e^{-2/3})^3\det\mathbf{B}_e = 1 \, , \\ \det\mathbf{B}_e' &= \frac{\mathbf{B}_e'\mathbf{m}^{1\prime} \times \mathbf{B}_e'\mathbf{m}^{2\prime} \cdot \mathbf{B}_e'\mathbf{m}^{3\prime}}{\mathbf{m}^{1\prime} \times \mathbf{m}^{2\prime} \cdot \mathbf{m}^{3\prime}} = \mathbf{m}_1' \times \mathbf{m}_2' \cdot \mathbf{m}_3' = 1 \, . \end{aligned} \tag{3.11.24}$$

Moreover, using the evolution equations (3.11.15) it can be shown that $\mathbf{B}_e'$ satisfies the evolution equation

$$\dot{\mathbf{B}}_e' = \mathbf{L}''\mathbf{B}_e' + \mathbf{B}_e'\mathbf{L}''^T \, , \tag{3.11.25}$$

which with the help of (3.11.21) ensures that $\mathbf{B}_e'$ remains unimodular since

$$\dot{\mathbf{B}}_e' \cdot \mathbf{B}_e'^{-1} = 0 \, . \tag{3.11.26}$$

Summary
For anisotropic response, the elastic deformations can be characterized by a right-handed triad of linearly independent microstructural vectors $\mathbf{m}_i$, which satisfy the evolution equations

$$\dot{\mathbf{m}}_i = \mathbf{L}\,\mathbf{m}_i \,. \tag{3.11.27}$$

Alternatively, the elastic deformations can be characterized by the elastic dilatation J_e and the elastic distortional deformation vectors $\mathbf{m}'_i$, which satisfy the evolution equations

$$\dot{J}_e = J_e\,\mathbf{D}\cdot\mathbf{I}\,, \quad \dot{\mathbf{m}}'_i = \mathbf{L}''\,\mathbf{m}'_i\,, \tag{3.11.28}$$

where $\mathbf{L}''$ is the deviatoric part of $\mathbf{L}$.

For isotropic response the elastic deformations can be characterized by the elastic deformation tensor $\mathbf{B}_e$, which satisfies the evolution equation

$$\dot{\mathbf{B}}_e = \mathbf{L}\mathbf{B}_e + \mathbf{B}_e\mathbf{L}^T\,, \tag{3.11.29}$$

or, alternatively, by the elastic dilatation J_e and the elastic distortional deformation $\mathbf{B}'_e$, which satisfy the evolution equations

$$\dot{J}_e = J_e\,\mathbf{D}\cdot\mathbf{I}\,, \quad \dot{\mathbf{B}}'_e = \mathbf{L}''\mathbf{B}'_e + \mathbf{B}'_e\mathbf{L}''^T\,. \tag{3.11.30}$$

Equations (3.11.27)–(3.11.30) represent Eulerian formulations of evolution equations because they depend only on quantities that can be determined in the current state of the material.

Transformations Under SRBM
Under Superposed Rigid Body Motions SRBM the quantities $\mathbf{m}_i$, m_{ij}, $\mathbf{m}^i$, J_e, $\mathbf{m}'_i$, m'_{ij}, $\mathbf{m}^{i\prime}$, $\mathbf{B}_e$ and $\mathbf{B}'_e$ transform to $\mathbf{m}^+_i$, m^+_{ij}, $\mathbf{m}^{i+}$, J^+_e, $\mathbf{m}'^+_i$, m'^+_{ij}, $\mathbf{m}^{i\prime+}$, $\mathbf{B}^+_e$, and $\mathbf{B}'^+_e$, such that

$$\begin{aligned}
&\mathbf{m}^+_i = \mathbf{Q}\mathbf{m}_i\,, && m^+_{ij} = m_{ij}\,, && \mathbf{m}^{i+} = \mathbf{Q}\mathbf{m}^i\,,\\
&J^+_e = J_e\,, && \mathbf{m}'^+_i = \mathbf{Q}\mathbf{m}'_i\,, && m'^+_{ij} = m'_{ij}\,,\\
&\mathbf{m}^{i\prime+} = \mathbf{Q}\mathbf{m}^{i\prime}\,, && \mathbf{B}^+_e = \mathbf{Q}\mathbf{B}_e\mathbf{Q}^T\,, && \mathbf{B}'^+_e = \mathbf{Q}\mathbf{B}'_e\mathbf{Q}^T\,.
\end{aligned} \tag{3.11.31}$$

These transformation relations make the evolution equations form-invariant under SRBM, so for examples (3.11.6) and (3.11.20) are consistent with the evolution equations

$$\dot{\mathbf{m}}^+_i = \mathbf{L}^+\,\mathbf{m}^+_i\,, \quad \dot{\mathbf{B}}^+_e = \mathbf{L}^+\mathbf{B}^+_e + \mathbf{B}^+_e\mathbf{L}^{+T}\,, \tag{3.11.32}$$

where under SRBM $\mathbf{L}$ transforms to $\mathbf{L}^+$.

Additional Eulerian Strain Measures
Using the condition (3.11.9) it is convenient to introduce elastic strains e_{ij} measured relative to zero-stress material states by

$$e_{ij} = \frac{1}{2}(m_{ij} - \delta_{ij}) , \tag{3.11.33}$$

which in view of (3.11.10) satisfy the evolution equations

$$\dot{e}_{ij} = (\mathbf{m}_i \otimes \mathbf{m}_j) \cdot \mathbf{D} . \tag{3.11.34}$$

Similarly, using the elastic distortional deformation (3.11.16), the elastic distortional strains e'_{ij} relative to zero-stress material states are defined by

$$e'_{ij} = \frac{1}{2}(m'_{ij} - \delta_{ij}) , \tag{3.11.35}$$

which in view of (3.11.17) satisfy the evolution equations

$$\dot{e}'_{ij} = (\mathbf{m}'_i \otimes \mathbf{m}'_j - \frac{1}{3} m'_{ij}\mathbf{I}) \cdot \mathbf{D} . \tag{3.11.36}$$

In addition, for elastically isotropic response, the elastic distortional strain $\mathbf{g}'_e$ and its deviatoric part $\mathbf{g}''_e$ can be defined by

$$\mathbf{g}'_e = \frac{1}{2}(\mathbf{B}'_e - \mathbf{I}) , \quad \mathbf{g}''_e = \frac{1}{2}\mathbf{B}''_e , \tag{3.11.37}$$

where $\mathbf{B}''_e$ is the deviatoric part of $\mathbf{B}'_e$

$$\mathbf{B}''_e = \mathbf{B}'_e - \frac{1}{3}(\mathbf{B}'_e \cdot \mathbf{I})\,\mathbf{I} . \tag{3.11.38}$$

3.12 Compatibility

Since the velocity gradient $\mathbf{L}$ is defined by the gradient of a velocity field $\mathbf{v}$, it follows that if $\mathbf{v}$ is continuously differentiable with respect to $\mathbf{x}$ then the total deformations are compatible in the sense that a motion $\hat{\mathbf{x}}(\mathbf{X}, t)$ exists and the deformation gradient $\mathbf{F}$ defined in (3.3.1c) is consistent with the value of $\mathbf{F}$ obtained by integrating the evolution equation (3.5.4).

Within the context of the Eulerian formulation for anisotropic elastic solids, the microstructural vectors $\mathbf{m}_i$ obtained by integrating the evolution equations (3.11.6) will also be compatible in the sense that a motion can be characterized by the invertible mapping

$$\mathbf{x} = \mathbf{x}(\theta^i, t) , \tag{3.12.1}$$

where θ^i are convected coordinates. Moreover, since for the elastic case $\mathbf{m}_i$ can be identified as material line elements these convected coordinates can be defined so that

$$\mathbf{m}_i = \frac{\partial \mathbf{x}}{\partial \theta^i}\,. \tag{3.12.2}$$

For a continuously differentiable motion

$$\frac{\partial^2 \mathbf{x}}{\partial \theta^i \partial \theta^j} = \frac{\partial^2 \mathbf{x}}{\partial \theta^j \partial \theta^i}\,, \tag{3.12.3}$$

with $\mathbf{m}_i$ satisfying the integrability conditions

$$\frac{\partial \mathbf{m}_i}{\partial \theta^j} = \frac{\partial \mathbf{m}_j}{\partial \theta^i}\,. \tag{3.12.4}$$

Now, expressing $\mathbf{m}_i$ as functions of $\mathbf{x}$, t and using the fact that

$$\frac{\partial \mathbf{m}_i}{\partial \theta^j} = (\partial \mathbf{m}_i / \partial \mathbf{x})\, \mathbf{m}_j\,, \tag{3.12.5}$$

it is convenient to define the three vectors $\mathbf{c}_k$ by

$$\mathbf{c}_k = \varepsilon_{kij} (\partial \mathbf{m}_i / \partial \mathbf{x})\, \mathbf{m}_j\,. \tag{3.12.6}$$

Then, for elastic response the integrability conditions (3.12.4) require

$$\mathbf{c}_k = 0\,. \tag{3.12.7}$$

For inelastic material response that will be discussed later, the vectors $\mathbf{m}_i$ will be obtained by integrating evolution equations which include an inelastic deformation rate, with $\mathbf{m}_i$ still characterizing elastic deformations. For the general case when $\mathbf{L}$ depends on $\mathbf{x}$ the total deformations and the inelastic deformation rate will be inhomogeneous so the elastic deformations need not be compatible in the sense that $\mathbf{c}_k$ in (3.12.6) need not satisfy the compatibility conditions (3.12.7).

3.13 Strongly Objective, Robust Numerical Integration Algorithms

Since the general equations of continuum mechanics are nonlinear, it is necessary to use numerical methods to obtain solutions to challenging problems. Computational mechanics is a field of mechanics that develops computational methods and applies them to analyze fundamental and practical problems in continuum mechanics.

To this end, a numerical algorithm must be proposed to integrate the Eulerian formulations of the evolution equations for the internal state variables, discussed in the previous sections, over a typical time step that begins at time $t = t_n$ and ends at time $t = t_{n+1}$, with time increment $\Delta t = t_{n+1} - t_n$. Specifically, given the values

$$\mathbf{m}_i(t_n)\,, \quad J_e(t_n)\,, \quad \mathbf{m}_i'(t_n)\,, \quad \mathbf{B}_e(t_n)\,, \quad \mathbf{B}_e'(t_n) \tag{3.13.1}$$

of these internal state variables at the beginning of the time step, it is necessary to develop a numerical algorithm to determine their values

$$\mathbf{m}_i(t_{n+1})\,, \quad J_e(t_{n+1})\,, \quad \mathbf{m}_i'(t_{n+1})\,, \quad \mathbf{B}_e(t_{n+1})\,, \quad \mathbf{B}_e'(t_{n+1}) \tag{3.13.2}$$

at the end of the time step.

Following the work of Simo [12], it is convenient to introduce the relative deformation gradient $\mathbf{F}_r(t)$ from the beginning of a time step, which satisfies the evolution equation and initial condition

$$\dot{\mathbf{F}}_r = \mathbf{L}\mathbf{F}_r\,, \quad \mathbf{F}_r(t_n) = \mathbf{I}\,. \tag{3.13.3}$$

The associated relative dilatation $J_r(t)$ from the beginning of the time step is defined by

$$J_r = \det\mathbf{F}_r\,, \tag{3.13.4}$$

which with the help of (3.5.12) can be seen to satisfy the evolution equation and initial condition

$$\dot{J}_r = J_r\mathbf{F}_r^{-T} \cdot \dot{\mathbf{F}}_r = J_r\,\mathbf{D} \cdot \mathbf{I}\,, \quad J_r(t_n) = 1\,. \tag{3.13.5}$$

Also, the unimodular part $\mathbf{F}_r'$ of $\mathbf{F}_r$ is defined by

$$\mathbf{F}_r' = J_r^{-1/3}\,\mathbf{F}_r\,, \quad \det \mathbf{F}_r' = 1\,, \tag{3.13.6}$$

which satisfies the evolution equation and initial condition

$$\dot{\mathbf{F}}_r' = \mathbf{L}''\,\mathbf{F}_r'\,, \quad \mathbf{F}_r'(t_n) = \mathbf{I}\,, \tag{3.13.7}$$

where $\mathbf{L}''$ is the deviatoric part of $\mathbf{L}$.

These relative deformation quantities J_r, $\mathbf{F}_r$ and $\mathbf{F}_r'$ are independent of arbitrary choices of a reference configuration and therefore can used to integrate Eulerian forms of evolution equations for internal state variables that are themselves independent of arbitrariness of the reference configuration. Also, under SRBM J_r, $\mathbf{F}_r$ and $\mathbf{F}_r'$ transform to J_r^+, $\mathbf{F}_r^+$ and $\mathbf{F}_r'^+$ according to the transformation relations

$$J_r^+ = J_r\,, \quad \mathbf{F}_r^+ = \mathbf{Q}\mathbf{F}_r \quad \mathbf{F}_r'^+ = \mathbf{Q}\mathbf{F}_r'\,. \tag{3.13.8}$$

Specifically, the elastic trial quantities

$$\begin{aligned}
&\mathbf{m}_i^*(t) = \mathbf{F}_r(t)\mathbf{m}_i(t_n)\,, \qquad J_e^*(t) = J_r(t)J_e(t_n)\,,\\
&\mathbf{m}_i'^*(t) = \mathbf{F}_r'(t)\mathbf{m}_i'(t_n)\,, \qquad \mathbf{B}_e^*(t) = \mathbf{F}_r(t)\mathbf{B}_e(t_n)\mathbf{F}_r^T(t)\,,\\
&\mathbf{B}_e'^*(t) = \mathbf{F}_r'(t)\mathbf{B}_e'(t_n)\mathbf{F}_r'^T(t)
\end{aligned} \tag{3.13.9}$$

satisfy the evolution equations and initial conditions

$$\begin{aligned}
&\dot{\mathbf{m}}_i^* = \mathbf{L}\mathbf{m}_i^*\,, && \mathbf{m}_i^*(t_n) = \mathbf{m}_i(t_n)\,,\\
&\dot{J}_e^* = J_e^*\mathbf{D}\cdot\mathbf{I}\,, && J_e^*(t_n) = J_e(t_n)\,,\\
&\dot{\mathbf{m}}_i'^* = \mathbf{L}''\mathbf{m}_i'^*\,, && \mathbf{m}_i'^*(t_n) = \mathbf{m}_i'(t_n)\,.\\
&\dot{\mathbf{B}}_e^* = \mathbf{L}\mathbf{B}_e^* + \mathbf{B}_e^*\mathbf{L}^T\,, && \mathbf{B}_e^*(t_n) = \mathbf{B}_e(t_n)\,,\\
&\dot{\mathbf{B}}_e'^* = \mathbf{L}''\mathbf{B}_e'^* + \mathbf{B}_e'^*\mathbf{L}''^T\,, && \mathbf{B}_e'^*(t_n) = \mathbf{B}_e'(t_n)\,.
\end{aligned} \tag{3.13.10}$$

Also, for later reference it is noted that the deviatoric part $\mathbf{B}_e''^*$

$$\mathbf{B}_e''^* = \mathbf{B}_e'^* - \frac{1}{3}(\mathbf{B}_e'^*\cdot\mathbf{I})\,\mathbf{I}\,, \tag{3.13.11}$$

of the elastic trial $\mathbf{B}_e'^*$ satisfies the evolution equation and initial condition

$$\dot{\mathbf{B}}_e''^* = \mathbf{L}''\mathbf{B}_e'^* + \mathbf{B}_e'^*\mathbf{L}''^T - \frac{2}{3}(\mathbf{B}_e'\cdot\mathbf{D}'')\,\mathbf{I}\,, \quad \mathbf{B}_e''^*(t_n) = \mathbf{B}_e''(t_n)\,. \tag{3.13.12}$$

Consequently, the elastic trial values (3.13.9) and (3.13.11) are exact solutions of the evolution equations (3.11.6), (3.11.5), (3.11.15), (3.11.20), (3.11.25) and (3.13.12), respectively. A fundamental feature of these elastic trial values is that they satisfy the same transformation relations under SRBM as the exact values

$$\begin{aligned}
&\mathbf{m}_i^{*+} = \mathbf{Q}\mathbf{m}_i^*\,, && J_e^{*+} = J_e^*\,, && \mathbf{m}_i'^{*+} = \mathbf{Q}\mathbf{m}_i'^*\,,\\
&\mathbf{B}_e^{*+}(t) = \mathbf{Q}\mathbf{B}_e^*\mathbf{Q}^T\,, && \mathbf{B}_e'^{*+} = \mathbf{Q}\mathbf{B}_e'^*\mathbf{Q}^T\,, && \mathbf{B}_e''^{*+} = \mathbf{Q}\mathbf{B}_e''^*\mathbf{Q}^T\,.
\end{aligned} \tag{3.13.13}$$

In particular, robust, strongly objective numerical algorithms can be developed using these elastic trial values (e.g., [2, 3, 5, 7–10]).

Average Total Deformation Rate
Following the work in [11] the average deformation rate $\tilde{\mathbf{D}}$ in a time step $t_n \le t \le t_{n+1}$ is expressed in the form

$$\tilde{\mathbf{D}} = \frac{1}{3}(\tilde{\mathbf{D}}\cdot\mathbf{I})\,\mathbf{I} + \tilde{\mathbf{D}}''\,, \tag{3.13.14}$$

where $\tilde{\mathbf{D}}''$ is the deviatoric part of $\tilde{\mathbf{D}}$. Integration of the evolution equation (3.13.5) for J_r yields an expression for the average total dilatational rate

$$\tilde{\mathbf{D}}\cdot\mathbf{I} = \frac{1}{\Delta t}\,\ln[J_r(t_{n+1})]\,. \tag{3.13.15}$$

To develop an expression for the average total distortional deformation rate tensor $\tilde{\mathbf{D}}''$ it is convenient to define the unimodular relative deformation tensors

$$\mathbf{C}'_r = \mathbf{F}'^T_r \mathbf{F}'_r\,, \quad \mathbf{B}'_r = \mathbf{F}'_r \mathbf{F}'^T_r\,. \tag{3.13.16}$$

Then, with the help of (3.13.7) it can be shown that

$$\dot{\mathbf{C}}'_r = 2\mathbf{F}'^T_r \mathbf{D}'' \mathbf{F}'_r\,, \quad \mathbf{D}'' = \frac{1}{2}\mathbf{F}'^{-T}_r \dot{\mathbf{C}}'_r \mathbf{F}'^{-1}_r\,, \tag{3.13.17}$$

where $\mathbf{D}''$ is the deviatoric part of $\mathbf{D}$. Moreover, since $\mathbf{C}'_r$ is unimodular, it follows that

$$\dot{\mathbf{C}}'_r \cdot \mathbf{C}'^{-1}_r = 0\,. \tag{3.13.18}$$

This property is satisfied when the derivative $\dot{\mathbf{C}}'_r$ is approximated by

$$\dot{\mathbf{C}}'_r \approx \frac{1}{\Delta t}[\mathbf{C}'_r(t_{n+1}) - \{\frac{3}{\mathbf{C}'^{-1}_r(t_{n+1}) \cdot \mathbf{I}}\}\mathbf{I}]\,. \tag{3.13.19}$$

Then, using the fact that $\mathbf{C}'^{-1}_r \cdot \mathbf{I} = \mathbf{B}'^{-1}_r \cdot \mathbf{I}$, the average total distortional deformation rate $\tilde{\mathbf{D}}''$ during the time step can be approximated by

$$\tilde{\mathbf{D}}'' = \frac{1}{2\Delta t}\left[\mathbf{I} - \{\frac{3}{\mathbf{B}'^{-1}_r(t_{n+1}) \cdot \mathbf{I}}\}\mathbf{B}'^{-1}_r(t_{n+1})\right]\,, \tag{3.13.20}$$

with $\tilde{\mathbf{D}}$ given by

$$\tilde{\mathbf{D}} = \frac{1}{3}\frac{1}{\Delta t}\ln[J_r(t_{n+1})]\mathbf{I} + \frac{1}{2\Delta t}\left[\mathbf{I} - \{\frac{3}{\mathbf{B}'^{-1}_r(t_{n+1}) \cdot \mathbf{I}}\}\mathbf{B}'^{-1}_r(t_{n+1})\right]\,. \tag{3.13.21}$$

3.14 The Total Deformation Gradient Used to Parameterize Specific Solutions

The objective of this section is to discuss differences between an elastic deformation variable that characterizes material response and a total deformation measure which is used to parameterize the solution of a particular problem for an elastic material.

Recall from Sect. 3.13 that the deformation gradient $\mathbf{F}_r(t)$ relative to the initial configuration at $t = 0$ satisfies the evolution equation and initial condition (3.13.3)

$$\dot{\mathbf{F}}_r = \mathbf{L}\mathbf{F}_r\,, \quad \mathbf{F}_r(0) = \mathbf{I}\,, \tag{3.14.1}$$

where $\mathbf{L}$ is the velocity gradient. Also, the relative dilatation J_r and the unimodular part $\mathbf{F}_r'$ of $\mathbf{F}_r$, defined by (3.13.4) and (3.13.6)

$$J_r = \det\mathbf{F}_r\,, \quad \mathbf{F}_r' = J_r^{-1/3}\,\mathbf{F}_r\,, \tag{3.14.2}$$

satisfy the evolution equations and initial conditions (3.13.5) and (3.13.7)

$$\begin{aligned} \dot{J}_r &= J_r\,\mathbf{D}\cdot\mathbf{I}\,, \quad J_r(0) = 1\,, \\ \dot{\mathbf{F}}_r' &= \mathbf{L}''\,\mathbf{F}_r'\,, \quad \mathbf{F}_r'(0) = \mathbf{I}\,, \end{aligned} \tag{3.14.3}$$

where $\mathbf{L}''$ is the deviatoic part of $\mathbf{L}$.

It has been shown in Sect. 3.11 that the microstructural vectors $\mathbf{m}_i$ in the Eulerian formulation are internal state variables in the sense of Onat [4], since their values are measurable by experiments on identical samples of the material in its current state. These vectors characterize elastic deformations and orientations of anisotropy and satisfy the evolution equations (3.11.27)

$$\dot{\mathbf{m}}_i = \mathbf{L}\mathbf{m}_i\,. \tag{3.14.4}$$

It then follows that the elastic dilatation J_e and the elastic distortional deformation vectors $\mathbf{m}_i'$, which satisfy the evolution equations (3.11.28)

$$\dot{J}_e = J_e\,\mathbf{D}\cdot\mathbf{I}\,, \quad \dot{\mathbf{m}}_i' = \mathbf{L}''\,\mathbf{m}_i'\,, \tag{3.14.5}$$

are also internal state variables. Next, let $\mathbf{m}_i(0)$, $J_e(0)$ and $\mathbf{m}_i'(0)$ be the measured values of $\mathbf{m}_i$, J_e and $\mathbf{m}_i'$, respectively, in the initial configuration. Then, using $\mathbf{F}_r$ in (3.14.1) and J_r and $\mathbf{F}_r'$ in (3.14.3), the evolution equations (3.14.4) and (3.14.5) can be integrated to obtain

$$\mathbf{m}_i(t) = \mathbf{F}_r(t)\mathbf{m}_i(0)\,, \quad J_e(t) = J_r(t)J_e(0)\,, \quad \mathbf{m}_i'(t) = \mathbf{F}_r'(t)\mathbf{m}_i'(0)\,. \tag{3.14.6}$$

Next, recall from (3.5.4) that the total deformation gradient $\mathbf{F}$ from the reference configuration satisfies the evolution equation

$$\dot{\mathbf{F}} = \mathbf{L}\mathbf{F}\,. \tag{3.14.7}$$

Consequently, with the help of $\mathbf{F}_r$ in (3.14.1) the total deformation gradient $\mathbf{F}$ is given by

$$\mathbf{F}(t) = \mathbf{F}_r(t)\mathbf{F}(0)\,, \tag{3.14.8}$$

where $\mathbf{F}(0)$ is the initial value of $\mathbf{F}$. Furthermore, $\mathbf{F}_r$ can be written in the form

$$\mathbf{F}_r(t) = \mathbf{F}(t)\mathbf{F}^{-1}(0) = [\mathbf{F}(t)\mathbf{A}][\mathbf{F}(0)\mathbf{A}]^{-1}\,, \quad \det\mathbf{A} > 0\,, \tag{3.14.9}$$

where $\mathbf{A}$ is an arbitrary, time independent, second-order tensor with positive determinant which can be a function of the position of material points. This expression shows that the relative deformation tensor $\mathbf{F}_r$ is insensitive to arbitrariness of the choice of the reference configuration for defining the total deformation gradient $\mathbf{F}$.

For the solution of a particular problem, the initial configuration can be specified to be the reference configuration with

$$\mathbf{F}(t) = \mathbf{F}_r(t) \quad \text{for} \quad \mathbf{F}(0) = \mathbf{I}\,, \tag{3.14.10}$$

so that $\mathbf{F}$ or $\mathbf{F}_r$ can be used to parameterize the solution of a particular problem. However, neither of these tensors can be used to determine the microstructural vectors $\mathbf{m}_i(t)$, which determine elastic deformations and the orientation of directions of anisotropy, without measuring $\mathbf{m}_i(0)$ in the initial configuration.

In summary, although $\mathbf{F}$ or $\mathbf{F}_r$ can be used to parameterize the solution of a particular problem, they are not internal state variables and cannot be used by themselves to characterize the material response of an elastic material.

References

1. Flory PJ (1961) Thermodynamic relations for high elastic materials. Trans Faraday Soc 57:829–838
2. Hollenstein M, Jabareen M, Rubin MB (2013) Modeling a smooth elastic-inelastic transition with a strongly objective numerical integrator needing no iteration. Comput Mech 52:649–667
3. Hollenstein M, Jabareen M, Rubin MB (2015) Erratum to: Modeling a smooth elastic-inelastic transition with a strongly objective numerical integrator needing no iteration. Comput Mech 55:649–667
4. Onat ET (1968) The notion of state and its implications in thermodynamics of inelastic solids. In: Irreversible aspects of continuum mechanics and transfer of physical characteristics in moving fluids, pp 292–314
5. Papes O (2013) Nonlinear continuum mechanics in modern engineering applications. Ph.D. dissertation DISS ETH NO 19956
6. Rubin MB (1994) Plasticity theory formulated in terms of physically based microstructural variables - Part I. Theory. Int J Solids Struct 31:2615–2634
7. Rubin MB, Papes O (2011) Advantages of formulating evolution equations for elastic-viscoplastic materials in terms of the velocity gradient instead of the spin tensor. J Mech Mater Struct 6:529–543
8. Rubin MB (2012) Removal of unphysical arbitrariness in constitutive equations for elastically anisotropic nonlinear elastic-viscoplastic solids. Int J Eng Sci 53:38–45
9. Rubin MB (2019) An Eulerian formulation of Inelasticity - From metal plasticity to growth of biological tissues. Trans R Soc A 377:20180071
10. Rubin MB, Cardiff P (2017) Advantages of formulating an evolution equation directly for elastic distortional deformation in finite deformation plasticity. Comput Mech 60:703–707
11. Rubin MB (2020) A strongly objective expression for the average deformation rate with application to numerical integration algorithms. Finite Elem Anal Des 175:103409
12. Simo JC (1988) A framework for finite strain elastoplasticity based on maximum plastic dissipation and the multiplicative decomposition. Part II: Computational aspects. Comput Methods Appl Mech Eng 68:1–31

Chapter 4
Balance Laws for the Purely Mechanical Theory

Abstract The objective of this chapter is to discuss the balance laws in the purely mechanical theory. Specifically, the conservation of mass and the balances of linear and angular momentum are presented in both global and local forms. The properties of the Cauchy stress tensor are derived and the rate of material dissipation is proposed. Invariance under Superposed Rigid Body Motions (SRBM) is discussed along with the development of the transformation relations for specific tensors. It is shown that the local forms of the balance laws can be derived by using invariance under SRBM of the rate of material dissipation and these transformation relations. Also, linearization of the kinematic quantities and balance laws are discussed.

4.1 Conservation of Mass

The conservation of mass states that mass of a material region remains constant. Since the material region P_0 in the reference configuration is mapped into the material region P in the current configuration, it follows that the conservation of mass requires

$$\int_P \rho dv = \int_{P_0} \rho_0 dV \tag{4.1.1}$$

to be valid for every part P (or P_0) of the body. In this equation, $\rho(\mathbf{x}, t)$ is the mass density (mass per unit current volume) in the current configuration, $\rho_0(\mathbf{X})$ is the reference value of the mass density (mass per unit reference volume) in the reference configuration and dv and dV are the elemental volumes in the current and reference configurations, respectively. Since P_0 and ρ_0 are independent of time, it follows that the conservation of mass requires

$$\frac{d}{dt} \int_P \rho dv = 0\,. \tag{4.1.2}$$

Equations (4.1.1) and (4.1.2) are called global equations because they are formulated by integrating over a finite region of space. To derive the local form of (4.1.1),

M. B. Rubin, *Continuum Mechanics with Eulerian Formulations of Constitutive Equations*, Solid Mechanics and Its Applications 265,
https://doi.org/10.1007/978-3-030-57776-6_4

use is made of (3.3.21) to convert the integral over P to an integral over P_0 to obtain

$$\int_{P_0} (\rho J - \rho_0) dV = 0 \,. \tag{4.1.3}$$

Now, assuming that the integrand in (4.1.3) is a continuous function of space and assuming that (4.1.3) holds for all arbitrary parts P_0 of the body and for all times, and using the theorem proved in Appendix B, it follows that

$$\rho J = \rho_0 \tag{4.1.4}$$

must hold at every point of the body and for all time. The form (4.1.4) is the Lagrangian representation of the local form of conservation of mass. It is considered a local form because it holds at every point in the body.

Alternatively, use can be made of the transport theorem (3.10.4) to rewrite (4.1.2) in the form

$$\int_P (\dot{\rho} + \rho \, \mathrm{div}\mathbf{v}) dv = 0 \,. \tag{4.1.5}$$

Now, assuming that the integrand in (4.1.5) is a continuous function of space and assuming that (4.1.5) holds for all arbitrary parts P of the body and for all times, and using the theorem proved in Appendix B, it follows that

$$\dot{\rho} + \rho \, \mathrm{div}\mathbf{v} = 0 \tag{4.1.6}$$

must hold at every point of the body and for all time. The form (4.1.6) is the Eulerian representation of the local form of conservation of mass. Moreover, using (3.5.3) the conservation of mass can be rewritten in the form

$$\frac{d}{dt}[\ln \rho] = -\mathrm{div}\mathbf{v} = -\mathbf{D} \cdot \mathbf{I} \,. \tag{4.1.7}$$

For later convenience the transport theorem (3.10.4) is used with $\phi = \rho f$ to deduce that

$$\frac{d}{dt} \int_P \rho f dv = \int_P (\dot{\overline{\rho f}} + \rho f \mathrm{div}\mathbf{v}) dv = 0 = \int_P [\rho \dot{f} + (\dot{\rho} + \rho \, \mathrm{div}\mathbf{v}) f] dv = 0 \,. \tag{4.1.8}$$

Thus, with the help of the local form of the conservation of mass (4.1.6), it follows that this equation reduces to

$$\frac{d}{dt} \int_P \rho f dv = \int_P \rho \dot{f} dv = 0 \,. \tag{4.1.9}$$

Letting $dm = \rho dv$ and using the fact that dm is conserved, this result can be rewritten in the form

$$\frac{d}{dt}\int_P \rho f\,dv = \frac{d}{dt}\int f\,dm = \int \dot{f}\,dm = \int_P \rho \dot{f}\,dv = 0\,, \tag{4.1.10}$$

which helps understand the result (4.1.9) that was proved in (4.1.8).

Next, recall the purely kinematical evolution equation (3.5.16) for the total dilatation J

$$\dot{J} = J(\mathbf{D}\cdot\mathbf{I})\,, \tag{4.1.11}$$

and rewrite the conservation of mass (4.1.7) in the form

$$\frac{d}{dt}[\ln(\rho J)] = 0\,. \tag{4.1.12}$$

This equation can be integrated using the values $J = 1$ and $\rho = \rho_0$ in the reference configuration to deduce that

$$J = \frac{\rho_0}{\rho}\,. \tag{4.1.13}$$

Similarly, using the evolution equation (3.11.28) for the elastic dilatation J_e

$$\dot{J}_e = J_e(\mathbf{D}\cdot\mathbf{I})\,, \tag{4.1.14}$$

the conservation of mass (4.1.7) can be rewritten in the form

$$\frac{d}{dt}[\ln(\rho J_e)] = 0\,, \tag{4.1.15}$$

which can be integrated using the condition (1.2.5) that $J_e = 1$ in any zero-stress material state to deduce that

$$J_e = \frac{\rho_z}{\rho}\,, \tag{4.1.16}$$

where ρ_z is the constant mass density of the purely mechanical elastic material in any zero-stress material state.

Although the total dilatation J and the elastic dilatation J_e satisfy the same forms of evolution equations, J is not a state variable since its initial value is not measurable. Specifically, J depends on arbitrariness of the reference configuration through the associated arbitrariness of the value of ρ_0. In contrast, J_e is a state variable since its initial value depends on the current value of the density ρ and on the zero-stress density ρ_z, both of which are measurable. Therefore, the elastic dilatation J_e will be used as a state variable in constitutive equations that are discussed in the next chapter.

4.2 Balances of Linear and Angular Momentum

In the previous section, the conservation of mass equation was discussed, which can be thought of as an equation to determine the mass density ρ. For the purely mechanical theory it is necessary to add two additional balance laws called the balances of linear and angular momentum.

Balance of Linear Momentum
In words, the balance of linear momentum states that the rate of change of the linear momentum of an arbitrary part P of a body is equal to the total external force applied to that part of the body. The total external force is due to two types of forces: body forces which are applied to each point of the part P and surface tractions which are applied to each point of the surface ∂P of P. The body force per unit mass is denoted by the vector $\mathbf{b}$ and the surface traction (i.e., force per unit current area da) is denoted by the traction vector $\mathbf{t}(\mathbf{n})$, which depends explicitly on the unit outward normal $\mathbf{n}$ to the surface ∂P. Then, the global form of the balance of linear momentum can be expressed as

$$\frac{d}{dt}\int_P \rho\mathbf{v}dv = \int_P \rho\mathbf{b}dv + \int_{\partial P} \mathbf{t}(\mathbf{n})da\,, \tag{4.2.1}$$

where the velocity $\mathbf{v}$ is the linear momentum per unit mass.

Balance of Angular Momentum
In words, the balance of angular momentum states that the rate of change of the angular momentum of an arbitrary part P of a body is equal to the total external moment applied to that part of the body by the body forces and surface tractions. In this statement the angular momentum and the moment are referred to the same arbitrary but fixed (inertial) origin. Letting $\mathbf{x}$ be the position vector of an arbitrary material point in P relative to the fixed origin, the global form of the balance of angular momentum can be expressed as

$$\frac{d}{dt}\int_P \mathbf{x}\times\rho\mathbf{v}dv = \int_P \mathbf{x}\times\rho\mathbf{b}dv + \int_{\partial P} \mathbf{x}\times\mathbf{t}(\mathbf{n})da\,. \tag{4.2.2}$$

4.3 Existence of the Stress Tensor

The procedure for proving the existence of the Cauchy stress tensor is called Cauchy's theorem. For this theorem, consider an arbitrary part P of the body with closed boundary ∂P and let P be divided by a material surface S into two parts P_1 and P_2 with closed boundaries ∂P_1 and ∂P_2, respectively. Furthermore, let the intersection of ∂P_1 with ∂P, be denoted by $\partial P'$ and the intersection of ∂P_2 with ∂P be denoted by $\partial P''$ (see Fig. 4.2). Mathematically, this separation is summarized by

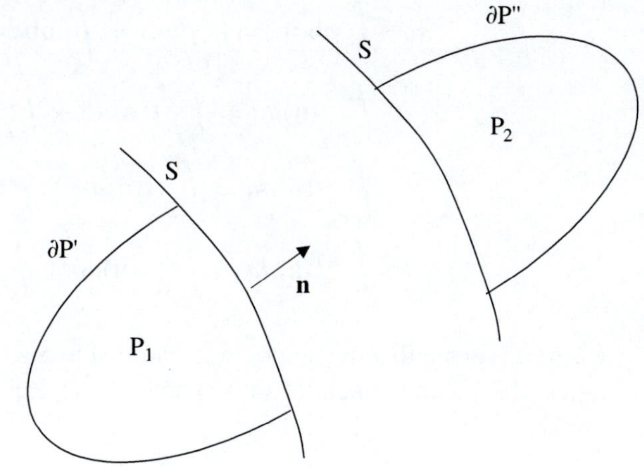

Fig. 4.1 Parts P_1 and P_2 of an arbitrary part P of a body

$$
\begin{aligned}
&P = P_1 \cup P_2\,, \quad \partial P' = \partial P_1 \cap \partial P\,, \quad \partial P'' = \partial P_2 \cap \partial P\,, \\
&\partial P = \partial P' \cup P''\,, \quad \partial P_1 = \partial P' \cup S\,, \quad \partial P_2 = \partial P'' \cup S\,.
\end{aligned} \tag{4.3.1}
$$

Also, let $\mathbf{n}$ be the unit normal to the surface s measured outward from the part P_1 (see Fig. 4.1).

Now recall that the balance of linear momentum is assumed to apply to an arbitrary material part of the body so its application to the parts P, P_1 and P_2 yields

$$
\begin{aligned}
&\frac{d}{dt}\int_P \rho\mathbf{v}dv - \int_P \rho\mathbf{b}dv - \int_{\partial P} \mathbf{t}(\mathbf{n})da = 0\,, \\
&\frac{d}{dt}\int_{P_1} \rho\mathbf{v}dv - \int_{P_1} \rho\mathbf{b}dv - \int_{\partial P_1} \mathbf{t}(\mathbf{n})da = 0\,, \\
&\frac{d}{dt}\int_{P_2} \rho\mathbf{v}dv - \int_{P_2} \rho\mathbf{b}dv - \int_{\partial P_2} \mathbf{t}(\mathbf{n})da = 0\,,
\end{aligned} \tag{4.3.2}
$$

where $\mathbf{n}$ in these equations is considered to be the unit outward normal to the boundary of each part and should not be confused with the specific definition of $\mathbf{n}$ associated with the surface S. Since the regions P, P_1 and P_2 are material and since the local form (4.1.6) of the conservation of mass is assumed to hold in each of these parts, the result (4.1.9) can be used to deduce that

$$
\begin{aligned}
&\frac{d}{dt}\int_P \rho\mathbf{v}dv = \int_P \rho\dot{\mathbf{v}}dv = \int_{P_1} \rho\dot{\mathbf{v}}dv + \int_{P_2} \rho\dot{\mathbf{v}}dv\,, \\
&\frac{d}{dt}\int_{P_1} \rho\mathbf{v}dv = \int_{P_1} \rho\dot{\mathbf{v}}dv\,, \quad \frac{d}{dt}\int_{P_2} \rho\mathbf{v}dv = \int_{P_2} \rho\dot{\mathbf{v}}dv\,.
\end{aligned} \tag{4.3.3}
$$

Also, using (4.3.1), it follows that

$$
\begin{aligned}
\int_P \rho \mathbf{b} dv &= \int_{P_1} \rho \mathbf{b} dv + \int_{P_2} \rho \mathbf{b} dv \,, \\
\int_{\partial P} \mathbf{t}(\mathbf{n}) da &= \int_{\partial P'} \mathbf{t}(\mathbf{n}) da + \int_{\partial P''} \mathbf{t}(\mathbf{n}) da \,, \\
\int_{\partial P_1} \mathbf{t}(\mathbf{n}) da &= \int_{\partial P'} \mathbf{t}(\mathbf{n}) da + \int_S \mathbf{t}(\mathbf{n}) da \,, \\
\int_{\partial P_2} \mathbf{t}(\mathbf{n}) da &= \int_{\partial P''} \mathbf{t}(\mathbf{n}) da + \int_S \mathbf{t}(-\mathbf{n}) da \,,
\end{aligned}
\tag{4.3.4}
$$

where it is noted that the unit outward normal to S associated the boundary of P_2 is $(-\mathbf{n})$. Thus, with the help of (4.3.3) and (4.3.4), Eq. (4.3.2) can be rewritten in the forms

$$
\begin{aligned}
&\Big[\int_{P_1} \rho \dot{\mathbf{v}} dv - \int_{P_1} \rho \mathbf{b} dv - \int_{\partial P'} \mathbf{t}(\mathbf{n}) da\Big] \\
&+ \Big[\int_{P_2} \rho \dot{\mathbf{v}} dv - \int_{P_2} \rho \mathbf{b} dv - \int_{\partial P''} \mathbf{t}(\mathbf{n}) da\Big] = 0 \,,
\end{aligned}
\tag{4.3.5a}
$$

$$
\Big[\int_{P_1} \rho \dot{\mathbf{v}} dv - \int_{P_1} \rho \mathbf{b} dv - \int_{\partial P'} \mathbf{t}(\mathbf{n}) da - \int_S \mathbf{t}(\mathbf{n}) da\Big] = 0 \,,
\tag{4.3.5b}
$$

$$
\Big[\int_{P_2} \rho \dot{\mathbf{v}} dv - \int_{P_2} \rho \mathbf{b} dv - \int_{\partial P''} \mathbf{t}(\mathbf{n}) da - \int_S \mathbf{t}(-\mathbf{n}) da\Big] = 0 \,.
\tag{4.3.5c}
$$

Next, (4.3.5b) and (4.3.5c) can be subtracted from (4.3.5a) to deduce that

$$
\int_S [\mathbf{t}(\mathbf{n}) + \mathbf{t}(-\mathbf{n})] da = 0 \,.
\tag{4.3.6}
$$

Since (4.3.6) must hold for arbitrary material surfaces S and assuming that the integrand is a continuous function of points on S, it follows by a generalization of the result developed in Appendix B that

$$
\mathbf{t}(-\mathbf{n}) = -\mathbf{t}(\mathbf{n})
\tag{4.3.7}
$$

must hold for all points on S and all times. Note that this result, which is called Cauchy's lemma, is the analogue of Newton's law of action and reaction because it states that the traction vector applied by part P_2 on part P_1 is equal in magnitude and opposite in direction to the traction vector applied by part P_1 on part P_2 at a material point on their common boundary S.

In general, the traction vector (or traction vector) $\mathbf{t}$ is a function of position $\mathbf{x}$, time t, and the unit outward normal $\mathbf{n}$ to the surface on which it is applied

$$
\mathbf{t} = \mathbf{t}(\mathbf{x}, t; \mathbf{n}) \,.
\tag{4.3.8}
$$

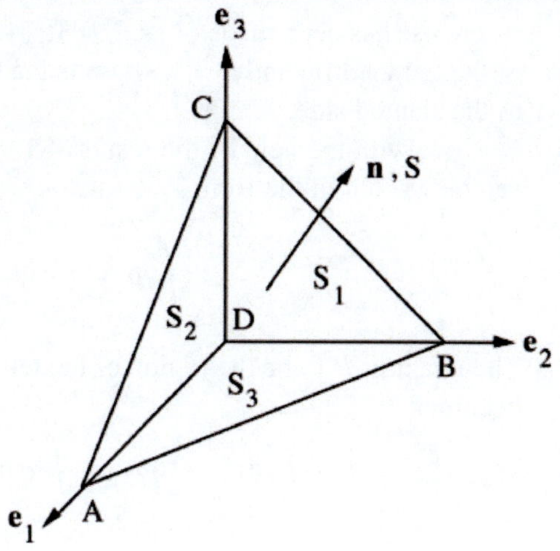

Fig. 4.2 An elemental tetrahedron

The state of stress at a point $\mathbf{x}$ and at time t must be determined by the infinite number of traction vectors obtained by considering all possible orientations $\mathbf{n}$ of planes passing through $\mathbf{x}$ at time t. However, it is not necessary to consider all possible orientations. To verify this statement, it is noted that the simplest polyhedron is a tetrahedron with four faces and that any three-dimensional region of space can be approximated to any finite degree of accuracy using a finite collection of tetrahedrons. Therefore, by analyzing the state of stress in a simple tetrahedron it is possible to analyze the stress at a point in an arbitrary body.

To this end, consider the tetrahedron with three faces that are perpendicular to the Cartesian base vectors $\mathbf{e}_i$, and whose fourth face is defined by the unit outward normal vector $\mathbf{n}$ (see Fig. 4.2). Specifically, let the vertex D (Fig. 4.2) be located at an arbitrary material point in the part P of the body, the surfaces perpendicular to $\mathbf{e}_i$ have surface areas S_i, respectively, and the slanted surface whose normal is $\mathbf{n}$ has surface area S. Denoting $\mathbf{x}_{AD}$, $\mathbf{x}_{BD}$ and $\mathbf{x}_{CD}$ as the vectors from the vertex D to the vertices A, B and C, respectively, it follows by vector algebra that

$$\begin{aligned}
2\,S\mathbf{n} &= (\mathbf{x}_{BD} - \mathbf{x}_{AD}) \times (\mathbf{x}_{CD} - \mathbf{x}_{AD})\,, \\
2\,S\mathbf{n} &= (\mathbf{x}_{BD} \times \mathbf{x}_{CD}) + (\mathbf{x}_{CD} \times \mathbf{x}_{AD}) + (\mathbf{x}_{AD} \times \mathbf{x}_{BD})\,, \\
2\,S\mathbf{n} &= 2\,S_1\mathbf{e}_1 + 2\,S_2\mathbf{e}_2 + 2\,S_3\mathbf{e}_3\,,
\end{aligned} \tag{4.3.9}$$

so the areas S_j are related to S and $\mathbf{n}$ by the formula

$$S_j = \mathbf{e}_j \cdot S\mathbf{n} = Sn_j\,, \tag{4.3.10}$$

where $n_j = \mathbf{n} \cdot \mathbf{e}_j$ are the Cartesian components of $\mathbf{n}$. Also, the volume of the tetrahedron is given by

$$V_{\text{tet}} = \frac{1}{6}(\mathbf{x}_{BD} - \mathbf{x}_{AD}) \times (\mathbf{x}_{CD} - \mathbf{x}_{AD}) \cdot \mathbf{x}_{CD} = \frac{1}{6}(2\,S\mathbf{n}) \cdot \mathbf{x}_{CD} = \frac{1}{3}Sh\,, \tag{4.3.11}$$

where use has been made of (4.3.9). In (4.3.11) S is the area of the slanted side ABC of the tetrahedron and $h = \mathbf{x}_{CD} \cdot \mathbf{n}$ is the height of the tetrahedron measured normal to the slanted side.

Now, with the help of the result (4.1.9), the balance of linear momentum (4.2.1) can be written in the form

$$\int_P \rho(\dot{\mathbf{v}} - \mathbf{b})dv = \int_{\partial P} \mathbf{t}(\mathbf{n})da \,. \tag{4.3.12}$$

Then, taking P to be the region of the tetrahedron, this balance of linear momentum becomes

$$\int_P \rho(\dot{\mathbf{v}} - \mathbf{b})dv = \int_S \mathbf{t}(\mathbf{n})da + \sum_{j=1}^{3} \int_{S_j} \mathbf{t}(-\mathbf{e}_j)da \,. \tag{4.3.13}$$

However, by Cauchy's lemma (4.3.7) yields

$$\mathbf{t}(-\mathbf{e}_j) = -\mathbf{t}(\mathbf{e}_j) \,. \tag{4.3.14}$$

Next, defining the three vectors $\mathbf{T}_j$ to be the traction vectors applied to the surfaces whose outward normals are $\mathbf{e}_j$

$$\mathbf{T}_j = \mathbf{t}(\mathbf{e}_j) \,, \tag{4.3.15}$$

and equation (4.3.13) becomes

$$\int_P \rho(\dot{\mathbf{v}} - \mathbf{b})dv = \int_S \mathbf{t}(\mathbf{n})da - \sum_{j=1}^{3} \int_{S_j} \mathbf{T}_j da \,. \tag{4.3.16}$$

Assuming that the term $\rho(\dot{\mathbf{v}} - \mathbf{b})$ is bounded and recalling that

$$\left| \int_P f dv \right| \leq \int_P |f| dv \,, \tag{4.3.17}$$

it follows that there exists a positive constant K such that

$$\left| \int_P \rho(\dot{\mathbf{v}} - \mathbf{b})dv \right| \leq \int_P |\rho(\dot{\mathbf{v}} - \mathbf{b})|dv \leq \int_P K dv = K \int_P dv = K\frac{1}{3}Sh \,. \tag{4.3.18}$$

Furthermore, assuming that the traction vector is a continuous function of the position $\mathbf{x}$ and the normal $\mathbf{n}$, the mean value theorem for integrals states that there exist points on the surfaces S, S_j for which the values $\mathbf{t}^*(\mathbf{n})$ and $\mathbf{T}_j^*$ of the quantities $\mathbf{t}(\mathbf{n})$ and $\mathbf{T}_j$, respectively, evaluated at these points are related to the integrals of the quantities by the expressions

$$\int_S \mathbf{t}(\mathbf{n})da = \mathbf{t}^*(\mathbf{n})S\,, \quad \sum_{j=1}^{3} \int_{S_j} \mathbf{T}_j da = \mathbf{T}_j^* S_j = \mathbf{T}_j^* n_j S\,, \tag{4.3.19}$$

where use has been made of the result (4.3.10). Then, with the help of (4.3.13) and (4.3.19), Eq. (4.3.18) yields

$$\left|\mathbf{t}^*(\mathbf{n}) - \mathbf{T}_j^* n_j\right| \leq \frac{1}{3} K h\,, \tag{4.3.20}$$

where the result has been divided by the positive area S. Now, considering the set of similar tetrahedrons with the same vertex D and with diminishing heights h, it follows that in the limit that h approaches zero

$$\mathbf{t}^*(\mathbf{n}) = \mathbf{T}_j^* n_j\,. \tag{4.3.21}$$

However, in this limit all functions are evaluated at the same point $\mathbf{x}$ so the star notation can be surpressed to deduce that

$$\mathbf{t}(\mathbf{n}) = \mathbf{T}_j n_j\,. \tag{4.3.22}$$

Also, since $\mathbf{x}$ was an arbitrary point in the above argument, it follows that this equation must hold for all points $\mathbf{x}$, all normals $\mathbf{n}$ and all times.

In words, the result (4.3.22) states that the traction vector on an arbitrary surface can be expressed as a linear combination of the traction vectors applied to the surfaces whose unit normals are in the coordinate directions $\mathbf{e}_j$ and that the coefficients are the components of the normal $\mathbf{n}$ in the directions $\mathbf{e}_j$. Notice that by introducing the definition

$$\mathbf{T} = \mathbf{T}_j \otimes \mathbf{e}_j\,, \tag{4.3.23}$$

equation (4.3.15) and the expression (4.3.22) for the traction vector (or stress vector) can be written in the alternative forms

$$\mathbf{T}_j = \mathbf{T}\mathbf{e}_j\,, \quad \mathbf{t}(\mathbf{n}) = \mathbf{T}\mathbf{n}\,. \tag{4.3.24}$$

It follows from (4.3.24) that since $\mathbf{T}$ transforms an arbitrary vector $\mathbf{n}$ into a vector $\mathbf{t}$, $\mathbf{T}$ must be a second-order tensor. This tensor $\mathbf{T}$ is called the Cauchy stress tensor and its Cartesian components T_{ij} are given by

$$T_{ij} = \mathbf{T} \cdot (\mathbf{e}_i \otimes \mathbf{e}_j) = \mathbf{e}_i \cdot \mathbf{T}_j\,, \tag{4.3.25}$$

so the component form of (4.3.24) becomes

$$t_i = T_{ij} n_j\,, \tag{4.3.26}$$

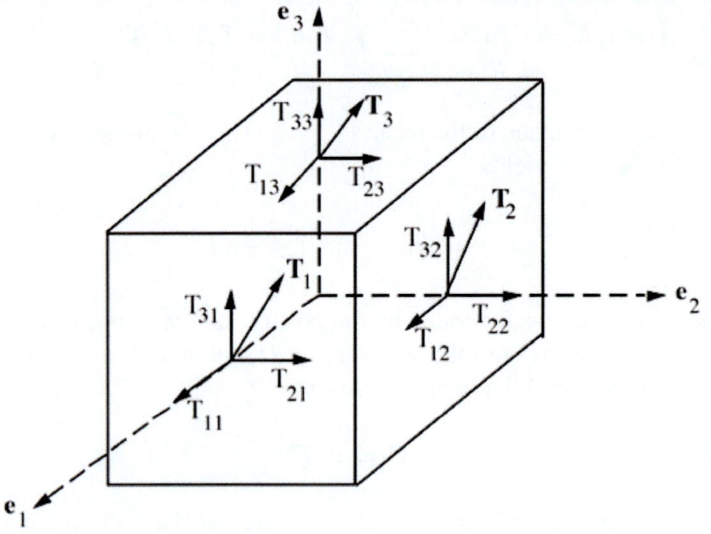

Fig. 4.3 Components of the stress tensor

where t_i are the Cartesian components of $\mathbf{t}$. Furthermore, in view of (4.3.15), it follows that components T_{ij} of $\mathbf{T}_j$ are the components of the traction vectors on the surfaces whose outward normals are $\mathbf{e}_j$ (see Fig. 4.3) and that the first index i of T_{ij} refers to the direction of the component of the traction vector and the second index j of T_{ij} refers to the plane on which the traction vector acts.

The existence of the Cauchy stress tensor $\mathbf{T}$ and the linear relationship (4.3.24) between the traction vector $\mathbf{t}$ and the unit normal $\mathbf{n}$ is called Cauchy's theorem. It is important to emphasize that the Cauchy stress tensor $\mathbf{T}(\mathbf{x}, t)$ is a function of position $\mathbf{x}$ and time t and in particular is independent of the normal $\mathbf{n}$. Therefore the state of stress at a point is characterized by the stress tensor $\mathbf{T}$. On the other hand, the traction vector $\mathbf{t}(\mathbf{x}, t; \mathbf{n})$ includes explicit dependence on the normal $\mathbf{n}$ and characterizes the force per unit present area acting on the particular plane defined by $\mathbf{n}$ that passes through the point $\mathbf{x}$ at time t.

The traction vector $\mathbf{t}$ on any surface can be separated into a component $\mathbf{t}_n$ normal to the surface and a component $\mathbf{t}_s$ parallel to the surface, such that

$$\mathbf{t} = \mathbf{t}_n + \mathbf{t}_s \,, \quad \mathbf{t}_n = \sigma \mathbf{n} \,, \quad \mathbf{t}_s = \tau \mathbf{s} \,, \tag{4.3.27}$$

where the normal stress σ, the magnitude τ of the shearing component and the shearing direction $\mathbf{s}$ are defined by

$$\begin{aligned} &\sigma = \mathbf{t} \cdot \mathbf{n} \,, && \tau = |\mathbf{t}_s| = (\mathbf{t} \cdot \mathbf{t} - \sigma^2)^{1/2} \,, \\ &\mathbf{s} = \tfrac{\mathbf{t}_s}{\tau} = \tfrac{\mathbf{t} - \sigma \mathbf{n}}{\tau} \,, && \mathbf{s} \cdot \mathbf{s} = 1 \,. \end{aligned} \tag{4.3.28}$$

It is important to note that σ and τ are functions of the state of the material through the value of the stress tensor $\mathbf{T}$ at the point of interest and are functions of the normal $\mathbf{n}$ to the plane of interest.

Sometimes a failure criterion for a brittle material is formulated in terms of a critical value of tensile stress whereas a failure criterion (like the Tresca condition) for a metal is formulated in terms of a critical value of the shear stress. Consequently, it is natural to determine the maximum values of the normal stress σ and the shear stress τ. It will be shown in the next section that the stress tensor $\mathbf{T}$ must be symmetric. Here, it is convenient to use the symmetry of $\mathbf{T}$ to rewrite terms in Eq. (4.3.28) in the forms

$$\sigma = \mathbf{T} \cdot (\mathbf{n} \otimes \mathbf{n}), \quad \tau^2 = \mathbf{T}^2 \cdot (\mathbf{n} \otimes \mathbf{n}) - \sigma^2. \tag{4.3.29}$$

Then, it is necessary to search for critical values of σ and τ as functions $\mathbf{n}$. However, it is important to remember that the components of $\mathbf{n}$ are not independent because $\mathbf{n}$ must be a unit vector

$$\mathbf{n} \cdot \mathbf{n} = 1. \tag{4.3.30}$$

Appendix C reviews the method of Lagrange Multipliers, which is used to determine critical values of functions subject to constraints, and Appendix D determines the critical values of σ and τ. In particular, it is recalled that the critical values of σ occur on the planes whose normals are parallel to the principal directions of the stress tensor $\mathbf{T}$. Also, letting σ_1, σ_2 and σ_3 be the ordered principal values of $\mathbf{T}$ and $\mathbf{p}_1$, $\mathbf{p}_2$ and $\mathbf{p}_3$ be the associated orthonormal vectors in the principal directions of $\mathbf{T}$

$$\mathbf{T}\mathbf{p}_1 = \sigma_1 \mathbf{p}_1, \quad \mathbf{T}\mathbf{p}_2 = \sigma_2 \mathbf{p}_2, \quad \mathbf{T}\mathbf{p}_3 = \sigma_3 \mathbf{p}_3, \quad \sigma_1 \geq \sigma_2 \geq \sigma_3, \tag{4.3.31}$$

it can be shown that σ is bounded by the values σ_1 and σ_3

$$\sigma_1 \geq \sigma \geq \sigma_3. \tag{4.3.32}$$

Therefore, the maximum value of tensile stress σ equals σ_1 and it occurs on the plane whose normal is in the direction $\mathbf{p}_1$. Moreover, it can be shown that the traction vector acting on this critical plane has no shearing component

$$\mathbf{t} = \sigma_1 \mathbf{n}, \quad \sigma = \sigma_1, \quad \tau = 0 \quad \text{for} \quad \mathbf{n} = \pm \mathbf{p}_1. \tag{4.3.33}$$

In Appendix D it is also shown that the maximum shear stress τ_{max} occurs on a plane which bisects the planes defined by the plane of maximum tensile stress σ_1 with normal $\mathbf{p}_1$ and the plane of minimum tensile stress σ_3 with normal $\mathbf{p}_3$, such that

$$\sigma = \frac{1}{2}(\sigma_1 + \sigma_3), \quad \tau_{max} = \frac{1}{2}(\sigma_1 - \sigma_3) \quad \text{for} \quad \mathbf{n} = \pm \frac{1}{\sqrt{2}}(\mathbf{p}_1 \pm \mathbf{p}_3). \tag{4.3.34}$$

Notice that on this plane the normal stress σ does not necessarily vanish so the traction vector $\mathbf{t}$ does not apply a pure shear stress on the plane where τ is maximum.

4.4 Local Forms of the Balance Laws

Assuming sufficient continuity and using the local form of conservation of mass (4.1.6) together with the result (4.1.9), it follows that

$$\begin{aligned} \frac{d}{dt}\int_P \rho\mathbf{v}dv &= \int_P \rho\dot{\mathbf{v}}dv\,, \\ \frac{d}{dt}\int_P \mathbf{x}\times\rho\mathbf{v}dv &= \int_P \rho\,\overline{\mathbf{x}\times\mathbf{v}}^{\,\cdot}\,dv = \int_P \mathbf{x}\times\rho\dot{\mathbf{v}}dv\,. \end{aligned} \tag{4.4.1}$$

Also, using the relationship (4.3.24) between the traction vector $\mathbf{t}$, the stress tensor $\mathbf{T}$ and the unit normal $\mathbf{n}$ together with the divergence theorem (2.5.10) yields

$$\begin{aligned} \int_{\partial P} \mathbf{t}da &= \int_{\partial P} \mathbf{T}\mathbf{n}da = \int_P \mathrm{div}\mathbf{T}dv\,, \\ \int_{\partial P} \mathbf{x}\times\mathbf{t}da &= \int_{\partial P} \mathbf{x}\times\mathbf{T}\mathbf{n}da = \int_P \mathrm{div}(\mathbf{x}\times\mathbf{T})dv\,, \\ \int_{\partial P} \mathbf{x}\times\mathbf{t}da &= \int_P (\mathbf{e}_j\times\mathbf{T}_j+\mathbf{x}\times\mathrm{div}\mathbf{T})dv\,, \end{aligned} \tag{4.4.2}$$

where (2.5.4) and (4.3.24) have been used to deduce that

$$\begin{aligned} \mathrm{div}(\mathbf{x}\times\mathbf{T}) &= (\mathbf{x}\times\mathbf{T}),_j\cdot\mathbf{e}_j = (\mathbf{x},_j\times\mathbf{T}+\mathbf{x}\times\mathbf{T},_j)\cdot\mathbf{e}_j\,, \\ \mathrm{div}(\mathbf{x}\times\mathbf{T}) &= \mathbf{e}_j\times\mathbf{T}\mathbf{e}_j+\mathbf{x}\times(\mathbf{T},_j\cdot\mathbf{e}_j) = \mathbf{e}_j\times\mathbf{T}\mathbf{e}_j+\mathbf{x}\times\mathrm{div}\mathbf{T}\,. \end{aligned} \tag{4.4.3}$$

Now, the balance of linear momentum (4.2.1) can be rewritten in the form

$$\int_{\partial P} (\rho\dot{\mathbf{v}}-\rho\mathbf{b}-\mathrm{div}\mathbf{T})dv = 0\,. \tag{4.4.4}$$

Assuming that the integrand in (4.4.4) is a continuous function and that (4.4.4) must hold for arbitrary regions P, it follows from the results of Appendix B that

$$\rho\dot{\mathbf{v}} = \rho\mathbf{b}+\mathrm{div}\mathbf{T} \tag{4.4.5}$$

must hold for each point of P and for all time. Letting v_i, b_i and T_{ij} be the Cartesian components of $\mathbf{v}$, $\mathbf{b}$ and $\mathbf{T}$, respectively, the component form of balance of linear momentum becomes

$$\rho\dot{v}_i = \rho b_i + T_{ij},_j\,. \tag{4.4.6}$$

Similarly, the balance of angular momentum (4.2.2) can be rewritten in the form

$$\int_{\partial P} [\mathbf{x}\times(\rho\dot{\mathbf{v}}-\rho\mathbf{b}-\mathrm{div}\mathbf{T})-\mathbf{e}_j\times\mathbf{T}_j]dv = 0\,. \tag{4.4.7}$$

Assuming that the integrand in (4.4.7) is a continuous function, using the local form (4.4.5) of balance of linear momentum and assuming that (4.4.7) must hold for arbitrary regions P, it follows from the results of Appendix B that

$$\mathbf{e}_j \times \mathbf{T}_j = 0 \tag{4.4.8}$$

must hold for each point of P and for all time. Then, using (2.4.19) and (4.3.24) this equation can be rewritten in the form

$$\mathbf{e}_j \times \mathbf{T}\mathbf{e}_j = \boldsymbol{\varepsilon} \cdot (\mathbf{e}_j \otimes \mathbf{T}\mathbf{e}_j) = \boldsymbol{\varepsilon} \cdot (\mathbf{e}_j \otimes \mathbf{e}_j \mathbf{T}^T) = \boldsymbol{\varepsilon} \cdot (\mathbf{I}\,\mathbf{T}^T) = \boldsymbol{\varepsilon} \cdot \mathbf{T}^T \,. \tag{4.4.9}$$

Since $\boldsymbol{\varepsilon}$ is skew-symmetric in any two of its indices, it follows that the local form of angular momentum requires the stress tensor to be symmetric

$$\mathbf{T}^T = \mathbf{T}\,, \quad T_{ji} = T_{ij} \,. \tag{4.4.10}$$

4.5 Rate of Material Dissipation

Within the context of the purely mechanical theory, it is possible to define the rate of material dissipation $\mathcal{D}$ per unit current volume by the equation

$$\int_P \mathcal{D} dv = \mathcal{W} - \dot{\mathcal{K}} - \dot{\mathcal{U}} \geq 0\,, \tag{4.5.1}$$

where the rate of work $\mathcal{W}$ done on the body, the kinetic energy $\mathcal{K}$, the total strain energy $\mathcal{U}$ are defined by

$$\mathcal{W} = \int_P \rho \mathbf{b} \cdot \mathbf{v} dv + \int_{\partial P} \mathbf{t} \cdot \mathbf{v} da\,, \quad \mathcal{K} = \int_P \frac{1}{2} \rho \mathbf{v} \cdot \mathbf{v} dv\,, \quad \mathcal{U} = \int_P \rho \Sigma dv\,, \tag{4.5.2}$$

and Σ is the strain energy function per unit mass. From these expressions it can be seen that $\mathcal{D}$ measures the rate of work done on the body which is not stored in kinetic or strain energy. When $\mathcal{D}$ is positive this excess rate of work is dissipative. Next, using the divergence theorem (2.5.10) and (4.3.24), the rate of work can be expressed in the form

$$\mathcal{W} = \int_P [(\rho \mathbf{b} + \text{div}\mathbf{T}) \cdot \mathbf{v} + \mathbf{T} \cdot \mathbf{L}] dv \,. \tag{4.5.3}$$

Also, using (4.1.8) the rates of change of the kinetic and strain energies are given by

$$\dot{\mathcal{K}} = \int_P [\rho \dot{\mathbf{v}} \cdot \mathbf{v} + \frac{1}{2} (\dot{\rho} + \rho\, \text{div}\mathbf{v})(\mathbf{v} \cdot \mathbf{v})] dv\,, \quad \dot{\mathcal{U}} = \int_P [\rho \dot{\Sigma} dv + (\dot{\rho} + \rho\, \text{div}\mathbf{v}) \Sigma] dv\,, \tag{4.5.4}$$

so that (4.5.1) can be rewritten as

$$\int_P [\mathcal{D} - (\rho \mathbf{b} + \text{div}\mathbf{T} - \rho \dot{\mathbf{v}}) \cdot \mathbf{v} - \mathbf{T} \cdot \mathbf{L} + \rho \dot{\Sigma} - (\dot{\rho} + \rho\, \text{div}\mathbf{v})(\frac{1}{2}\mathbf{v} \cdot \mathbf{v} + \Sigma)]dv = 0\,. \tag{4.5.5}$$

Then, assuming that this equation is valid for all parts P of the body and that the integrand is a continuous function, it follows that

$$\mathcal{D} = \mathbf{T} \cdot \mathbf{L} - \rho \dot{\Sigma} + (\rho \mathbf{b} + \text{div}\mathbf{T} - \rho \dot{\mathbf{v}}) \cdot \mathbf{v} - (\dot{\rho} + \rho\, \text{div}\mathbf{v})(\frac{1}{2}\mathbf{v} \cdot \mathbf{v} + \Sigma) \tag{4.5.6}$$

must hold at each point in P and for all time. Moreover, using the local forms (4.1.6) and (4.4.5) and (4.4.10) of the conservation of mass and the balances of linear and angular momentum, respectively, the rate of material dissipation requires

$$\mathcal{D} = \mathbf{T} \cdot \mathbf{D} - \rho \dot{\Sigma} \geq 0\,. \tag{4.5.7}$$

4.6 Referential Forms of the Equations of Motion

In the previous sections the traction vector $\mathbf{t}$ has been defined as the force per unit area in the current configuration. This leads to a definition of stress which is sometimes referred to as the true stress. Alternatively, since the material surface ∂P in the current configuration maps to the material surface ∂P_0 in the reference configuration, it is possible to define the Piola-Kirchhoff traction vector $\boldsymbol{\pi}$ as the force acting in the current configuration but measured per unit area in the reference configuration. This leads to a definition of stress which is sometimes referred to as engineering stress.

Recalling that the traction vector $\mathbf{t}$ depends on position $\mathbf{x}$, time t, and the unit outward normal $\mathbf{n}$ to the surface ∂P, it follows that the traction vector $\boldsymbol{\pi}$ depends on position $\mathbf{X}$, time t, and the unit outward normal $\mathbf{N}$ to the surface ∂P_0. Thus, the force acting in the current configuration on an arbitrary material part S of the present surface ∂P or the associated material part S_0 of the reference surface ∂P_0 of the body can be expressed in the equivalent forms

$$\int_S \mathbf{t}(\mathbf{n})da = \int_{S_0} \boldsymbol{\pi}(\mathbf{N})dA\,, \tag{4.6.1}$$

where dA is the element of area in the reference configuration. Similarly, the quantities $\mathbf{v}$, $\mathbf{b}$, $\mathbf{x} \times \mathbf{v}$ and $\mathbf{x} \times \mathbf{b}$, respectively, represent the linear momentum, body force, angular momentum and moment of body force per unit mass. Therefore, since ρ_0 is the reference mass density (i.e., mass per unit reference volume), it follows that

$$\begin{aligned}
&\int_P \rho \mathbf{v} dv = \int_{P_0} \rho_0 \mathbf{v} dV\,,\\
&\int_P \rho \mathbf{b} dv = \int_{P_0} \rho_0 \mathbf{b} dV\,,\\
&\int_P \mathbf{x} \times \rho \mathbf{v} dv = \int_{P_0} \mathbf{x} \times \rho_0 \mathbf{v} dV\,,\\
&\int_P \mathbf{x} \times \rho \mathbf{b} dv = \int_{P_0} \mathbf{x} \times \rho_0 \mathbf{b} dV\,,
\end{aligned} \tag{4.6.2}$$

where dV is the element of volume in the reference configuration.

Then, with the help of the results (4.6.1) and (4.6.2), the balances of linear momentum (4.2.1) and angular momentum (4.2.2) can be rewritten in the forms

$$\begin{aligned}
&\frac{d}{dt}\int_{P_0} \rho \mathbf{v} dV = \int_{P_0} \rho_0 \mathbf{b} dV + \int_{\partial P_0} \boldsymbol{\pi}(\mathbf{N}) dA\,,\\
&\frac{d}{dt}\int_{P_0} \mathbf{x} \times \rho \mathbf{v} dV = \int_{P_0} \mathbf{x} \times \rho_0 \mathbf{b} dV + \int_{\partial P_0} \mathbf{x} \times \boldsymbol{\pi}(\mathbf{N}) dA\,.
\end{aligned} \tag{4.6.3}$$

Following similar arguments to those in Sect. 4.3, it can be proved that the traction vector $\boldsymbol{\pi}(\mathbf{N})$ is a linear function of $\mathbf{N}$ which can be represented in the form

$$\boldsymbol{\pi}(\mathbf{X}, t; \mathbf{N}) = \boldsymbol{\Pi}(\mathbf{X}, t)\mathbf{N}\,, \quad \pi_i(\mathbf{X}, t; N_A) = \Pi(\mathbf{X}, t) N_A\,, \quad \boldsymbol{\Pi} = \Pi_{iA} \mathbf{e}_i \otimes \mathbf{e}_A\,, \tag{4.6.4}$$

where π_i are the components of $\boldsymbol{\pi}$, and $\boldsymbol{\Pi}$, with components Π_{iA}, is a second-order tensor called the first Piola-Kirchhoff stress tensor.

With the help of (4.6.4) the local form of balance of linear momentum becomes

$$\rho_0 \dot{\mathbf{v}} = \rho_0 \mathbf{b} + \mathrm{Div}\boldsymbol{\Pi}\,, \quad \rho_0 \dot{v}_i = \rho_0 b_i + \Pi_{iA,A}\,, \tag{4.6.5}$$

where Div denotes the divergence with respect to $\mathbf{X}$ and $(,_A)$ denotes partial differentiation with respect to X_A. To obtain the local form of angular momentum, consider

$$\begin{aligned}
\mathrm{Div}(\mathbf{x} \times \boldsymbol{\Pi}) &= (\mathbf{x} \times \boldsymbol{\Pi}),_A \cdot \mathbf{e}_A = (\mathbf{x},_A \times \boldsymbol{\Pi}) \cdot \mathbf{e}_A + \mathbf{x} \times (\boldsymbol{\Pi},_A \cdot \mathbf{e}_A)\\
&= (\mathbf{F}\mathbf{e}_A) \times (\boldsymbol{\Pi}\mathbf{e}_A) + \mathbf{x} \times (\mathrm{Div}\boldsymbol{\Pi})\,.
\end{aligned} \tag{4.6.6}$$

Thus, with the help of (4.6.5) the local form of the balance of angular momentum yields

$$(\mathbf{F}\mathbf{e}_A) \times (\boldsymbol{\Pi}\mathbf{e}_A) = 0\,. \tag{4.6.7}$$

Next, using (2.4.19) this equation can be written in the form

$$0 = (\mathbf{F}\mathbf{e}_A) \times (\boldsymbol{\Pi}\mathbf{e}_A) = \boldsymbol{\varepsilon} \cdot (\mathbf{F}\mathbf{e}_A \otimes \boldsymbol{\Pi}\mathbf{e}_A) = \boldsymbol{\varepsilon} \cdot (\mathbf{F}\mathbf{e}_A \otimes \mathbf{e}_A \boldsymbol{\Pi}^T) = \boldsymbol{\varepsilon} \cdot (\mathbf{F}\,\mathbf{I}\,\boldsymbol{\Pi}^T)\,,$$

$$\boldsymbol{\varepsilon} \cdot (\mathbf{F}\boldsymbol{\Pi}^T) = 0 . \tag{4.6.8}$$

Thus, since $\boldsymbol{\varepsilon}$ is skew-symmetric in any two of its indices, it follows that the tensor $\mathbf{F}\boldsymbol{\Pi}^T$ must be symmetric

$$\mathbf{F}\boldsymbol{\Pi}^T = (\mathbf{F}\boldsymbol{\Pi}^T)^T = \boldsymbol{\Pi}\mathbf{F}^T , \quad F_{iA}\Pi_{jA} = \Pi_{iA}F_{jA} . \tag{4.6.9}$$

This means that the first Piola-Kirchhoff stress tensor $\boldsymbol{\Pi}$ is not necessarily symmetric.

Since the traction vector $\mathbf{t}$ is related to the Cauchy stress $\mathbf{T}$ by the formula $\mathbf{t}(\mathbf{n}) = \mathbf{Tn}$ and since Eq. (4.6.1) relates the force acting on the material part S of the surface ∂P to the force acting on the material part S_0 of the surface ∂P_0, it should be possible to relate the Cauchy stress $\mathbf{T}$ to the first Piola-Kirchhoff stress $\boldsymbol{\Pi}$. To this end, use is made of Nanson's formula (3.3.35) to deduce that

$$\boldsymbol{\pi}(\mathbf{N})dA = \boldsymbol{\Pi}\mathbf{N}dA = J^{-1}\boldsymbol{\Pi}\mathbf{F}^T\mathbf{n}da , \tag{4.6.10}$$

so that (4.6.1) can be rewritten in the form

$$\int_S (\mathbf{T} - J^{-1}\boldsymbol{\Pi}\mathbf{F}^T)\,\mathbf{n}da = 0 . \tag{4.6.11}$$

Assuming that the integrand is continuous and that S is arbitrary, it follows that

$$(\mathbf{T} - J^{-1}\boldsymbol{\Pi}\mathbf{F}^T)\,\mathbf{n} = 0 \tag{4.6.12}$$

at each point on S. However, since the tensor in the brackets is independent of the normal $\mathbf{n}$, and $\mathbf{n}$ is arbitrary, the Cauchy stress tensor $\mathbf{T}$ must be related to the first Piola-Kirchhoff stress tensor $\boldsymbol{\Pi}$ by

$$\mathbf{T} = J^{-1}\boldsymbol{\Pi}\mathbf{F}^T , \quad T_{ij} = J^{-1}\Pi_{iB}F_{jB} . \tag{4.6.13}$$

Notice that (4.6.9) and (4.6.13) ensure that the Cauchy stress $\mathbf{T}$ is symmetric, which is the same result that was obtained from the balance of angular momentum referred to the current configuration.

The first Piola-Kirchhoff stress $\boldsymbol{\Pi}$, with components Π_{iA}, is a two-point tensor referred to both the current configuration and the reference configuration and it is also called the nonsymmetric Piola-Kirchhoff stress tensor. Sometimes it is convenient to introduce the second Piola-Kirchhoff stress tensor $\mathbf{S}$, with components S_{AB}, which is referred to the reference configuration only and is defined by

$$\boldsymbol{\Pi} = \mathbf{FS} , \quad \Pi_{iB} = F_{iA}S_{AB} . \tag{4.6.14}$$

It then follows from (4.6.9) that $\mathbf{S}$ is a symmetric tensor

$$\mathbf{S}^T = \mathbf{S}\,, \quad S_{BA} = S_{AB}\,. \tag{4.6.15}$$

For this reason **S** is also called the symmetric Piola-Kirchhoff stress tensor. Moreover, it is noted from (4.6.13) and (4.6.14) that the Cauchy stress **T** is related to **S** by the formula

$$\mathbf{T} = J^{-1}\mathbf{F}\mathbf{S}\mathbf{F}^T\,, \quad T_{ij} = J^{-1}F_{iA}S_{AB}F_{jB}\,. \tag{4.6.16}$$

Furthermore, recall that the Cauchy stress **T** can be separated into its spherical part $-p\mathbf{I}$ and its deviatoric part $\mathbf{T}''$, such that

$$\mathbf{T} = -p\,\mathbf{I} + \mathbf{T}''\,, \quad p = -\frac{1}{3}\mathbf{T}\cdot\mathbf{I}\,, \quad \mathbf{T}''\cdot\mathbf{I} = 0\,, \tag{4.6.17}$$

where p denotes the pressure. It follows from (4.6.16) and (4.6.17) that the symmetric Piola-Kirchhoff stress **S** admits an analogous separation

$$\begin{aligned} &\mathbf{S} = J\mathbf{F}^{-1}\mathbf{T}\mathbf{F}^{-T}\,, \quad \mathbf{S} = -p\,\mathbf{C}^{-1} + \mathbf{S}'\,, \\ &p = -\tfrac{1}{3}J^{-1}\mathbf{S}\cdot\mathbf{C}\,, \quad \mathbf{S}' = J\mathbf{F}^{-1}\mathbf{T}''\mathbf{F}^{-T}\,, \quad \mathbf{S}'\cdot\mathbf{C} = 0\,. \end{aligned} \tag{4.6.18}$$

It is important to emphasize that although $\mathbf{T}''$ is a deviatoric tensor, the associated quantity $\mathbf{S}'$ is not a standard deviatoric tensor. Instead, $\mathbf{S}'$ is orthogonal to **C** and not **I**. In addition, using (3.5.5), (4.1.4) and (4.6.16), the rate of material dissipation (4.5.7) requires

$$J\mathcal{D} = 2\mathbf{S}\cdot\dot{\mathbf{C}} - \rho_0\dot{\Sigma} \geq 0\,. \tag{4.6.19}$$

Since the Cauchy stress **T** is a measure of the force per unit present area it is referred to as the *true* stress. Also, since the Piola-Kirchhoff stress $\mathbf{\Pi}$ is a measure of the force per unit reference area it is referred to as the *engineering* stress. The notion of *true* stress somehow infers that it is the correct stress. However, a failure criterion for some materials might be more physical using the Cauchy stress **T** whereas a failure criterion for other materials might be more physical using the Piola-Kirchhoff stress $\mathbf{\Pi}$.

For example, to determine the value of uniaxial stress when a metal bar breaks it is reasonable to attempt to determine the force required to break atomic bonds and the number of atomic bonds that need to be broken to fracture the bar. In this regard, it is noted that plasticity in metal is characterized by dislocations that move through the atomic lattice. Since the elastic distortions required for these dislocations to move are small, the angles and lengths of the atomic lattice remain relatively constant during plastic flow. Consequently, the current area of the necked cross section of the bar when it breaks can be used to estimate the number of atomic bonds that are breaking. This means that the Cauchy stress, which depends on the current deformed cross-sectional area of the bar, might be a more physical measure of the stress needed to break a metal bar.

Alternatively, consider a bar made of a polymer with long coiled polymer chains. To estimate the stress when the bar breaks in uniaxial stress it is necessary to determine the force required to break a polymer chain as well as the number of chains that are being broken. Since the polymer chains are typically coiled, the effective cross-sectional area of the chain decreases as it is stretc.hed and the coils straighten. This means that the deformed cross section of the bar contains about the same number of polymer chains as it did when the chains were coiled and unstretc.hed in a zero-stress reference configuration. Consequently, it might be more physical to use the first Piola-Kirchhoff stress $\mathbf{\Pi}$ to formulate a failure criterion for a polymer bar.

4.7 Invariance Under Superposed Rigid Body Motions (SRBM)

As motivation, consider the simple case of a nonlinear massless elastic spring. Let one end of the spring be attached to a fixed point in a room and attach a ball with mass on the other end. When the ball is thrown in the room it can bounce off the walls. During this motion the force $\mathbf{f}$ applied by the spring on the ball remains parallel to the changing orientation of the spring and its tension depends only on the extension of the spring relative to its zero-stress length. In particular, this tension does not depend on the orientation of the spring. This section uses the kinematics of SRBM discussed in Sect. 3.8 to develop restrictions on constitutive equations which generalize the notion that the constitutive response of the spring relative to its orientation is insensitive to SRBM. Specifically, the force $\mathbf{f}$ in the spring rotates with the changing orientation of the spring and the tension in the spring depends only on the deformation of the spring.

From Sect. 3.8 it is recalled that under SRBM the point $\mathbf{x}$ in the current configuration at time t moves to the point $\mathbf{x}^+$ in the superposed configuration at time $t^+ = t + c$ such that $\mathbf{x}^+$ and $\mathbf{x}$ are related by the mapping

$$\mathbf{x}^+ = \mathbf{c}(t) + \mathbf{Q}(t)\mathbf{x}\,, \quad \mathbf{Q}\mathbf{Q}^T = \mathbf{I}\,, \quad \det\mathbf{Q} = +1\,, \tag{4.7.1}$$

where $\mathbf{c}$ is a vector, and $\mathbf{Q}$ is a second-order proper orthogonal tensor, both functions of time only, and c is a constant time shift. Furthermore, it is recalled from Sect. 3.8 that the mapping (4.7.1) was used to derive a number of expressions for the values of various kinematic quantities associated with the superposed configuration P^+. In this section, expressions will be developed for the superposed values of a number of kinetic quantities that include: the mass density ρ, the traction vector $\mathbf{t}$, the Cauchy stress tensor $\mathbf{T}$, and the body force $\mathbf{b}$. Specifically, expressions will be developed for

$$(\rho^+, \mathbf{t}^+, \mathbf{T}^+, \mathbf{b}^+)\,. \tag{4.7.2}$$

Following the work in [1], the notion of invariance under SRBM is based on the following two restrictions

(R-1): The balance laws must be form-invariant under SRBM, (4.7.3a)

(R-2): The constitutive response of the material relative to its orientation is the same for all SRBM. (4.7.3b)

The first restriction (R-1) requires the balance laws in the superposed configuration with a superscript $()^+$ added to each variable to be valid for all SRBM and the second restriction (R-2) states in what sense the constitutive responses of a material under SRBM are equivalent.

Specifically, (R-1) requires the global forms of the balance laws to be form-invariant in the superposed configuration P^+ with all kinematic and kinetic quantities taking their superposed values in P^+. Therefore, the conservation of mass and balances of linear and angular momentum can be stated relative to P^+ in the forms

$$\frac{d}{dt}\int_{P^+} \rho^+ dv^+ = 0\,, \tag{4.7.4a}$$

$$\frac{d}{dt}\int_{P^+} \rho^+ \mathbf{v}^+ dv^+ = \int_{P^+} \rho^+ \mathbf{b}^+ dv^+ + \int_{\partial P^+} \mathbf{t}^+ da^+\,, \tag{4.7.4b}$$

$$\frac{d}{dt}\int_{P^+} \mathbf{x}^+ \times \rho^+ \mathbf{v}^+ dv^+ = \int_{P^+} \mathbf{x}^+ \times \rho^+ \mathbf{b}^+ dv^+ + \int_{\partial P^+} \mathbf{x}^+ \times \mathbf{t}^+ da^+\,, \tag{4.7.4c}$$

where ∂P^+ is the closed boundary of P^+, with unit outward normal $\mathbf{n}^+$ and the time derivative with respect to t^+ has been replaced by the time derivative with respect to t since $t^+ = t + c$ and c is constant. Using the arguments in Sect. 4.3 it can be shown that a Cauchy tensor $\mathbf{T}^+$ exists which is a function of position and time only, such that

$$\mathbf{t}^+ = \mathbf{T}^+ \mathbf{n}^+\,. \tag{4.7.5}$$

Then, with the help of the divergence theorem (2.5.10) and the transport theorem (4.1.9), the local forms of the conservation of mass (4.7.4a) and the balance laws (4.7.4b) and (4.7.4c) can be written as

$$\frac{d}{dt}[\ln \rho^+] = -(\mathbf{D}^+ \cdot \mathbf{I})\,, \quad \rho^+ \dot{\mathbf{v}}^+ = \rho^+ \mathbf{b}^+ + \mathrm{div}^+ \mathbf{T}^+\,, \quad \mathbf{T}^{+T} = \mathbf{T}^+\,, \tag{4.7.6}$$

where div^+ denotes the divergence operation with respect to $\mathbf{x}^+$ and use has been made of (4.1.7).

Next, with the help of the conservation of mass (4.1.7) and the transformation relation (3.8.24) for $\mathbf{D}$

$$\frac{d}{dt}[\ln \rho] = -\mathbf{D} \cdot \mathbf{I}\,, \quad \mathbf{D}^+ = \mathbf{Q}\mathbf{D}\mathbf{Q}^T\,, \tag{4.7.7}$$

it follows that

$$\frac{d}{dt}\Big[\ln(\frac{\rho^+}{\rho})\Big] = \mathbf{D}\cdot\mathbf{I} - \mathbf{D}^+\cdot\mathbf{I} = 0\,. \tag{4.7.8}$$

Since this equation is valid for all SRBM it can be integrated for the trivial superposed motion with $\mathbf{x}^+ = \mathbf{x}$ to deduce that

$$\rho^+ = \rho\,. \tag{4.7.9}$$

Next, with the help of (2.5.4) and (4.7.1) it can be shown that

$$\begin{aligned}&\text{div}^+\mathbf{T}^+ = (\partial\mathbf{T}^+/\partial\mathbf{x}^+)\cdot\mathbf{I} = (\partial\mathbf{T}^+/\partial\mathbf{x})(\partial\mathbf{x}/\partial\mathbf{x}^+)\cdot\mathbf{I} = (\partial\mathbf{T}^+/\partial\mathbf{x})\mathbf{Q}^T\cdot\mathbf{I}\,,\\ &\text{div}^+\mathbf{T}^+ = \text{div}(\mathbf{T}^+\mathbf{Q})\,,\quad \mathbf{Q}\text{div}\mathbf{T} = \text{div}(\mathbf{Q}\mathbf{T})\,.\end{aligned} \tag{4.7.10}$$

It then follows that the balance of linear momentum in $(4.7.6)_2$ can be solved for the body force $\mathbf{b}^+$ to obtain

$$\mathbf{b}^+ = \dot{\mathbf{v}}^+ - \frac{1}{\rho}\text{div}(\mathbf{T}^+\mathbf{Q})\,. \tag{4.7.11}$$

Moreover, the balance of linear momentum in the current configuration yields an expression for the body force given by

$$\mathbf{b} = \dot{\mathbf{v}} - \frac{1}{\rho}\text{div}(\mathbf{T})\,. \tag{4.7.12}$$

Now, taking the juxtaposition of $\mathbf{Q}$ on the left-hand side of (4.7.12) and subtracting the result from (4.7.11) yields

$$\mathbf{b}^+ = \dot{\mathbf{v}}^+ + \mathbf{Q}(\mathbf{b} - \dot{\mathbf{v}}) - \frac{1}{\rho}\text{div}(\mathbf{T}^+\mathbf{Q} - \mathbf{Q}\mathbf{T})\,. \tag{4.7.13}$$

In this regard, it is noted that the restriction (R-1) tacitly assumes that the balance of linear momentum is valid for any specified body force. Consequently, it is also valid for the body force (4.7.13), which enforces SRBM.

Since the traction vector $\mathbf{t}$ characterizes the constitutive response of the material and the unit outward normal vector $\mathbf{n}$ rotates with the material under SRBM, it follows that the restriction (R-2) in (4.7.3b) requires[1]

$$\mathbf{t}^+\cdot\mathbf{n}^+ = \mathbf{t}\cdot\mathbf{n} \tag{4.7.14}$$

to be valid for all material points, all unit normals $\mathbf{n}$ and all SRBM. Now, using Cauchy's theorem (4.3.24) for the current configuration, (4.7.5) for the superposed configuration and the kinematic result (3.8.20) that $\mathbf{n}$ rotates under SRBM

[1]This proposal was presented in a class that I took from Prof. P.M. Naghdi in the fall of 1973.

$$\mathbf{n}^+ = \mathbf{Q}\mathbf{n}\,, \tag{4.7.15}$$

the expression (4.7.14) can be rewritten in the form

$$(\mathbf{T}^+ - \mathbf{Q}\mathbf{T}\mathbf{Q}^T) \cdot \mathbf{n}^+ \otimes \mathbf{n}^+ = 0\,. \tag{4.7.16}$$

Since the coefficient in the brackets is symmetric and independent of $\mathbf{n}^+$ and since $\mathbf{n}^+$ is an arbitrary unit vector, the Cauchy stress tensor must satisfy the restriction that

$$\mathbf{T}^+ = \mathbf{Q}\mathbf{T}\mathbf{Q}^T\,, \tag{4.7.17}$$

is valid for all SRBM. This transformation relation automatically satisfies the reduced form (4.7.6) of the balance of angular momentum when $\mathbf{T}$ satisfies the reduced form (4.4.10) of the balance of angular momentum. Moreover, it follows from (4.7.5), (4.7.15) and (4.7.17) that under SRBM the traction vector transforms by

$$\mathbf{t}^+ = \mathbf{Q}\mathbf{t}\,. \tag{4.7.18}$$

Furthermore, from (4.7.17) and (4.7.18) it can be shown that

$$\mathbf{t}^+ \cdot \mathbf{t}^+ = \mathbf{t} \cdot \mathbf{t}\,, \quad \mathbf{T}^+ \cdot \mathbf{T}^+ = \mathbf{T} \cdot \mathbf{T}\,, \tag{4.7.19}$$

which means that the magnitudes of the traction vector and the Cauchy stress tensor remain unchanged by SRBM. Consequently, the traction vector and stress tensor, which characterize the response of the material, are merely rotated by SRBM.

Also, using (4.7.17) the expression (4.7.13) for the body force $\mathbf{b}^+$ reduces to

$$\mathbf{b}^+ = \dot{\mathbf{v}}^+ + \mathbf{Q}(\mathbf{b} - \dot{\mathbf{v}})\,. \tag{4.7.20}$$

This is the body force that must be applied to ensure that the superposed motion remains rigid. In this regard, consider an isotropic disk of an elastic material which is spinning about its axis of symmetry at constant angular velocity. In the absence of body force, this disk expands to balance the centripetal accelerations of material points but in the presence of material dissipation it eventually spins like a rigid body with constant angular velocity. This motion should not be confused with a SRBM. If this disk is subjected to a SRBM which increases or decreases the angular velocity of the disk, it is necessary to apply a body force of the type (4.7.20) to ensure that the superposed motion is rigid with the radial position of each material point remaining constant.

For later convenience, these transformation relations for the kinetic quantities are summarized

$$\rho^+ = \rho\,, \quad \mathbf{T}^+ = \mathbf{Q}\mathbf{T}\mathbf{Q}^T\,, \quad \mathbf{b}^+ = \dot{\mathbf{v}}^+ + \mathbf{Q}(\mathbf{b} - \dot{\mathbf{v}})\,. \tag{4.7.21}$$

In addition, since the strain energy Σ also characterizes the constitutive response of the material and it does not depend on the orientation of the material, the restriction (R-2) in (4.7.3b) requires Σ to be unaltered by SRBM

$$\Sigma^+ = \Sigma \,, \tag{4.7.22}$$

so with the help of (3.8.24) and (4.7.21) it can be shown that the rate of material dissipation $\mathcal{D}$ in (4.5.7) is also unaltered by SRBM

$$\mathcal{D}^+ = \mathcal{D} \,. \tag{4.7.23}$$

Also, it is emphasized that the restriction (R-2) applies only to constitutive response. For example, the kinetic energy $\frac{1}{2}\mathbf{v} \cdot \mathbf{v}$ per unit mass is a scalar which is not specified by a constitutive equation so it need not remain unaltered by SRBM.

Furthermore, with the help of (4.6.4), (4.6.14), (4.6.16) and (4.7.21), it can be shown that the Piola-Kirchhoff traction vector $\boldsymbol{\pi}$, the nonsymmetric Piola-Kirchhoff stress tensor $\boldsymbol{\Pi}$, and the symmetric Piola-Kirchhoff stress tensor $\mathbf{S}$ transform under SRBM by

$$\boldsymbol{\pi}^+ = \mathbf{Q}\boldsymbol{\pi} \,, \quad \boldsymbol{\Pi}^+ = \mathbf{Q}\boldsymbol{\Pi} \,, \quad \mathbf{S}^+ = \mathbf{S} \,. \tag{4.7.24}$$

4.8 An Alternative Derivation of the Local Balance Laws

This section shows that the conservation of mass (4.1.6), the balance of linear momentum (4.4.5) and the reduced form (4.4.10) of the balance of angular momentum can be derived directly from the rate of material dissipation (4.5.1) and the invariance requirements (4.7.21)–(4.7.23) under SRBM. This interrelationship is an example of the fundamental nature of invariance requirements under SRBM in the general theory of a continuum.

Specifically, in view of the restriction (4.7.3a), the rate of material dissipation (4.5.1) is required to be form-invariant under SRBM so that

$$\int_{P^+} \mathcal{D}^+ dv^+ = \mathcal{W}^+ - \dot{\mathcal{K}}^+ - \dot{\mathcal{U}}^+ \geq 0 \,, \tag{4.8.1}$$

where the rate of work $\mathcal{W}^+$ done on the body, the kinetic energy $\mathcal{K}^+$ and the total strain energy $\mathcal{U}^+$ are defined by

$$\begin{aligned} \mathcal{W}^+ &= \int_{P^+} \rho^+ \mathbf{b}^+ \cdot \mathbf{v}^+ dv^+ + \int_{\partial P^+} \mathbf{t}^+ \cdot \mathbf{v}^+ da^+ \,, \\ \mathcal{K}^+ &= \int_{P^+} \frac{1}{2} \rho^+ \mathbf{v}^+ \cdot \mathbf{v}^+ dv^+ \,, \quad \mathcal{U}^+ = \int_{P^+} \rho^+ \Sigma^+ dv^+ \,. \end{aligned} \tag{4.8.2}$$

Using a development similar to that in Sect. 4.5, the local form of (4.8.1) is given by

$$\begin{aligned}\mathcal{D}^+ &= \mathbf{T}^+ \cdot \mathbf{L}^+ - \rho^+\dot{\Sigma}^+ + (\rho^+\mathbf{b}^+ + \mathrm{div}^+\mathbf{T}^+ - \rho^+\dot{\mathbf{v}}^+) \cdot \mathbf{v}^+ \\ &- (\dot{\rho}^+ + \rho^+ \,\mathrm{div}^+\mathbf{v}^+)(\frac{1}{2}\mathbf{v}^+ \cdot \mathbf{v}^+ + \Sigma^+) .\end{aligned} \tag{4.8.3}$$

Now, with the help of the invariance conditions (3.8.24), (4.7.1), (4.7.10) and (4.7.21)–(4.7.23), as well as the results

$$\begin{aligned}&\mathrm{div}^+\mathbf{v}^+ = \mathbf{L}^+ \cdot \mathbf{I} = (\mathbf{QLQ}^T + \mathbf{\Omega}) \cdot \mathbf{I} = \mathbf{L} \cdot \mathbf{I} = \mathrm{div}\mathbf{v} , \\ &\mathbf{T}^+ \cdot \mathbf{L}^+ = \mathbf{QTQ}^T \cdot (\mathbf{QLQ}^T + \mathbf{\Omega}) = \mathbf{T} \cdot \mathbf{L} + \mathbf{T} \cdot \mathbf{Q}^T \mathbf{\Omega}\mathbf{Q} .\end{aligned} \tag{4.8.4}$$

Equation (4.8.3) reduces to

$$\begin{aligned}\mathcal{D} &= \mathbf{T} \cdot \mathbf{L} - \rho\dot{\Sigma} + \mathbf{T} \cdot \mathbf{Q}^T \mathbf{\Omega}\mathbf{Q} + \mathbf{v}^+ \cdot \mathbf{Q}(\rho\mathbf{b} + \mathrm{div}\mathbf{T} - \rho\dot{\mathbf{v}}) \\ &- (\dot{\rho} + \rho \,\mathrm{div}\mathbf{v})(\frac{1}{2}\mathbf{v}^+ \cdot \mathbf{v}^+ + \Sigma) .\end{aligned} \tag{4.8.5}$$

Equation (4.8.5) must hold for all motions and all SRBM (3.8.27)

$$\mathbf{x}^+ = \mathbf{c}(t) + \mathbf{Q}(t)\mathbf{x} , \quad \mathbf{Q}^T\mathbf{Q} = \mathbf{I} , \quad \dot{\mathbf{Q}} = \mathbf{\Omega}\mathbf{Q} , \quad \mathbf{\Omega}^T = -\mathbf{\Omega} . \tag{4.8.6}$$

To develop the conservation of mass and the balances of linear and angular momentum, consider the following three special cases of SRBM.

Case 1: No SRBM
This case considers no SRBM with

$$\mathbf{c} = 0 , \quad \dot{\mathbf{c}} = 0 , \quad \mathbf{Q} = \mathbf{I} , \quad \dot{\mathbf{Q}} = 0 , \quad \mathbf{v}^+ = \mathbf{v} , \quad \dot{\mathbf{v}}^+ = \dot{\mathbf{v}} , \tag{4.8.7}$$

so (4.8.5) requires

$$\mathcal{D} = \mathbf{T} \cdot \mathbf{L} - \rho\dot{\Sigma} + \mathbf{v} \cdot (\rho\mathbf{b} + \mathrm{div}\mathbf{T} - \rho\dot{\mathbf{v}}) - (\dot{\rho} + \rho \,\mathrm{div}\mathbf{v})(\frac{1}{2}\mathbf{v} \cdot \mathbf{v} + \Sigma) , \tag{4.8.8}$$

for all motions.

Case 2: Superposed Constant Translational Velocity
This case considers superposed constant translational velocity with

$$\begin{aligned}&\mathbf{c} = u\mathbf{u}\,t , \qquad \dot{\mathbf{c}} = u\mathbf{u} , \qquad \mathbf{u} \cdot \mathbf{u} = 1 , \\ &\mathbf{Q} = \mathbf{I} , \qquad \dot{\mathbf{Q}} = 0 , \\ &\mathbf{v}^+ = \mathbf{v} + u\mathbf{u} , \qquad \dot{\mathbf{v}}^+ = \dot{\mathbf{v}} ,\end{aligned} \tag{4.8.9}$$

where u is an arbitrary constant scalar and $\mathbf{u}$ is an arbitrary constant unit vector, both characterizing the superposed constant translational velocity. It then follows that for this case (4.8.5) requires

$$\begin{aligned}\mathcal{D} &= \mathbf{T}\cdot\mathbf{L} - \rho\dot{\Sigma} + (\mathbf{v} + u\mathbf{u})\cdot(\rho\mathbf{b} + \text{div}\mathbf{T} - \rho\dot{\mathbf{v}}) \\ &- (\dot{\rho} + \rho\,\text{div}\mathbf{v})[\frac{1}{2}(u^2 + 2u\mathbf{u}\cdot\mathbf{v} + \mathbf{v}\cdot\mathbf{v}) + \Sigma]\,.\end{aligned} \tag{4.8.10}$$

Again, this equation must hold for all motions. Thus, subtracting (4.8.8) from (4.8.10) yields

$$\frac{1}{2}(u^2 + 2u\mathbf{u}\cdot\mathbf{v})(\dot{\rho} + \rho\,\text{div}\mathbf{v}) + u\mathbf{u}\cdot(\rho\dot{\mathbf{v}} - \rho\mathbf{b} - \text{div}\mathbf{T}) = 0\,, \tag{4.8.11}$$

which must hold for all motions and all u, $\mathbf{u}$. Now, dividing this equation by u requires

$$\frac{1}{2}(u + 2\mathbf{u}\cdot\mathbf{v})(\dot{\rho} + \rho\,\text{div}\mathbf{v}) + \mathbf{u}\cdot(\rho\dot{\mathbf{v}} - \rho\mathbf{b} - \text{div}\mathbf{T}) = 0\,. \tag{4.8.12}$$

Since the coefficient of u in this equation is independent of u, it follows that this equation can only be satisfied for arbitrary values of u if mass is conserved

$$\dot{\rho} + \rho\,\text{div}\mathbf{v} = 0\,. \tag{4.8.13}$$

Next, substituting the conservation of mass into (4.8.12) yields

$$\mathbf{u}\cdot(\rho\dot{\mathbf{v}} - \rho\mathbf{b} - \text{div}\mathbf{T}) = 0\,. \tag{4.8.14}$$

Since the coefficient of $\mathbf{u}$ in this equation is independent of the arbitrary unit vector $\mathbf{u}$ it can satisfied only if the balance of linear momentum is satisfied

$$\rho\dot{\mathbf{v}} = \rho\mathbf{b} + \text{div}\mathbf{T}\,. \tag{4.8.15}$$

Case 3: Superposed Rate of Rotation
Using the conservation of mass and the balance of linear momentum and subtracting (4.8.8) from (4.8.5) yields the restriction that

$$\mathbf{T}\cdot\mathbf{Q}^T\boldsymbol{\Omega}\mathbf{Q} = 0 \tag{4.8.16}$$

must hold for all motions and all SRBM. Specifically, for superposed rate of rotation $\boldsymbol{\Omega}$ is a nonzero skew-symmetric tensor and $\mathbf{T}$ is independent of the skew-symmetric tensor $\mathbf{Q}^T\boldsymbol{\Omega}\mathbf{Q}$ so the stress tensor must be symmetric

$$\mathbf{T}^T = \mathbf{T}\,, \tag{4.8.17}$$

which is the reduced form of the balance of angular momentum.

In the above analysis, the conservation of mass and the balances of linear and angular momentum have been shown to be necessary consequences of the assumption that the rate of material dissipation is form-invariant under SRBM (4.8.3). Although these results were obtained using special simple SRBM it is easy to see using the invariance conditions (4.7.1), (4.7.10) and (4.8.4) that these balance laws and the rate of material dissipation remain form-invariant

$$\begin{aligned} &\dot{\rho}^+ + \rho^+ \mathrm{div}^+ \mathbf{v}^+ = 0\,, \qquad \rho^+ \dot{\mathbf{v}}^+ = \rho^+ \mathbf{b}^+ + \mathrm{div}^+ \mathbf{T}^+\,, \\ &\mathbf{T}^{+T} = \mathbf{T}^+\,, \qquad \mathcal{D}^+ = \mathbf{T}^+ \cdot \mathbf{D}^+ - \rho^+ \dot{\Sigma}^+\,, \end{aligned} \tag{4.8.18}$$

for all motions and all SRBM.

Furthermore, it is noted that not all scalars remain form-invariant under SRBM. For example, the kinetic energy $\mathcal{K}^+$ in (4.8.2) can be written in the form

$$\mathcal{K}^+ = \mathcal{K} + \int_P \frac{1}{2}\rho\big[2(\dot{\mathbf{c}} + \dot{\mathbf{Q}}\mathbf{x}) \cdot \mathbf{Q}\mathbf{v} + (\dot{\mathbf{c}} + \dot{\mathbf{Q}}\mathbf{x}) \cdot (\dot{\mathbf{c}} + \dot{\mathbf{Q}}\mathbf{x})\big] dv\,, \tag{4.8.19}$$

which is influenced by superposed translational rate $\dot{\mathbf{c}}$ and superposed rotation $\mathbf{Q}$ and rotation rate $\dot{\mathbf{Q}}$.

4.9 Initial and Boundary Conditions

In this section, attention is confined to the discussion of initial and boundary conditions for the purely mechanical theory. In general, the number of initial conditions required and the type of boundary conditions required will depend on the specific type of material under consideration. However, it is possible to make some general observations that apply to all materials.

To this end, it is recalled that the local forms of conservation of mass (4.1.6) and balance of linear momentum (4.4.5) are partial differential equations which require both initial and boundary conditions. Specifically, the conservation of mass (4.1.6) is first order in time with respect to density ρ so it is necessary to specify the initial value of density at each point of the body

$$\rho(\mathbf{x}, 0) = \bar{\rho}(\mathbf{x}) \text{ on } P \text{ for } t = 0\,. \tag{4.9.1}$$

Also, the balance of linear momentum (4.4.5) is second order in time with respect to position $\mathbf{x}$ so that it is necessary to specify the initial value of $\mathbf{x}$ and the initial value of the velocity $\mathbf{v}$ at each point of the body

$$\begin{aligned} &\hat{\mathbf{x}}(\mathbf{X}, 0) = \bar{\mathbf{x}}(\mathbf{X}) \qquad \text{on } P \text{ for } t = 0\,, \\ &\hat{\mathbf{v}}(\mathbf{X}, 0) = \tilde{\mathbf{v}}(\mathbf{x}, 0) = \bar{\mathbf{v}}(\mathbf{x}) \qquad \text{on } P \text{ for } t = 0\,. \end{aligned} \tag{4.9.2}$$

Guidance for determining the appropriate form of boundary conditions is usually obtained by considering the rate of work done by the traction vector. From (4.5.2) it is observed that $\mathbf{t} \cdot \mathbf{v}$ is the rate of work per unit current area done by the traction vector. At each point of the surface ∂P it is possible to define a right-handed orthonormal coordinate system with base vectors $\mathbf{s}_1, \mathbf{s}_2$ and $\mathbf{n}$, such that $\mathbf{n}$ is the unit outward normal to ∂P and $\mathbf{s}_1$ and $\mathbf{s}_2$ are orthogonal vectors tangent to ∂P. Then, with reference to this coordinate system, it follows that

$$\mathbf{t} \cdot \mathbf{v} = (\mathbf{t} \cdot \mathbf{s}_1)(\mathbf{v} \cdot \mathbf{s}_1) + (\mathbf{t} \cdot \mathbf{s}_2)(\mathbf{v} \cdot \mathbf{s}_2) + (\mathbf{t} \cdot \mathbf{n})(\mathbf{v} \cdot \mathbf{n}) \text{ on } \partial P\,. \tag{4.9.3}$$

Using this representation, it is possible to define three types of boundary conditions:
Kinematic
For kinematic boundary conditions all three components of the velocity are specified

$$(\mathbf{v} \cdot \mathbf{s}_1)\,,\ (\mathbf{v} \cdot \mathbf{s}_2)\,,\ (\mathbf{v} \cdot \mathbf{n}) \text{ specified on } \partial P \text{ for all } t \geq 0\,. \tag{4.9.4}$$

Kinetic For kinetic boundary conditions all three components of the traction vector are specified

$$(\mathbf{t} \cdot \mathbf{s}_1)\,,\ (\mathbf{t} \cdot \mathbf{s}_2)\,,\ (\mathbf{t} \cdot \mathbf{n}) \text{ specified on } \partial P \text{ for all } t \geq 0\,. \tag{4.9.5}$$

Mixed For mixed boundary conditions conjugate components of both the velocity and the traction vector are specified

$$\begin{array}{lllll} (\mathbf{v} \cdot \mathbf{s}_1) & \text{or} & (\mathbf{t} \cdot \mathbf{s}_1) & \text{specified on } \partial P & \text{for all } t \geq 0\,, \\ (\mathbf{v} \cdot \mathbf{s}_2) & \text{or} & (\mathbf{t} \cdot \mathbf{s}_2) & \text{specified on } \partial P & \text{for all } t \geq 0\,, \\ (\mathbf{v} \cdot \mathbf{n}) & \text{or} & (\mathbf{t} \cdot \mathbf{n}) & \text{specified on } \partial P & \text{for all } t \geq 0\,. \end{array} \tag{4.9.6}$$

Essentially, the conjugate components $(\mathbf{t} \cdot \mathbf{s}_1)$, $(\mathbf{t} \cdot \mathbf{s}_2)$ and $(\mathbf{t} \cdot \mathbf{n})$ are the responses to the motions $(\mathbf{v} \cdot \mathbf{s}_1)$, $(\mathbf{v} \cdot \mathbf{s}_2)$ and $(\mathbf{v} \cdot \mathbf{n})$, respectively. Therefore, it is important to emphasize that, for example, both $(\mathbf{v} \cdot \mathbf{n})$ and $(\mathbf{t} \cdot \mathbf{n})$ cannot be specified at the same point of ∂P because this would mean that both the motion and the stress response can be specified independently of the material properties and geometry of the body. Notice also, that since the initial position of points on the boundary ∂P are specified by the kinematic initial condition in (4.9.2), the velocity boundary conditions (4.9.4) can be used to determine the position of the boundary for all time. This means that the kinematic boundary conditions (4.9.4) can also be characterized by specifying the position of points on the boundary for all time.

4.10 Linearization

The previous sections have considered the exact formulation of the theory of simple continua. The resulting equations are nonlinear so they are quite difficult to solve analytically. However, often it is possible to obtain relevant physical insight about the solution of a problem by considering a simpler approximate theory. This section develops the linearized equations associated with this nonlinear theory.

First it is noted that a tensor $\mathbf{u}$ is said to be of order ε^n [denoted by $O(\varepsilon^n)$] if there exists a real finite number C, independent of ε, such that

$$|\mathbf{u}| < C\varepsilon^n \text{ as } \varepsilon \to 0\,. \tag{4.10.1}$$

In what follows various kinematic quantities as well as the conservation of mass and the balance of linear momentum and boundary conditions are linearized by considering small deviations from a reference configuration in which the body is in a zero-stress material state, at rest and free of body force. To this end, it is assumed that the density ρ is of order $O(\varepsilon^0)$

$$\rho = \rho_z + O(\varepsilon)\,, \tag{4.10.2}$$

where ρ_z is the zero-stress reference density. Also, it is assumed that the displacement $\mathbf{u}$, body force $\mathbf{b}$, Cauchy stress $\mathbf{T}$, nonsymmetric Piola-Kirchhoff stress $\boldsymbol{\Pi}$ and symmetric Piola-Kirchhoff stress $\mathbf{S}$ are of order $O(\varepsilon)$

$$(\,\mathbf{u}\,,\,\mathbf{b}\,,\,\mathbf{T}\,,\,\boldsymbol{\Pi}\,,\,\mathbf{S}\,) = O(\varepsilon)\,. \tag{4.10.3}$$

The resulting theory will be a linear theory if ε is small enough

$$\varepsilon << 1 \tag{4.10.4}$$

that quadratic and higher order terms in ε can be neglected relative to terms of order $O(\varepsilon)$. It is tacitly assumed that all derivatives with respect to time and space of a quantity of order $O(\varepsilon)$ are also order $O(\varepsilon)$.

Kinematics

Recalling from (3.3.31) that the position $\mathbf{x}$ of a material point in the current configuration can be represented by

$$\mathbf{x} = \mathbf{X} + \mathbf{u}\,, \tag{4.10.5}$$

with deformation gradient $\mathbf{F}$ given by

$$\mathbf{F} = \partial\mathbf{x}/\partial\mathbf{X} = \mathbf{I} + \partial\mathbf{u}/\partial\mathbf{X}\,. \tag{4.10.6}$$

In what follows, use is made of this expression to derive a number of kinematical results. For this purpose it is convenient to separate the displacement gradient into its symmetric part $\boldsymbol{\varepsilon}$ and its skew-symmetric part $\boldsymbol{\omega}$, such that

$$\begin{aligned}
&\partial\mathbf{u}/\partial\mathbf{X} = \boldsymbol{\varepsilon} + \boldsymbol{\omega}\,,\\
&\boldsymbol{\varepsilon} = \frac{1}{2}[\partial\mathbf{u}/\partial\mathbf{X} + (\partial\mathbf{u}/\partial\mathbf{X})^T] = \boldsymbol{\varepsilon}^T\,,\\
&\boldsymbol{\omega} = \frac{1}{2}[\partial\mathbf{u}/\partial\mathbf{X} - (\partial\mathbf{u}/\partial\mathbf{X})^T] = \boldsymbol{\omega}^T\,,
\end{aligned} \tag{4.10.7}$$

where it is noted that both $\boldsymbol{\varepsilon}$ and $\boldsymbol{\omega}$ are of order $O(\varepsilon)$. Now, with the help of (4.10.6) and (4.10.7), it follows that

$$\begin{aligned}
&\mathbf{F} = \mathbf{I} + \partial\mathbf{u}/\partial\mathbf{X} = \mathbf{I} + \boldsymbol{\varepsilon} + \boldsymbol{\omega}\,, \\
&\mathbf{F}^{-1} = \mathbf{I} - \partial\mathbf{u}/\partial\mathbf{X} + O(\varepsilon^2) = \mathbf{I} - \boldsymbol{\varepsilon} - \boldsymbol{\omega} + O(\varepsilon^2) \\
&\mathbf{C} = \mathbf{F}^T\mathbf{F} = \mathbf{I} + 2\boldsymbol{\varepsilon} + O(\varepsilon^2)\,, \\
&\mathbf{E} = \frac{1}{2}(\mathbf{C} - \mathbf{I}) = \boldsymbol{\varepsilon} + O(\varepsilon^2)\,, \\
&\mathbf{U} = \mathbf{C}^{1/2} = \mathbf{I} + \boldsymbol{\varepsilon} + O(\varepsilon^2)\,, \\
&\mathbf{U}^{-1} = \mathbf{I} - \boldsymbol{\varepsilon} + O(\varepsilon^2)\,, \\
&\mathbf{R} = \mathbf{F}\mathbf{U}^{-1} = \mathbf{I} + \boldsymbol{\omega} + O(\varepsilon^2)\,,
\end{aligned} \tag{4.10.8}$$

which indicates that $\boldsymbol{\varepsilon}$ is the linearized strain measure and $\boldsymbol{\omega}$ is the linearized rotation measure. Furthermore, these approximations can be used to deduce that

$$\begin{aligned}
&\partial\mathbf{u}/\partial\mathbf{x} = (\partial\mathbf{u}/\partial\mathbf{X})\mathbf{F}^{-1} = \partial\mathbf{u}/\partial\mathbf{x} + O(\varepsilon^2)\,, \\
&\mathbf{e} = \frac{1}{2}(\mathbf{I} - \mathbf{F}^{-T}\mathbf{F}^{-1}) = \boldsymbol{\varepsilon} + O(\varepsilon^2)\,,
\end{aligned} \tag{4.10.9}$$

so for the linear theory where terms of order $O(\varepsilon^2)$ and higher are neglected, there is no distinction between the Lagrangian strain $\mathbf{E}$ and the Almansi strain $\mathbf{e}$.

To determine the linearized expression for the change in volume it is recalled that

$$J^2 = \det\mathbf{C} \approx \det[2\boldsymbol{\varepsilon} - (-1)\,\mathbf{I}]\,. \tag{4.10.10}$$

Now, the characteristic equation of this determinant is given by

$$\begin{aligned}
\det[2\boldsymbol{\varepsilon} - (-1)\,\mathbf{I}] &= -(-1)^3 + (2\boldsymbol{\varepsilon}\cdot\mathbf{I})(-1)^2 \\
&\quad - \frac{1}{2}[(2\boldsymbol{\varepsilon}\cdot\mathbf{I})^2 - (2\boldsymbol{\varepsilon}\cdot 2\boldsymbol{\varepsilon})](-1) + \det(2\boldsymbol{\varepsilon})\,.
\end{aligned} \tag{4.10.11}$$

However,

$$(2\boldsymbol{\varepsilon}\cdot\mathbf{I})^2 = O(\varepsilon^2)\,,\quad (2\boldsymbol{\varepsilon}\cdot 2\boldsymbol{\varepsilon}) = O(\varepsilon^2)\,,\quad \det(2\boldsymbol{\varepsilon}) = O(\varepsilon^3)\,, \tag{4.10.12}$$

so neglecting quadratic terms in ε yields

$$J^2 = 1 + 2\boldsymbol{\varepsilon}\cdot\mathbf{I} + O(\varepsilon^2)\,,\quad J = 1 + \boldsymbol{\varepsilon}\cdot\mathbf{I} + O(\varepsilon^2)\,. \tag{4.10.13}$$

Thus, the trace of the linearized strain $\boldsymbol{\varepsilon}$ is the relative increase in volume

$$\frac{dv}{dV} - 1 = \frac{dv - dV}{dV} = \boldsymbol{\varepsilon} \cdot \mathbf{I}\,,$$
$$\frac{\dot{j}}{J} = \text{div}\,\mathbf{v} = \mathbf{D} \cdot \mathbf{I} = \dot{\boldsymbol{\varepsilon}} \cdot \mathbf{I} + O(\varepsilon^2)\,. \tag{4.10.14}$$

Kinetics

It follows from (4.10.2) and (4.10.14) that the conservation of mass (4.1.7) for the linear theory becomes

$$\dot{\rho} + \rho_z(\dot{\boldsymbol{\varepsilon}} \cdot \mathbf{I}) = 0\,, \tag{4.10.15}$$

where terms of order $O(\varepsilon^2)$ and higher have been neglected. Also, since the stresses $\mathbf{T}$, $\boldsymbol{\Pi}$ and $\mathbf{S}$ are related by Eqs. (4.6.13), (4.6.14) and (4.6.16), it follows that

$$\boldsymbol{\Pi} = \mathbf{FS} = \mathbf{S} + O(\varepsilon^2)\,, \quad \mathbf{T} = J^{-1}\boldsymbol{\Pi}\mathbf{F}^T = \mathbf{S} + O(\varepsilon^2)\,. \tag{4.10.16}$$

This means that for the linear theory where terms of order $O(\varepsilon^2)$ and higher are neglected, there is no distinction between the three types of stresses

$$\mathbf{T} = \boldsymbol{\Pi} = \mathbf{S}\,. \tag{4.10.17}$$

This is consistent with the fact that for the linear theory, the geometry of the current configuration is only slightly different from the geometry of the reference configuration. Further in this regard, it is noted that

$$\begin{aligned} \text{div}\mathbf{T} &= \mathbf{T}_{,i} \cdot \mathbf{e}_i = (\partial \mathbf{T}/\partial \mathbf{X})(\partial \mathbf{X}/\partial \mathbf{x})\,\mathbf{e}_i \cdot \mathbf{e}_i \\ &= (\partial \mathbf{T}/\partial \mathbf{X})\mathbf{I} \cdot (\mathbf{e}_i \otimes \mathbf{e}_i) + O(\varepsilon^2) \\ &= (\partial \mathbf{T}/\partial \mathbf{X}) \cdot \mathbf{I} + O(\varepsilon^2)\,, \\ \text{div}\mathbf{T} &= \text{Div}\mathbf{T} + O(\varepsilon^2)\,. \end{aligned} \tag{4.10.18}$$

Also, it can be shown that

$$\text{Div}\boldsymbol{\Pi} = \text{Div}\mathbf{S} + O(\varepsilon^2)\,, \tag{4.10.19}$$

so that the balance of linear momentum (4.4.5) or (4.6.5) yield

$$\rho_z \ddot{\mathbf{u}} = \rho_x \mathbf{b} + \text{Div}\mathbf{T}\,, \tag{4.10.20}$$

where again terms of order $O(\varepsilon^2)$ and higher have been neglected.

Boundary Conditions

The boundary conditions (4.9.4)–(4.9.6) are expressed in terms of values of functions of order $O(\varepsilon)$ that are evaluated at points on the boundary ∂P of the surface in the current configuration. The linearized forms of these boundary conditions can be determined by considering an arbitrary function f of order $O(\varepsilon)$ and using a Taylor series expansion to deduce that

$$f(\mathbf{x},t) = f(\mathbf{X}+\mathbf{u},t) = f(\mathbf{X},t) + \partial f/\partial \mathbf{x} \cdot \mathbf{u} + O(\varepsilon^3) = f(\mathbf{X},t) + O(\varepsilon^2)\,. \tag{4.10.21}$$

This means that for the linear theory the distinction between the Lagrangian and Eulerian representations of any function of order $O(\varepsilon)$ vanishes. Thus, to within the order of accuracy of the linear theory the boundary conditions can be evaluated at points on the reference boundary ∂P_0 instead of on the present boundary ∂P.

Next, it is emphasized that the linear theory derived from a given nonlinear theory is unique but not vice versa. This means that an infinite number of nonlinear theories exist which when linearized yield the same linear theory. Consequently, a linear theory provides little guidance for developing a physical nonlinear theory.

Linearization of the Kinematics in the Eulerian Formulation
To linearize the Eulerian formulation, it is convenient to consider an initial zero-stress material state with the vectors $\mathbf{m}_i$ specified by the orthonormal triad $\mathbf{M}_i$, such that

$$\mathbf{m}_i(0) = \mathbf{M}_i\,, \quad \mathbf{M}_i \cdot \mathbf{M}_j = \delta_{ij}\,, \quad \mathbf{M}_1 \times \mathbf{M}_2 \cdot \mathbf{M}_3 = 1\,. \tag{4.10.22}$$

Now, let $\tilde{\mathbf{X}}_i$ be the position of a material point in this initial configuration and $\tilde{\mathbf{u}}$ be the displacement of this material point, such that

$$\mathbf{x} = \tilde{\mathbf{X}} + \tilde{\mathbf{u}}\,. \tag{4.10.23}$$

Then, taking $t_n = 0$ in (3.13.3), the relative deformation gradient $\mathbf{F}_r$ can be expressed in the form

$$\mathbf{F}_r = \partial \mathbf{x}/\partial \tilde{\mathbf{x}} = \mathbf{I} + \partial \tilde{\mathbf{u}}/\partial \tilde{\mathbf{x}}\,, \tag{4.10.24}$$

so the vectors $\mathbf{m}_i$ in (3.13.9) are given by

$$\mathbf{m}_i = (\mathbf{I} + \partial \tilde{\mathbf{u}}/\partial \tilde{\mathbf{x}})\,\mathbf{M}_i\,, \tag{4.10.25}$$

where the superscirpt $()^*$ has been omitted for notational simplicity.

Furthermore, the displacement gradient $\partial \tilde{\mathbf{u}}/\partial \tilde{\mathbf{x}}$ can be separated into its symmetric part $\tilde{\mathbf{e}}$ and its skew-symmtric part $\tilde{\boldsymbol{\omega}}$

$$\begin{aligned} \partial \tilde{\mathbf{u}}/\partial \tilde{\mathbf{x}} &= \tilde{\mathbf{e}} + \tilde{\boldsymbol{\omega}}\,, \\ \tilde{\mathbf{e}} &= \frac{1}{2}[\partial \tilde{\mathbf{u}}/\partial \tilde{\mathbf{x}} + (\partial \tilde{\mathbf{u}}/\partial \tilde{\mathbf{x}})^T] = \tilde{\mathbf{e}}^T\,, \\ \tilde{\boldsymbol{\omega}} &= \frac{1}{2}[\partial \tilde{\mathbf{u}}/\partial \tilde{\mathbf{x}} - (\partial \tilde{\mathbf{u}}/\partial \tilde{\mathbf{x}})^T] = -\tilde{\boldsymbol{\omega}}^T\,. \end{aligned} \tag{4.10.26}$$

Next, expressing these quantities in terms of their components relative to $\mathbf{M}_i$

$$\tilde{\mathbf{x}} = \tilde{x}_i \mathbf{M}_i\,, \quad \tilde{\mathbf{u}} = \tilde{u}_i \mathbf{M}_i\,, \quad \tilde{\mathbf{e}} = \tilde{e}_{ij} \mathbf{M}_i \otimes \mathbf{M}_j\,, \quad \tilde{\boldsymbol{\omega}} = \tilde{\omega}_{ij} \mathbf{M}_i \otimes \mathbf{M}_j\,, \tag{4.10.27}$$

it follows that

$$\tilde{e}_{ij} = \frac{1}{2}(\tilde{u}_{i},_{j} + \tilde{u}_{j},_{i}) \,, \quad \tilde{\omega}_{ij} = \frac{1}{2}(\tilde{u}_{i},_{j} - \tilde{u}_{j},_{i}) \,, \tag{4.10.28}$$

where here a comma denotes partial differentiation with respect to $\tilde{x}_i$.

Now, with the help of (4.10.25)–(4.10.27) the vectors $\mathbf{m}_i$ can be expressed in the forms

$$\mathbf{m}_i = (\mathbf{I} + \tilde{\mathbf{e}} + \tilde{\boldsymbol{\omega}})\,\mathbf{M}_i \,, \quad \mathbf{m}_i = (\delta_{ij} + \tilde{e}_{ij} + \tilde{\omega}_{ij})\,\mathbf{M}_i \,. \tag{4.10.29}$$

Thus, neglecting quadratic terms in the derivatives of $\tilde{u}_i$ and using the definition (3.11.33) of the elastic strains e_{ij} it can be shown that

$$e_{ij} = \tilde{e}_{ij} \,, \tag{4.10.30}$$

which indicates that $\tilde{e}_{ij}$ characterize the linearized strains and $\tilde{\omega}_{ij}$ characterize the linearized rotations.

Moreover, using (3.11.7) and neglecting quadratic terms in $\tilde{e}_{ij}$ and $\tilde{\omega}_{ij}$, the expression for the elastic dilatation can be approximated by

$$J_e = 1 + \tilde{e}_{mm} \,, \tag{4.10.31}$$

so the distortional strains e'_{ij} (3.11.35) can be approximated by

$$e'_{ij} = \tilde{e}''_{ij} = \tilde{e}_{ij} - \frac{1}{3}\tilde{e}_{mm}\delta_{ij} \,, \tag{4.10.32}$$

where $\tilde{e}''_{ij}$ is the deviatoric part of $\tilde{e}_{ij}$.

Reference

1. Rubin MB (2020) Invariance under superposed rigid body motions with constraints. To appear in the International Journal of Elasticity

Chapter 5
Purely Mechanical Constitutive Equations

Abstract The objective of this chapter is to discuss purely mechanical constitutive equations. After identifying unphysical arbitrariness of the classical Lagrangian formulation of constitutive equations, an Eulerian formulation for nonlinear elastic materials is developed using evolution equations for microstructural vectors $\mathbf{m}_i$. The influence of kinematic constraints on constitutive equations is discussed and specific nonlinear constitutive equations are presented for a number of materials including: elastic solids, viscous fluids and elastic–inelastic materials.

5.1 The Classical Lagrangian Formulation for Nonlinear Elastic Solids

In general, a constitutive equation is an equation that characterizes the response of a given material to deformations, deformation rates, thermal, electrical, magnetic or mechanobiological loads. An elastic material is a very special material because it exhibits ideal behavior in the sense that it has no material dissipation. One of the most important features of an elastic material is that it is characterized by a total strain energy $\mathcal{U}$ and a strain energy function Σ per unit mass defined in (4.5.2)

$$\mathcal{U} = \int_P \rho \Sigma dv \,. \tag{5.1.1}$$

Generalizing the notions of a simple nonlinear elastic spring, an elastic material is characterized by the following four assumptions:

Assumption 5.1 The material response is ideal in the sense that the rate of material dissipation $\mathcal{D}$ in (4.5.7) vanishes

$$\mathcal{D} = \mathbf{T} \cdot \mathbf{D} - \rho \dot{\Sigma} = 0 \,, \tag{5.1.2}$$

M. B. Rubin, *Continuum Mechanics with Eulerian Formulations of Constitutive Equations*, Solid Mechanics and Its Applications 265,
https://doi.org/10.1007/978-3-030-57776-6_5

for all motions. This generalizes the notion that the elastic spring is non-dissipative.

Assumption 5.2 Within the context of the Lagrangian formulation, the strain energy Σ is a function of the total deformation gradient $\mathbf{F}$ and the reference position $\mathbf{X}$ only

$$\Sigma = \tilde{\Sigma}(\mathbf{F}; \mathbf{X}) , \tag{5.1.3}$$

where dependence on the reference position $\mathbf{X}$ has been included to allow for the possibility that the material can be inhomogeneous in the reference configuration. This generalizes the notion that the elastic energy in an elastic spring depends only on extension of the spring from its zero-stress length.

Assumption 5.3 The strain energy Σ is form-invariant under SRBM

$$\Sigma^+ = \Sigma . \tag{5.1.4}$$

With regard to a spring, this follows directly from the fact that every member of the group of SRBM has the same length of the spring at each time.

Assumption 5.4 The Cauchy stress $\mathbf{T}$ is independent of the rate of deformation $\mathbf{L}$. This is consistent with the fact that the force in an elastic spring does not depend on the rate of extension of the spring.

To explore the physical consequences of Assumption 5.1 (5.1.2), use is made of global form (4.5.1) of the rate of material dissipation to obtain

$$\mathcal{W} = \dot{\mathcal{K}} + \dot{\mathcal{U}} , \tag{5.1.5}$$

which states that for an elastic material the rate of work done on the body due to body forces and contact forces equals the rate of change of kinetic and strain energies. In particular, the total work $W_{2/1}$ done on the body during the time interval $t_1 \leq t \leq t_2$ is given by

$$\begin{gathered} W_{2/1} = \int_{t=t_1}^{t=t_2} \mathcal{W} dt = \Delta \mathcal{K} + \Delta U , \\ \Delta \mathcal{K} = \mathcal{K}(t_2) - \mathcal{K}(t_1) , \quad \Delta \mathcal{U} = \mathcal{U}(t_2) - \mathcal{U}(t_1) . \end{gathered} \tag{5.1.6}$$

In view of Assumption 5.2 (5.1.3), the strain energy Σ depends on the current configuration through the current value of $\mathbf{F}$ only. Similarly, the value of the kinetic energy $\mathcal{K}$ depends only on the values of the density ρ and the velocity $\mathbf{v}$ at the beginning and ends of the time interval. Moreover, the values of ρ at the beginning and end of the time interval are connected by the conservation of mass (4.1.13) which requires $\rho \det \mathbf{F} =$ constant. Consequently, $\Delta\mathcal{K}$, $\Delta\mathcal{U}$ and the work done $W_{2/1}$ during the time interval depend only on the values of $\mathbf{v}$ and $\mathbf{F}$ at the beginning and end of the time interval. In particular, this means that the work $W_{2/1}$ done on the body between any two states defined by $\mathbf{v}(t_1)$ and $\mathbf{F}(t_1)$ and $\mathbf{v}(t_2)$ and $\mathbf{F}(t_2)$ is independent of the path of the deformation between these two states. This is consistent with the notion that the work done on an elastic spring between any two states is path independent. Also, it follows that the work $W_{2/1}$ done on the body vanishes for an arbitrary closed cycle for which the values of $\mathbf{v} \cdot \mathbf{v}$ and the deformation gradient $\mathbf{F}$ are the same at the beginning and end of the cycle

$$W_{2/1} = 0\,, \quad \Delta\mathcal{K} = 0\,, \quad \Delta\mathcal{U} = 0\,. \tag{5.1.7}$$

In this regard, it is noted that $\mathbf{v}$ and $\mathbf{F}$ are functions of position and time so the notion of a closed cycle implies that each point starts and ends with the same values of $\mathbf{v} \cdot \mathbf{v}$ and the same values of $\mathbf{F}$.

Assumption 5.3 (5.1.4) places restrictions on the functional form (5.1.3) of the strain energy. Using the fact that $\mathbf{F}^+ = \mathbf{QF}$ under SRBM, it follows that

$$\Sigma^+ = \tilde{\Sigma}(\mathbf{F}^+; \mathbf{X}) = \tilde{\Sigma}(\mathbf{QF}; \mathbf{X}) = \tilde{\Sigma}(\mathbf{F}; \mathbf{X}) \tag{5.1.8}$$

must hold for arbitrary proper orthogonal tensors $\mathbf{Q}$ and all times. Since the deformation can be inhomogeneous, the rotation tensor $\mathbf{R}$ can be a function of position $\mathbf{X}$. However, for an arbitrary but specified value $\mathbf{X}_1$ of $\mathbf{X}$, choose $\mathbf{Q}(t) = \mathbf{R}^T(\mathbf{X}_1, t)$ so that this equation requires

$$\tilde{\Sigma}(\mathbf{F}; \mathbf{X}) = \tilde{\Sigma}(\mathbf{R}^T(\mathbf{X}_1)\mathbf{RU}; \mathbf{X})\,, \tag{5.1.9}$$

where the dependence of $\mathbf{R}(\mathbf{X}_1, t)$ on time has been suppressed for notational simplicity. Now, evaluating this expression at $\mathbf{X} = \mathbf{X}_1$, it follows that locally

$$\tilde{\Sigma}(\mathbf{F}; \mathbf{X}) = \tilde{\Sigma}(\mathbf{U}; \mathbf{X}_1) = \hat{\Sigma}(\mathbf{C}; \mathbf{X}_1)\,. \tag{5.1.10}$$

Thus, a necessary condition for the strain energy Σ to be locally invariant under SRBM is that the strain energy function Σ be dependent on the deformation gradient $\mathbf{F}$ only through its dependence on the deformation tensor $\mathbf{C}$. It is easy to see that this condition is also a sufficient condition for the strain energy function to be form-invariant under SRBM since $\mathbf{C}^+ = \mathbf{C}$. Moreover, since $\mathbf{X}_1$ is an arbitrary material point, this restriction on Σ must hold for each point $\mathbf{X}$ so the strain energy Σ can depend on $\mathbf{F}$ only through its dependence on $\mathbf{C}$ for all material points $\mathbf{X}$

$$\Sigma = \hat{\Sigma}(\mathbf{C}; \mathbf{X})\,. \tag{5.1.11}$$

Next, with the help of (5.1.11) Assumption 5.1 (5.1.2) requires

$$\mathbf{T} \cdot \mathbf{D} = \rho \frac{\partial \Sigma}{\partial \mathbf{C}} \cdot \dot{\mathbf{C}} = \rho \frac{\partial \Sigma}{\partial \mathbf{C}} \cdot 2\mathbf{F}^T\mathbf{DF} = 2\rho \mathbf{F} \frac{\partial \Sigma}{\partial \mathbf{C}} \mathbf{F}^T \cdot \mathbf{D}\,, \tag{5.1.12a}$$

$$\left(\mathbf{T} - 2\rho \mathbf{F} \frac{\partial \Sigma}{\partial \mathbf{C}} \mathbf{F}^T\right) \cdot \mathbf{D} = 0\,. \tag{5.1.12b}$$

However, since the coefficient of $\mathbf{D}$ in (5.1.12b) is independent of the rate $\mathbf{D}$ and is symmetric, it follows that for any fixed values of $\mathbf{F}$ and $\mathbf{X}$ the coefficient of $\mathbf{D}$ is fixed and yet $\mathbf{D}$ can be an arbitrary symmetric tensor. Therefore, the necessary condition that (5.1.12b) be valid for arbitrary motions is that the Cauchy stress be determined by a derivative of the strain energy

$$\mathbf{T} = 2\rho \mathbf{F} \frac{\partial \Sigma}{\partial \mathbf{C}} \mathbf{F}^T . \tag{5.1.13}$$

Using the conservation of mass (4.1.4) and the relationship (4.6.16), the symmetric Piola-Kirchhoff stress $\mathbf{S}$ is also determined by a derivative of the strain energy

$$\mathbf{S} = 2\rho_0 \frac{\partial \Sigma}{\partial \mathbf{C}} . \tag{5.1.14}$$

Notice that the results (5.1.13) and (5.1.14) are automatically properly invariant under SRBM. Also, it can be seen that the result (5.1.14) is similar to the result for an elastic spring that the force is equal to the derivative of the potential (strain) energy.

Green Elasticity (Hyperelasticity)
The elastic response of the material described by (5.1.13) is called Green elasticity or hyperelasticity with all four assumptions satisfied. In particular, the stress $\mathbf{T}$ is independent of velocity gradient $\mathbf{L}$. Also, the stress is an explicit function of the deformation gradient $\mathbf{F}$ which is related to the derivative of a strain energy function Σ that depends only on $\mathbf{F}$ through the right Cauchy–Green deformation tensor $\mathbf{C} = \mathbf{F}^T\mathbf{F}$. This means that the stress is determined by the deformation state $\mathbf{F}$ and is independent of the path of deformation. Moreover, the work done between two states of deformation $\mathbf{F}_1$ and $\mathbf{F}_2$ is independent of the path.

Cauchy Elasticity
For Cauchy elasticity, only Assumption 5.4 is satisfied with the stress $\mathbf{T}$ being a function of $\mathbf{F}$ only

$$\mathbf{T} = \mathbf{T}(\mathbf{F}) . \tag{5.1.15}$$

This material has the property that the stress is determined by the deformation state $\mathbf{F}$ and is independent of the velocity gradient $\mathbf{L}$ and of the path of deformation. However, in general, the function in (5.1.15) does not satisfy integrability conditions necessary for a strain energy function to exist. This means that (5.1.13) is not valid and the work done between two states of deformation $\mathbf{F}_1$ and $\mathbf{F}_2$ can be path dependent. Moreover, since under SRBM the stress $\mathbf{T}$ must satisfy the transformation relations (4.7.17), it follows that the functional form for $\mathbf{T}(\mathbf{F})$ must satisfy the restriction that

$$\mathbf{T}^+ = \mathbf{T}(\mathbf{F}^+) = \mathbf{T}(\mathbf{QF}) = \mathbf{QT}(\mathbf{F})\mathbf{Q}^T \tag{5.1.16}$$

is satisfied for all proper orthogonal $\mathbf{Q}$. This restriction requires $\mathbf{T}$ to be an isotropic tensor function of the left Cauchy–Green deformation tensor $\mathbf{B} = \mathbf{FF}^T$

$$\mathbf{T} = \mathbf{T}(\mathbf{B})\,, \quad \mathbf{T}(\mathbf{Q}\mathbf{B}\mathbf{Q}^T) = \mathbf{Q}\mathbf{T}(\mathbf{B})\mathbf{Q}^T\,, \tag{5.1.17}$$

which can only characterize elastically isotropic response.

Hypoelasticity
For hypoelasticity, only Assumption 5.4 is satisfied and the stress is determined by integrating an evolution equation of the form

$$\overset{\nabla}{\mathbf{T}} = \mathbf{K}(\mathbf{T}) \cdot \mathbf{D}\,, \tag{5.1.18}$$

where $\mathbf{K}$ is a fourth-order tensor function of $\mathbf{T}$ having the symmetry properties that

$$^{LT}\mathbf{K} = \mathbf{K}^T = \mathbf{K}\,. \tag{5.1.19}$$

Also, the rate of stress $\overset{\nabla}{\mathbf{T}}$ and the stiffness tensor $\mathbf{K}$ transform under SRBM such that

$$\overset{\nabla}{\mathbf{T}}{}^{+} = \mathbf{Q}\overset{\nabla}{\mathbf{T}}\mathbf{Q}^T\,, \quad \mathbf{K}(\mathbf{T}^+) \cdot \mathbf{D}^+ = \mathbf{K}(\mathbf{Q}\mathbf{T}\mathbf{Q}^T) \cdot \mathbf{Q}\mathbf{D}\mathbf{Q}^T = \mathbf{Q}\,[\mathbf{K}(\mathbf{T}) \cdot \mathbf{D}]\,\mathbf{Q}^T\,, \tag{5.1.20}$$

so that the evolution equation (5.1.18) remains form-invariant under SRBM

$$\overset{\nabla}{\mathbf{T}}{}^{+} = \mathbf{K}(\mathbf{T}^+) \cdot \mathbf{D}^+\,. \tag{5.1.21}$$

Since this equation is homogeneous of order one in time, the predicted material response is rate independent. Stress rates which satisfy the restriction $(5.1.20)_1$ for all SRBM are called *objective*.

Truesdell Stress Rate
Recalling that

$$\dot{\mathbf{F}} = \mathbf{L}\mathbf{F}\,, \quad \dot{\rho} = -\rho\mathbf{D} \cdot \mathbf{I}\,, \tag{5.1.22}$$

it is possible to differentiate the hyperelastic constitutive equation (5.1.13) to deduce that

$$\dot{\mathbf{T}} = \mathbf{L}\mathbf{T} + \mathbf{T}\mathbf{L}^T - (\mathbf{D} \cdot \mathbf{I})\mathbf{T} + 2\rho\mathbf{F}\left(\frac{\partial^2 \Sigma}{\partial \mathbf{C} \otimes \partial \mathbf{C}} \cdot \dot{\mathbf{C}}\right)\mathbf{F}^T\,. \tag{5.1.23}$$

This equation can be rewritten in the form

$$\overset{T}{\mathbf{T}} = 2\rho\mathbf{F}\left(\frac{\partial^2 \Sigma}{\partial \mathbf{C} \otimes \partial \mathbf{C}} \cdot \dot{\mathbf{C}}\right)\mathbf{F}^T\,, \tag{5.1.24}$$

where the Truesdell stress rate is defined by

$$\overset{T}{\mathbf{T}} = \dot{\mathbf{T}} - \mathbf{L}\mathbf{T} - \mathbf{T}\mathbf{L}^T + (\mathbf{D} \cdot \mathbf{I})\mathbf{T}\,. \tag{5.1.25}$$

Next, recalling that under SRBM

$$\begin{aligned}
&\mathbf{Q}\mathbf{Q}^T = \mathbf{I}\,, \quad \dot{\mathbf{Q}} = \boldsymbol{\Omega}\mathbf{Q}\,, \quad \boldsymbol{\Omega}^T = -\boldsymbol{\Omega}\,,\\
&\mathbf{T}^+ = \mathbf{Q}\mathbf{T}\mathbf{Q}^T\,, \quad \mathbf{L}^+ = \mathbf{D}^+ + \mathbf{W}^+ = \mathbf{Q}\mathbf{L}\mathbf{Q}^T + \boldsymbol{\Omega}\,,\\
&\mathbf{D}^+ = \mathbf{Q}\mathbf{D}\mathbf{Q}^T\,, \quad \mathbf{W}^+ = \mathbf{Q}\mathbf{W}\mathbf{Q}^T + \boldsymbol{\Omega}\,,
\end{aligned} \tag{5.1.26}$$

it can be shown by differentiating the expression for $\mathbf{T}^+$ that the Truesdell is objective

$$\overset{T}{\mathbf{T}}{}^+ = \dot{\mathbf{T}}^+ - \mathbf{L}^+\mathbf{T}^+ - \mathbf{T}^+\mathbf{L}^{+T} + (\mathbf{D}^+ \cdot \mathbf{I})\mathbf{T}^+ = \mathbf{Q}\overset{T}{\mathbf{T}}\mathbf{Q}^T\,. \tag{5.1.27}$$

Thus, the evolution equation (5.1.24) based on the Truesdell stress rate satisfies the restriction $(5.1.20)_1$ so it is form-invariant under SRBM and can be used to formulate hypoelastic constitutive equations of the type (5.1.18).

Jaumann Stress Rate
The Jaumann stress rate defined by

$$\overset{J}{\mathbf{T}} = \dot{\mathbf{T}} - \mathbf{W}\mathbf{T} - \mathbf{T}\mathbf{W}^T \tag{5.1.28}$$

is also objective

$$\overset{J}{\mathbf{T}}{}^+ = \dot{\mathbf{T}}^+ - \mathbf{W}^+\mathbf{T}^+ - \mathbf{T}^+\mathbf{W}^{+T} = \mathbf{Q}\overset{J}{\mathbf{T}}\mathbf{Q}^T\,. \tag{5.1.29}$$

Consequently, it can be used for form-invariant hypoelastic constitutive equations of the type (5.1.18). Moreover, it follows from (5.1.25) and (5.1.28) that the Truesdell and Jaumann stress rates are related by

$$\overset{T}{\mathbf{T}} = \overset{J}{\mathbf{T}} - \mathbf{D}\mathbf{T} - \mathbf{T}\mathbf{D} + (\mathbf{D} \cdot \mathbf{I})\mathbf{T}\,, \tag{5.1.30}$$

for all SRBM.

In this regard, it is noted that there are an infinite number of stress rates that transform like $(5.1.20)_1$ under SRBM. For example, consider a generalized hypoelastic material specified by the evolution equation

$$\overset{J}{\mathbf{T}} = \mathbf{K}(\mathbf{T}, \mathbf{D})\,, \tag{5.1.31}$$

where $\mathbf{K}(\mathbf{T}, \mathbf{D})$ is a homogeneous function of order one in $\mathbf{D}$ which satisfies the restrictions

$$\mathbf{K}(\mathbf{T}, \alpha\mathbf{D}) = \alpha\mathbf{K}(\mathbf{T}, \mathbf{D})\,, \quad \mathbf{K}^T = \mathbf{K}\,, \quad \mathbf{K}(\mathbf{T}^+, \mathbf{D}^+) = \mathbf{Q}\mathbf{K}(\mathbf{T}, \mathbf{D})\mathbf{Q}^T\,, \tag{5.1.32}$$

for all scalars α and all proper orthogonal tensor functions $\mathbf{Q}$. Next, let n be an arbitrary positive integer and consider the stress rate $\overset{\nabla}{\mathbf{T}}$ in (5.1.18) to be specified by the form

$$\overset{\triangledown}{\mathbf{T}} = \overset{J}{\mathbf{T}} - \beta\,[\mathbf{D}\mathbf{T}^n + \mathbf{T}^n\mathbf{D} - (\mathbf{D}\cdot\mathbf{I})\mathbf{T}^n]\,, \tag{5.1.33}$$

where β is a constant scaling constant having the units $[(\text{stress})^{1-n}]$ so that (5.1.33) has the units of stress. It can be shown that

$$-\mathbf{D}^+\mathbf{T}^{+n} - \mathbf{T}^{+n}\mathbf{D}^+ + (\mathbf{D}^+\cdot\mathbf{I})\mathbf{T}^{+n} = \mathbf{Q}[-\mathbf{D}\mathbf{T}^n - \mathbf{T}^n\mathbf{D} + (\mathbf{D}\cdot\mathbf{I})\mathbf{T}^n]\mathbf{Q}^T\,, \tag{5.1.34}$$

for all SRBM. Consequently, since the Jaumann rate is objective, it follows that $\overset{\triangledown}{\mathbf{T}}$ in (5.1.33) is objective

$$\overset{\triangledown}{\mathbf{T}}{}^+ = \mathbf{Q}\overset{\triangledown}{\mathbf{T}}\mathbf{Q}^T\,. \tag{5.1.35}$$

Next, define the new function $\hat{\mathbf{K}}(\mathbf{T},\mathbf{D})$ by

$$\hat{\mathbf{K}}(\mathbf{T},\mathbf{D}) = \mathbf{K}(\mathbf{T},\mathbf{D}) - \beta\,[\mathbf{D}\mathbf{T}^n + \mathbf{T}^n\mathbf{D} - (\mathbf{D}\cdot\mathbf{I})\mathbf{T}^n]\,, \tag{5.1.36}$$

which satisfies restrictions similar to the forms (5.1.32)

$$\hat{\mathbf{K}}(\mathbf{T},\alpha\mathbf{D}) = \alpha\hat{\mathbf{K}}(\mathbf{T},\mathbf{D})\,,\quad \hat{\mathbf{K}}^T = \hat{\mathbf{K}}\,,\quad \hat{\mathbf{K}}(\mathbf{T}^+,\mathbf{D}^+) = \mathbf{Q}\hat{\mathbf{K}}(\mathbf{T},\mathbf{D})\mathbf{Q}^T\,. \tag{5.1.37}$$

It then follows that the stress $\mathbf{T}$, which satisfies the form-invariant evolution equation

$$\overset{\triangledown}{\mathbf{T}} = \hat{\mathbf{K}}(\mathbf{T},\mathbf{D})\,, \tag{5.1.38}$$

with $\overset{\triangledown}{\mathbf{T}}$ defined by (5.1.33), predicts the same hypoelastic material response as that predicted by (5.1.31). This means that for this general form of a hypoelastic material there is no fundamental physical significance of any of the infinite stress rates in (5.1.33) that satisfy under SRBM.

Summary

Equation (5.1.24) shows that any hyperelastic equation can be formulated in terms of an evolution equation for stress if the right-hand side of (5.1.18) is appropriately modified. However, in general, rate equations of the type (5.1.18) produce hypoelastic response since they do not satisfy integrability conditions necessary for a strain energy function to exist [3]. Due to the physical deficiencies of both Cauchy elasticity and hypoelasticity, the term *elastic material* is used here only for a material that exhibits Green elasticity (hyperelasticity).

5.2 Unphysical Arbitrariness of the Lagrangian Formulation of Constitutive Equations

The classical Lagrangian formulation of constitutive equations for hyperelastic materials specifies the strain energy function Σ to be a function of the deformation gradient $\mathbf{F}$ through the right Cauchy–Green deformation tensor $\mathbf{C}$, such that

$$\Sigma = \Sigma(\mathbf{C}), \quad \mathbf{C} = \mathbf{F}^T \mathbf{F}, \tag{5.2.1}$$

where $\mathbf{F}$ characterizes deformations from an arbitrary, but fixed, reference configuration. The only restriction on this reference configuration is that the mapping

$$\mathbf{x} = \mathbf{x}(\mathbf{X}, t) \tag{5.2.2}$$

be invertible, which requires

$$J = \det\mathbf{F} > 0, \tag{5.2.3}$$

for all material points in the material region under consideration and for all time.

The strain energy function Σ characterizes the response of a specific material, which should be independent of arbitrariness of the choice of the reference configuration. This means that Σ should be a function of internal state variables that can be measured by experiments on identical samples of the material in its current state. In this regard, it is recalled from Sect. 3.11 that $\mathbf{F}$ is not an internal state variable in the sense of Onat [31].

To be more specific, consider a homogeneous deformation of a homogeneous hyperelastic material from a uniform zero-stress material state in its reference configuration with $\mathbf{C} = \mathbf{I}$. It is always possible to unload this material to a zero-stress material state with $\mathbf{C} = \mathbf{I}$, which is satisfied whenever $\mathbf{F}$ is a proper orthogonal tensor. However, anisotropic response requires characterization of the deformation and orientation of material fibers relative to observable material orientations. This arbitrariness of $\mathbf{F}$ makes it impossible to use experiments on the material in its current configuration to determine the orientations of specific material fibers associated with the arbitrary choice of the reference configuration used to specify Σ in (5.2.1).

5.3 An Eulerian Formulation for Nonlinear Elastic Solids

The Eulerian formulation of constitutive equations for nonlinear elastic solids in this section removes the unphysical arbitrariness of the choice of a reference configuration and a total strain measure. For this formulation, use is made of the microstructural vectors $\mathbf{m}_i$ and the elastic metric m_{ij} introduced in Sect. 3.11, determined by the equations

$$\dot{\mathbf{m}}_i = \mathbf{L}\,\mathbf{m}_i\,, \quad m_{ij} = \mathbf{m}_i \cdot \mathbf{m}_j = m_{ji}\,, \quad \dot{m}_{ij} = 2(\mathbf{m}_i \otimes \mathbf{m}_j) \cdot \mathbf{D}\,, \tag{5.3.1}$$

with the strain energy function Σ and the stress proposed in the forms

$$\Sigma = \hat{\Sigma}(m_{ij})\,, \quad \mathbf{T} = \hat{\mathbf{T}}(\mathbf{m}_i)\,. \tag{5.3.2}$$

Using these expressions together with assumption (5.1.2) requires

$$\left[\mathbf{T} - 2\rho \frac{\partial \hat{\Sigma}}{\partial m_{ij}} (\mathbf{m}_i \otimes \mathbf{m}_j)\right] \cdot \mathbf{D} = 0 \tag{5.3.3}$$

for arbitrary motions and all times. Since m_{ij} is symmetric, it follows that $\partial \tilde{\Sigma}/\partial m_{ij}$ $(\mathbf{m}_i \otimes \mathbf{m}_j)$ is a symmetric tensor. Consequently the coefficient of $\mathbf{D}$ in (5.3.3) is symmetric and is independent of $\mathbf{D}$ so the necessary condition that (5.3.3) be valid for arbitrary motions is that the Cauchy stress be determined by a derivative of the strain energy

$$\mathbf{T} = \hat{\mathbf{T}} = 2\rho \frac{\partial \hat{\Sigma}}{\partial m_{ij}} (\mathbf{m}_i \otimes \mathbf{m}_j)\,. \tag{5.3.4}$$

In this formulation, the vectors $\mathbf{m}_i$ are defined so that they form an orthonormal triad in any zero-stress material state (1.2.13) with

$$m_{ij} = \delta_{ij} \quad \text{for any zero-stress material state}\,, \tag{5.3.5}$$

which requires the strain energy function to satisfy the restrictions

$$\frac{\partial \hat{\Sigma}}{\partial m_{ij}} = 0 \quad \text{for} \quad m_{ij} = \delta_{ij}\,. \tag{5.3.6}$$

Moreover, using the conservation of mass (4.1.7), the material derivative of (5.3.4) yields the evolution equation

$$\dot{\mathbf{T}} = \mathbf{L}\mathbf{T} + \mathbf{T}\mathbf{L}^T - (\mathbf{D} \cdot \mathbf{I})\,\mathbf{T} + 2\rho \frac{\partial^2 \hat{\Sigma}}{\partial m_{ij} \partial m_{mn}} (\mathbf{m}_i \otimes \mathbf{m}_j \otimes \mathbf{m}_m \otimes \mathbf{m}_n) \cdot \mathbf{D}\,. \tag{5.3.7}$$

The microstructural vectors $\mathbf{m}_i$ are internal state variables in the sense of Onat [31] with their values in the current configuration being determined by experiments on identical samples of the material. Specifically, use is made of measurements of the current state of stress $\mathbf{T}$ and the value of $\dot{\mathbf{T}}$ for different values of the loading rate $\mathbf{L}$. Any differences between the $\mathbf{m}_1$, $\mathbf{m}_2$ and $\mathbf{m}_3$ directions which cannot be determined by these experiments should be consistent with material symmetries of the strain energy function Σ which ensure that these differences do not influence the prediction of the material response to an arbitrary loading rate $\mathbf{L}$.

A Separation of Elastic Dilatation and Distortional Deformations
To introduce separate control over the response of the material to dilatation and distortional rates of deformation it is convenient to use the elastic dilatation J_e defined in (3.11.7), the distortional deformation vectors $\mathbf{m}'_i$ defined in (3.11.14) and the elastic distortional deformation metric m'_{ij} defined in (3.11.16), which satisfy the Eqs. (3.11.17) and (3.11.28)

$$\begin{aligned} &J_e = \mathbf{m}_1 \times \mathbf{m}_2 \cdot \mathbf{m}_3 > 0\,, && \dot{J}_e = J_e \mathbf{D} \cdot \mathbf{I}\,, \\ &\mathbf{m}'_i = J_e^{-1/3}\,\mathbf{m}_i\,, && \dot{\mathbf{m}}'_i = \mathbf{L}''\,\mathbf{m}'_i\,, \\ &m'_{ij} = \mathbf{m}'_i \cdot \mathbf{m}'_j = m'_{ji}\,, && \dot{m}'_{ij} = 2\left(\mathbf{m}'_i \otimes \mathbf{m}'_j - \tfrac{1}{3} m'_{ij}\mathbf{I}\right) \cdot \mathbf{D}\,. \end{aligned} \tag{5.3.8}$$

Then, the strain energy function and the stress are proposed in the forms

$$\Sigma = \tilde{\Sigma}(J_e, m'_{ij})\,, \quad \mathbf{T} = \tilde{\mathbf{T}}(J_e, \mathbf{m}'_i)\,, \tag{5.3.9}$$

and the condition (5.1.2) requires

$$\left[\mathbf{T} - \rho J_e \frac{\partial \tilde{\Sigma}}{\partial J_e}\mathbf{I} - 2\rho \frac{\partial \tilde{\Sigma}}{\partial m'_{ij}}(\mathbf{m}'_i \otimes \mathbf{m}'_j - \frac{1}{3} m'_{ij}\mathbf{I})\right] \cdot \mathbf{D} = 0\,. \tag{5.3.10}$$

Since m'_{ij} is symmetric, it follows that $\partial \tilde{\Sigma}/\partial m'_{ij}(\mathbf{m}_i \otimes \mathbf{m}_j)$ is a symmetric tensor. Consequently the coefficient of $\mathbf{D}$ in (5.3.10) is symmetric and is independent of $\mathbf{D}$ so the necessary condition that (5.3.10) be valid for arbitrary motions is that the Cauchy stress be determined by a derivative of the strain energy

$$\begin{aligned} &\mathbf{T} = -p\,\mathbf{I} + \mathbf{T}'' = \tilde{\mathbf{T}}\,, \quad p = \tilde{p} = -\rho J_e \frac{\partial \tilde{\Sigma}}{\partial J_e}\,, \\ &\mathbf{T}'' = \tilde{\mathbf{T}}'' = 2\rho \frac{\partial \tilde{\Sigma}}{\partial m'_{ij}}\left(\mathbf{m}'_i \otimes \mathbf{m}'_j - \frac{1}{3} m'_{ij}\mathbf{I}\right)\,, \end{aligned} \tag{5.3.11}$$

where p is the pressure and $\mathbf{T}''$ is the deviatoric part of $\mathbf{T}$.

Also, the constitutive equation for stress is assumed to be restricted so that a zero-stress material state is characterized by

$$\mathbf{T} = 0\,, \quad \frac{\partial \tilde{\Sigma}}{\partial J_e} = 0\,, \quad \frac{\partial \tilde{\Sigma}}{\partial m'_{ij}} = \frac{1}{3}\frac{\partial \tilde{\Sigma}}{\partial m'_{nn}}\delta_{ij} \text{ for } J_e = 1 \text{ and } m'_{ij} = \delta_{ij}\,. \tag{5.3.12}$$

This means that the triad $\mathbf{m}'_i$ has been defined so that $\mathbf{m}'_i$ are orthonormal vectors in a zero-stress material state.

This form for the strain energy function makes it easy to separate the effects of dilatation and distortion. For example, a class of materials can be considered for which the strain energy function separates into two additive parts

$$\rho_z \Sigma = f(J_e) + \rho_z \tilde{\Sigma}_d(m'_{ij}) , \tag{5.3.13}$$

where ρ_z is the constant zero-stress mass density, f controls the response to dilatation and $\tilde{\Sigma}_d$ controls the response to distortional deformations. It then follows that the Cauchy stress $\mathbf{T}$ for this strain energy function is given by

$$\begin{aligned} &\mathbf{T} = -p\,\mathbf{I} + \mathbf{T}'' = \tilde{\mathbf{T}} , \quad p = -\left(\frac{\rho J_e}{\rho_z}\right) \frac{df}{dJ_e} , \\ &\mathbf{T}'' = \tilde{\mathbf{T}}''(J_e, m'_{ij}) = 2\rho \frac{\partial \tilde{\Sigma}_d}{\partial m'_{ij}} \left(\mathbf{m}'_i \otimes \mathbf{m}'_j - \frac{1}{3} m'_{ij} \mathbf{I} \right) , \end{aligned} \tag{5.3.14}$$

with the restrictions that

$$\frac{df}{dJ_e} = 0 , \quad \frac{\partial \tilde{\Sigma}_d}{\partial m'_{ij}} = \frac{1}{3} \frac{\partial \tilde{\Sigma}_d}{\partial m'_{nn}} \delta_{ij} \text{ for } J_e = 1 \text{ and } m'_{ij} = \delta_{ij} . \tag{5.3.15}$$

Furthermore, using the conservation of mass in the form (4.1.16)

$$J_e = \frac{\rho_z}{\rho} , \tag{5.3.16}$$

it follows that the pressure

$$p = \tilde{p}(J_e) = -\frac{df}{dJ_e} , \tag{5.3.17}$$

for this class of materials depends on the elastic dilatation J_e only.

5.4 Difference Between the Microstructural Vectors $\mathbf{m}_i$ and the Deformation Gradient F

Recall that the deformation gradient $\mathbf{F}$ satisfies the evolution equation (3.5.4)

$$\dot{\mathbf{F}} = \mathbf{L}\mathbf{F} . \tag{5.4.1}$$

It has been stated in Sect. 3.11 that $\mathbf{F}$ is not an internal state variable in the sense of Onat [31] since its initial value depends explicitly on an arbitrary choice of the reference configuration which cannot be measured in the current configuration.

In contrast, the microstructural vectors $\mathbf{m}_i$ for elastic response satisfy the Eqs. (3.11.6)

$$\dot{\mathbf{m}}_i = \mathbf{L}\,\mathbf{m}_i , \tag{5.4.2}$$

and Sect. 5.3 explained how their initial conditions can be determined by experiments on identical samples of material in the current state so they are internal state variables in the sense of Onat [31].

To further explore the arbitrariness of $\mathbf{F}$, consider an initial zero-stress material state (3.11.9) for which the measured values $\mathbf{m}_i(0)$ of $\mathbf{m}_i$ form a right-handed orthonormal triad

$$m_{ij}(0) = \mathbf{m}_i(0) \cdot \mathbf{m}_j(0) = \delta_{ij} \,. \tag{5.4.3}$$

Furthermore, define the elastic deformation tensor $\mathbf{F}_e$ by

$$\mathbf{F}_e = \mathbf{m}_i(t) \otimes \mathbf{m}_i(0) \,. \tag{5.4.4}$$

By definition, this tensor satisfies the evolution equation and initial condition

$$\dot{\mathbf{F}}_e = \mathbf{L}\mathbf{F}_e \,, \quad \mathbf{F}(0) = \mathbf{I} \,. \tag{5.4.5}$$

Although $\mathbf{m}_i(0)$ are measurable in the initial state and $\mathbf{m}_i(t)$ are measurable in the current state, the tensor $\mathbf{F}_e$ is not a state variable since it is impossible to know the orientation of $\mathbf{m}_i(0)$ in the reference state from experiments on the material in its current state. In this regard, it is emphasized that there is no need for the second-order tensor $\mathbf{F}_e$ because the microstructural vectors $\mathbf{m}_i$ with their elastic deformation metric m_{ij} are sufficient to characterize constitutive equations for general anisotropic elastic response (5.3.4).

To be more specific, let $\mathbf{M}_i$ be an *arbitrary* right-handed orthonormal triad of constant vectors $\mathbf{M}_i$ and define $\mathbf{F}$ by

$$\mathbf{F} = \mathbf{m}_i \otimes \mathbf{M}_i \,. \tag{5.4.6}$$

It follows that $\mathbf{F}$ satisfies the evolution equation and initial condition

$$\dot{\mathbf{F}} = \mathbf{L}\mathbf{F} \,, \quad \mathbf{F}(0) = \mathbf{m}_i(0) \otimes \mathbf{M}_i \,. \tag{5.4.7}$$

However, since $\mathbf{M}_i$ are arbitrary orthonormal vectors and $\mathbf{m}_i(0)$ are orthonormal vectors, it also follows that the initial value of $\mathbf{F}$ is an arbitrary proper orthogonal rotation tensor

$$\mathbf{F}(0)^T \mathbf{F}(0) = \mathbf{I} \,, \tag{5.4.8}$$

with arbitrariness of the specification of $\mathbf{M}_i$, which represents an arbitrary orientation of the body in a reference configuration that cannot be determined by experiments on the material in its current state.

5.5 Homogeneity and Uniformity

This section presents a brief discussion of notions of homogeneous deformation, a body that is materially uniform, a homogeneous body and a uniform material state. More detailed discussion of these notions can be found in ([47], Sect. 27.3).

Homogeneous Deformation
A body is said to experience a homogenous deformation during the time period $t_1 \le t \le t_2$ if the velocity gradient $\mathbf{L}$ is independent of $\mathbf{x}$ during this time period

$$\mathbf{L} = \mathbf{L}(t)\,, \quad \partial\mathbf{L}/\partial\mathbf{x} = 0\,, \quad \text{for} \quad t_1 \le t \le t_2\,. \tag{5.5.1}$$

With the help of (3.13.3), it follows that the relative deformation gradient $\mathbf{F}_r$ from the time t_1 depends on time only and satisfies equations

$$\dot{\mathbf{F}}_r = \mathbf{L}\mathbf{F}_r\,, \quad \mathbf{F}_r(t_1) = \mathbf{I}\,, \quad \mathbf{F}_r = \mathbf{F}_r(t) \quad \text{for} \quad t_1 \le t \le t_2\,. \tag{5.5.2}$$

Moreover, with the help of (3.11.1), it follows that the total deformation gradient $\mathbf{F}$ satisfies the evolution equation and initial condition

$$\dot{\mathbf{F}} = \mathbf{L}\mathbf{F}\,, \quad \mathbf{F}(\mathbf{X}, t_1) = \bar{\mathbf{F}}(\mathbf{X}, t_1)\,, \tag{5.5.3}$$

where the value $\bar{\mathbf{F}}(\mathbf{X}, t_1)$ of $\mathbf{F}$ at time t_1 can be a function of position $\mathbf{X}$. Using the relative deformation gradient $\mathbf{F}_r(t)$, the exact solution of $\mathbf{F}$ during this time period is given by

$$\mathbf{F}(\mathbf{X}, t) = \mathbf{F}_r(t)\bar{\mathbf{F}}(\mathbf{X}, t_1) \quad \text{for} \quad t_1 \le t \le t_2\,. \tag{5.5.4}$$

In particular, it is noted that although the deformation is homogeneous during the time period $t_1 \le t \le t_2$ the total deformation gradient $\mathbf{F}$ is not necessarily independent of space $\mathbf{X}$.

A Materially Uniform Body
A body is said to be *materially uniform* if the material functions that characterize the response of the material are explicitly independent of space. For example, a body made of an elastic material characterized by the strain energy function (5.2.1) associated with the Lagrangian formulation

$$\Sigma = \hat{\Sigma}(\mathbf{C})\,, \tag{5.5.5}$$

is materially uniform if Σ depends on $\mathbf{X}$ only through the dependence of $\mathbf{C} = \mathbf{F}^T\mathbf{F}$ on $\mathbf{X}$

$$\partial\hat{\Sigma}/\partial\mathbf{X} = 0\,. \tag{5.5.6}$$

To be precise, a superposed $(\hat{\ })$ has been used to indicate a specific functional dependence of $\hat{\Sigma}(\mathbf{C})$ on $\mathbf{C}$ only so the dependence of Σ on $\mathbf{X}$ must be evaluated using the chain rule of differentiation

$$\partial \Sigma / \partial \mathbf{X} = \partial \hat{\Sigma} / \partial \mathbf{C} \cdot \partial \mathbf{C} / \partial \mathbf{X} \,. \tag{5.5.7}$$

Similarly, a body made of an elastic material characterized by the strain energy function (5.3.2) associated with the Eulerian formulation

$$\Sigma = \hat{\Sigma}(m_{ij}) \tag{5.5.8}$$

is materially uniform if Σ depends on $\mathbf{x}$ only through the dependence of elastic deformation metric m_{ij} on $\mathbf{x}$

$$\partial \hat{\Sigma} / \partial \mathbf{x} = 0 \,, \tag{5.5.9}$$

so that

$$\partial \Sigma / \partial \mathbf{x} = (\partial \hat{\Sigma} / \partial m_{ij}) \, (\partial m_{ij} / \partial \mathbf{x}) \,. \tag{5.5.10}$$

For more general material response, like that of elastic–inelastic materials discussed in Sect. 5.11 or thermoelastic materials and thermoelastic–inelastic materials discussed in Chap. 6, all constitutive functions, including those in evolution equations, must be explicitly independent of **x** for a body to be *materially uniform.*

A Uniform Material State
A body is said to be in a *uniform material state* if the body is materially uniform and each response function characterizing the material has a value that is independent of **x** for all points in the body. With regard to the Eulerian formulation, it is emphasized that the notion of a uniform material state need not be connected with any specification of a configuration of the body which places the body in space at a specified time.

Homogeneous Body
A body is said to be a *homogeneous body* if it is materially uniform and a configuration exits for which it is also in a uniform material state.

Examples
To better understand the difference between a body that is materially uniform and a homogeneous body, consider a cylindrical region that is materially uniform. Its solid cylindrical inner core is a homogeneous body that has a zero-stress uniform material state and its outer cylindrical shell is also a homogeneous body that has a zero-stress uniform material state. Moreover, consider the case when the outer radius of the zero-stress inner core is larger than the inner radius of the zero-stress outer cylindrical shell. By cooling the inner core or heating the outer cylindrical shell, it is possible to assemble the inner core inside of the outer cylindrical shell. Then, when the temperature in returned to a uniform value, the resulting body will have residual stresses in both of its inner and outer regions even if the outer surface of the cylindrical shell is traction free. The resulting body remains materially uniform but is not in a uniform material state. Moreover, it is no longer a homogeneous body since no configuration exists in which it can be in a uniform material state.

To better understand the notion of a uniform material state, consider a homogeneous elastic body which is in a uniform zero-stress material state. Then, load the body with a body force and traction vectors that cause inhomogeneous deformation. The body remains homogeneous but the deformed material state is not in a uniform material state.

Arbitrariness of the Reference Configuration
To examine the influence of arbitrariness of the choice of the reference configuration in the Lagrangian formulation, consider a homogeneous body made from an elastic material with the strain energy function

$$\Sigma = \hat{\Sigma}(\mathbf{C}) , \quad \mathbf{C} = \mathbf{F}^T \mathbf{F} . \tag{5.5.11}$$

In these expressions, the deformation gradient $\mathbf{F}$ is measured from a reference configuration where the body is in a zero-stress uniform material state with $\mathbf{F} = \mathbf{I}$. Next, consider an arbitrary change in the reference configuration with $\bar{\mathbf{F}}$ measured relative to the new reference configuration, such that

$$\mathbf{F} = \bar{\mathbf{F}}\mathbf{A} , \quad \det \mathbf{A}(\mathbf{X}) > 0 , \quad \mathbf{C} = \mathbf{A}^T \bar{\mathbf{C}} \mathbf{A} , \quad \bar{\mathbf{C}} = \bar{\mathbf{F}}^T \bar{\mathbf{F}} , \tag{5.5.12}$$

where $\mathbf{A}(\mathbf{X})$ is an arbitrary second-order tensor function of $\mathbf{X}$ only with positive determinant. It then follows that the strain energy function can be expressed in terms of $\bar{\mathbf{C}}$ and $\mathbf{A}$ in the form

$$\Sigma = \bar{\Sigma}(\bar{\mathbf{C}}, \mathbf{A}) = \hat{\Sigma}(\mathbf{A}^T \bar{\mathbf{C}} \mathbf{A}) . \tag{5.5.13}$$

Since $\bar{\mathbf{C}}$ is the deformation relative to the new reference configuration and since $\mathbf{A}$ can be an arbitrary function of $\mathbf{X}$, it follows that the strain energy function $\bar{\Sigma}(\bar{\mathbf{C}}, \mathbf{A})$ depends on $\mathbf{X}$ explicitly through the tensor $\mathbf{A}$. This means that the notions of the body being materially uniform or homogeneous depend on the variables being used to describe the response and on arbitrariness of the choice of the reference configuration.

In contrast, the Eulerian formulation for a homogeneous body with the strain energy function

$$\Sigma = \tilde{\Sigma}(m_{ij}) , \tag{5.5.14}$$

which is deformed from a zero-stress uniform material state is insensitive to changes in the reference configuration with associated changes in the total deformation from the reference configuration.

Influence of Inelasticity
Elastic–inelastic response will be discussed in detail in Sect. 5.11 and in Chap. 6. To discuss the influence of inelasticity on the motions of a body that is materially uniform, a homogeneous body and a uniform material state it is sufficient to consider a homogeneous body with the strain energy function (5.5.14) which is initially in a zero-stress uniform material state with $m_{ij} = \delta_{ij}$. Loading the body with a body force and surface tractions can cause inhomogeneous total deformation with nonzero inhomogeneous inelastic deformation rate. When all external loads are removed, this inhomogeneous inelastic deformation rate causes the body to attain a nonuniform material state with nonzero residual stresses. This unloaded body remains materially

uniform but is no longer homogeneous since a configuration no longer exists for which it is also in a uniform material state.

5.6 Material Symmetry

Consider a general nonlinear homogeneous elastic material which is initially in a uniform zero-stress material state with initial orthonormal values $\mathbf{M}_i$ of the vectors $\mathbf{m}_i$

$$\mathbf{m}_i(0) = \mathbf{M}_i\,, \quad \mathbf{M}_i \cdot \mathbf{M}_j = \delta_{ij}\,, \quad \mathbf{M}_1 \times \mathbf{M}_2 \cdot \mathbf{M}_3 = 1\,. \tag{5.6.1}$$

Also, consider a set of experiments where tension specimens are machined from the material with different orientations relative to $\mathbf{M}_i$. The dependence of the measured nonlinear response for specimens with different orientations characterizes the material symmetry of the material. If the measured nonlinear response for different specimens is different, then the material is denoted as anisotropic. Whereas, if the measured nonlinear response for specimens with all possible orientations is the same relative to the orientation of the specimen for all deformations, then the material is denoted as isotropic.

To analyze this notion of material symmetry, consider a tension specimen that has a fixed orientation relative to another orthonormal triad of vectors $\tilde{\mathbf{M}}_i$ defined by the orthonormal matrix H_{ij}, such that

$$\begin{aligned} \tilde{\mathbf{M}}_i &= H_{ij}\mathbf{M}_j\,, & \mathbf{M}_i &= H_{ji}\tilde{\mathbf{M}}_j\,, \\ H_{ij} &= \tilde{\mathbf{M}}_i \cdot \mathbf{M}_j\cdot\,, & H_{im}H_{jm} &= H_{mi}H_{mj} = \delta_{ij}\,. \end{aligned} \tag{5.6.2}$$

The deformation tensor

$$m_{ij}\mathbf{M}_i \otimes \mathbf{M}_j \tag{5.6.3}$$

applies the elastic deformation metric m_{ij} to a specimen which has a specific alignment relative to the vectors $\mathbf{M}_i$ and the deformation tensor

$$m_{ij}\tilde{\mathbf{M}}_i \otimes \tilde{\mathbf{M}}_j \tag{5.6.4}$$

applies the same elastic deformation metric m_{ij} to a specimen which has the same specific alignment relative to the vectors $\tilde{\mathbf{M}}_i$. These two deformation tensors (5.6.3) and (5.6.4) are different and the components $\bar{m}_{ij}$ of (5.6.4) relative to $\mathbf{M}_i$ are given by

$$\bar{m}_{ij} = m_{mn}\tilde{\mathbf{M}}_m \otimes \tilde{\mathbf{M}}_n \cdot \mathbf{M}_i \otimes \mathbf{M}_j = H_{mi}H_{nj}m_{mn}\,. \tag{5.6.5}$$

Consequently, the elastic deformation applied to a specimen taken in, say the $\mathbf{M}_1$ direction, will be the same as that applied to a specimen taken in the $\tilde{\mathbf{M}}_1$ direction for all values of m_{ij} and all orthogonal matrices H_{ij}.

Now, the response of a nonlinear elastic material to arbitrary identical nonlinear deformations m_{ij} with different material orientations will be the same provided that

$$\hat{\Sigma}(m_{ij}) = \hat{\Sigma}(\bar{m}_{ij}) = \hat{\Sigma}(H_{mi} H_{nj} m_{mn}) \tag{5.6.6}$$

or

$$\tilde{\Sigma}(J_e, m'_{ij}) = \tilde{\Sigma}(J_e, \bar{m}'_{ij}) = \tilde{\Sigma}(J_e, H_{mi} H_{nj} m'_{mn}) \tag{5.6.7}$$

hold for all possible deformations m_{ij}, J_e and m'_{ij}. In other words, the functional forms of the strain energies $\hat{\Sigma}$ and $\tilde{\Sigma}$ remain form-invariant to a group of orthogonal transformations H_{ij} which characterize the material symmetries exhibited by a given material. For the case of crystalline materials these symmetry groups can be related to the different crystal structures.

For the most general anisotropic elastic response, the material has no symmetry, so the group of H_{ij} contains only the identity δ_{ij}. Whereas an isotropic elastic material has complete symmetry, so the group of H_{ij} is the full orthogonal group. Furthermore, it is important to emphasize that the notion of material symmetry is necessarily referred to identifiable material directions which are naturally represented by the vectors $\mathbf{m}_i$.

Moreover, the dependence of the functional forms $\hat{\Sigma}(m_{ij})$ and $\tilde{\Sigma}(J_e, m'_{ij})$ on the material directions $\mathbf{m}_i$ is explicit and is used to determine the initial values of $\mathbf{m}_i$. In particular, any anisotropic response of the material is measured relative to the microstructural vectors $\mathbf{m}_i$, which causes the characterization of anisotropy to be independent of arbitrariness of a specification of a reference configuration. Furthermore, any indeterminacy of $\mathbf{m}_i$ in the current state must be compensated by the material symmetry of the strain energy function rendering this indeterminacy irrelevant for the response of the material.

5.7 Kinematic Constraints

Some materials have special properties that can be exploited to obtain approximate constitutive equations that simplify analytical solutions to problems. For example, rubber is a material with its resistance to volumetric deformation being much larger than its resistance to distortional deformations. This means that large changes in pressure occur for small changes in volume. From a mathematical point of view, it is convenient to consider a kinematic condition which constrains the material to be incompressible.

Using (3.11.5), it follows that an elastically incompressible material can only experience deformations which satisfy the kinematic constraint

$$G = J_e - 1 = 0 \Rightarrow \mathbf{I} \cdot \mathbf{D} = 0 \,. \tag{5.7.1}$$

Another example is a fiber reinforced composite with stiff fibers relative to the response of its matrix. For such a material it is possible to approximate the fibers as being inextensible. Using the microstructural vectors $\mathbf{m}_i$ in (3.11.6), it follows that a material fiber in the $\mathbf{m}_1$ direction will remain inextensible (in tension and compression) if the material satisfies the kinematic constraint

$$G = m_{11} - 1 = 0 \;\Rightarrow\; (\mathbf{m}_1 \otimes \mathbf{m}_1) \cdot \mathbf{D} = 0\,. \tag{5.7.2}$$

In general, consider a kinematic constraint of the form

$$G = G(m_{ij}) = 0 \;\Rightarrow\; \boldsymbol{\Gamma} \cdot \mathbf{D} = 0\,, \quad \boldsymbol{\Gamma} \equiv \frac{\partial G}{\partial m_{ij}} \mathbf{m}_i \otimes \mathbf{m}_j\,, \tag{5.7.3}$$

which can be rewritten in the form

$$\frac{\partial G}{\partial m_{ij}} D_{ij} = 0\,, \quad D_{ij} = \mathbf{D} \cdot \mathbf{m}_i \otimes \mathbf{m}_j\,. \tag{5.7.4}$$

In particular, it is noted that $\boldsymbol{\Gamma}$ is a symmetric second-order tensor that is independent of the rate $\mathbf{D}$

$$\boldsymbol{\Gamma}^T = \boldsymbol{\Gamma}\,, \tag{5.7.5}$$

and under SRBM it satisfies the transformation relation

$$\boldsymbol{\Gamma}^+ = \mathbf{Q}\boldsymbol{\Gamma}\mathbf{Q}^T\,. \tag{5.7.6}$$

Moreover, consider a general unconstrained material that is characterized by a constitutive equation $\hat{\mathbf{T}}$ for the Cauchy stress $\mathbf{T}$. Next, consider a model of a constrained material for which $\mathbf{T}$ is additively separated into the constitutive part $\hat{\mathbf{T}}$ and a part $\bar{\mathbf{T}}$, called the constraint response, which enforces the kinematic constraint (5.7.3)

$$\mathbf{T} = \hat{\mathbf{T}} + \bar{\mathbf{T}}\,. \tag{5.7.7}$$

Although $\hat{\mathbf{T}}$ characterizes the response to general deformations, its value in (5.7.7) is determined by evaluating $\hat{\mathbf{T}}$ only for deformations that satisfy the imposed kinematic constraint. Moreover, $\hat{\mathbf{T}}$ automatically satisfies the restriction

$$\hat{\mathbf{T}}^T = \hat{\mathbf{T}} \tag{5.7.8}$$

due to the balance of angular momentum and it transforms under SRBM, such that

$$\hat{\mathbf{T}}^+ = \mathbf{Q}\hat{\mathbf{T}}\mathbf{Q}^T\,. \tag{5.7.9}$$

Now, since the reduced form (4.4.10) of the balance of angular momentum requires $\mathbf{T}$ to be a symmetric tensor, the constraint response $\bar{\mathbf{T}}$ must also be a symmetric tensor

$$\bar{\mathbf{T}}^T = \bar{\mathbf{T}} \,. \tag{5.7.10}$$

In addition, $\bar{\mathbf{T}}$ is assumed to be workless

$$\bar{\mathbf{T}} \cdot \mathbf{D} = 0 \,, \tag{5.7.11}$$

and independent of the rate $\mathbf{D}$.

Next, multiplying (5.7.3) by an arbitrary scalar Γ and subtracting the result from (5.7.11) yields

$$(\bar{\mathbf{T}} - \Gamma\, \boldsymbol{\Gamma}) \cdot \mathbf{D} = 0 \,. \tag{5.7.12}$$

Now, it is noted that the coefficient of $\mathbf{D}$ in this equation is a symmetric tensor that is independent of $\mathbf{D}$ and that this equation must hold for arbitrary rates $\mathbf{D}$ that satisfy the constraint (5.7.4). Moreover, since the constraint (5.7.4) is nontrivial, at least one component of $\boldsymbol{\Gamma}$ is nonzero. For example, let $\partial G/\partial m_{33}$ be nonzero. This means that the component D_{33} can be used to satisfy the constraint (5.7.4) for arbitrary values of the other components D_{ij}. By choosing the value of Γ in (5.7.12) so that the coefficient of D_{33} vanishes, and choosing the other components of D_{ij} arbitrarily, it follows that the constraint response $\bar{\mathbf{T}}$ must be given by

$$\bar{\mathbf{T}} = \Gamma\, \boldsymbol{\Gamma} \,, \tag{5.7.13}$$

with Γ being an arbitrary function of $\mathbf{x}$ and t that is determined by the equations of motion and boundary conditions. Due to (5.7.5) it can be seen that this form for $\bar{\mathbf{T}}$ automatically satisfies the restriction (5.7.10) due to the balance of angular momentum. Moreover, since $\mathbf{T}$ in (5.7.7) appears in the balance of linear momentum and characterizes the response of the constrained material, the restriction (R-2) in (4.7.3b), which defines how the constitutive response of the material relative to its orientation is the same for all SRBM, requires the constraint response $\bar{\mathbf{T}}$ to satisfy the transformation relation

$$\bar{\mathbf{T}}^+ = \Gamma^+ \boldsymbol{\Gamma}^+ = \mathbf{Q}\bar{\mathbf{T}}\mathbf{Q}^T = \Gamma\, \mathbf{Q}\boldsymbol{\Gamma}\mathbf{Q}^T \,, \tag{5.7.14}$$

which with the help of (5.7.6) requires the arbitrary function Γ to be unaffected by SRBM

$$\Gamma^+ = \Gamma \,. \tag{5.7.15}$$

In addition, since the constraint response $\bar{\mathbf{T}}$ is workless (5.7.11), it follows that

$$\mathbf{T} \cdot \mathbf{D} = \hat{\mathbf{T}} \cdot \mathbf{D} \,, \tag{5.7.16}$$

so the constraint response does not influence the restriction (4.5.7) characterizing the rate of material dissipation.

For the special case of an incompressible material, the constraint response is given by

$$\bar{\mathbf{T}} = -\bar{p}\,\mathbf{I}\,, \tag{5.7.17}$$

where $\bar{p}$ is an arbitrary function of $\mathbf{x}$ and t that is determined by the equations of motion and boundary conditions.

Furthermore, it is noted that up to five independent kinematic constraints of the type (5.7.3) can be imposed simultaneously without causing $\mathbf{T}$ to be totally indeterminate.

5.8 Isotropic Nonlinear Elastic Materials

For an isotropic nonlinear elastic material the strain energy function (5.6.7) remains form-invariant for the full orthogonal group of H_{ij}. This means that Σ can depend on m'_{ij} only through its invariants. This also means that experiments on identical samples of the material in its current state cannot distinguish between the microstructural vectors $\mathbf{m}'_1$, $\mathbf{m}'_2$ and $\mathbf{m}'_3$ so the material response functions must be insensitive to this arbitrariness of $\mathbf{m}'_i$. Consequently, the symmetric, positive-definite, unimodular tensor $\mathbf{B}'_e$ defined in (3.11.19)

$$\mathbf{B}'_e = \mathbf{m}'_i \otimes \mathbf{m}'_i \tag{5.8.1}$$

can be used to characterize the response of an elastically isotropic material to elastic distortional deformations.

To discuss the invariants of m'_{ij} it is recalled from (3.3.17) and (3.11.24) that the unimodular elastic distortional deformation tensor $\mathbf{B}'_e$ satisfies the equations

$$\begin{aligned}
\det\mathbf{B}'_e &= \frac{\mathbf{B}'_e\mathbf{m}'_1 \times \mathbf{B}'_e\mathbf{m}'_2 \cdot \mathbf{B}'_e\mathbf{m}'_3}{\mathbf{m}'_1 \times \mathbf{m}'_2 \cdot \mathbf{m}'_3} = m'_{i1}\mathbf{m}'_i \times m'_{j2}\mathbf{m}'_j \cdot m'_{k3}\mathbf{m}'_k\,,\\
\det\mathbf{B}'_e &= \varepsilon_{ijk} m'_{i1} m'_{j2} m'_{k3}\,,\\
\det\mathbf{B}'_e &= \frac{1}{6}\varepsilon_{ijk}\varepsilon_{rst} m'_{ir} m'_{js} m'_{kt} = \det(m'_{ij}) = 1\,,
\end{aligned} \tag{5.8.2}$$

where use has been made of (3.11.14) to conclude that

$$\varepsilon_{ijk} = \mathbf{m}'_i \times \mathbf{m}'_j \cdot \mathbf{m}'_k\,. \tag{5.8.3}$$

Thus, the metric m'_{ij} of elastic distortional deformations has only two nontrivial independent invariants which can be specified by

$$\alpha_1 = m'_{ii} = \mathbf{m}'_i \cdot \mathbf{m}'_i = \mathbf{B}'_e \cdot \mathbf{I}\,, \quad \alpha_2 = m'_{ij} m'_{ij} = \mathbf{B}'_e \cdot \mathbf{B}'_e\,. \tag{5.8.4}$$

Consequently, for an isotropic elastic material the strain energy function takes the form

$$\Sigma = \Sigma(J_e, \alpha_1, \alpha_2)\,. \tag{5.8.5}$$

Next, using the evolution equation (3.11.5) for the elastic dilatation J_e

$$\dot{J}_e = J_e \mathbf{D} \cdot \mathbf{I} , \tag{5.8.6}$$

the evolution equations for the scalar measures α_1 and α_2 of elastic distortional deformation are given by

$$\dot{\alpha}_1 = 2\mathbf{B}_e'' \cdot \mathbf{D} , \quad \dot{\alpha}_2 = 4\left(\mathbf{B}_e'^2 - \frac{1}{3}\alpha_2 \mathbf{I}\right) \cdot \mathbf{D} , \tag{5.8.7}$$

where $\mathbf{B}_e''$ is the deviatoric part of $\mathbf{B}_e'$. Consequently, the material derivative of the strain energy function (5.8.5) is given by

$$\rho\dot{\Sigma} = \left[\rho J_e \frac{\partial \Sigma}{\partial J_e}\mathbf{I} + 2\rho\frac{\partial \Sigma}{\partial \alpha_2}\mathbf{B}_e'' + 4\rho\frac{\partial \Sigma}{\partial \alpha_2}\left(\mathbf{B}_e'^2 - \frac{1}{3}\alpha_2\mathbf{I}\right)\right] \cdot \mathbf{D} . \tag{5.8.8}$$

Then, the condition that the material response of an elastic material is non-dissipative for all motions

$$\mathbf{T} \cdot \mathbf{D} = \rho\dot{\Sigma} \tag{5.8.9}$$

requires the stress to be given in the form

$$\begin{aligned} &\mathbf{T} = -p\,\mathbf{I} + \mathbf{T}'' , \quad p = -\rho J_e \frac{\partial \Sigma}{\partial J_e} , \\ &\mathbf{T}'' = 2\rho\frac{\partial \Sigma}{\partial \alpha_2}\mathbf{B}_e'' + 4\rho\frac{\partial \Sigma}{\partial \alpha_2}\left(\mathbf{B}_e'^2 - \frac{1}{3}\alpha_2\mathbf{I}\right) . \end{aligned} \tag{5.8.10}$$

In particular, notice that the deviatoric stress $\mathbf{T}''$ vanishes whenever $\mathbf{B}_e' = \mathbf{I}$ so the condition (5.3.5) characterizing a zero-stress material state requires

$$\frac{\partial \Sigma}{\partial J_e} = 0 \quad \text{for } J_e = 1 \text{ and } \mathbf{B}_e' = \mathbf{I} . \tag{5.8.11}$$

A Compressible Neo-Hookean Material

Significant advances in the theory of finite elasticity were made by Rivlin and co-workers [2] studying the response of natural rubber, which is a material that can experience large distortional deformations and is relatively stiff to volumetric deformations. For such a material it is convenient to additively separate the strain energy function into a part that controls the response to elastic dilatation and depends only on J_e and another part that depends only on elastic distortional deformations through the invariants α_1, α_2. For the simplest compressible Neo-Hookean the strain energy function is specified by

$$\rho_z \Sigma = f(J_e) + \frac{1}{2}\mu(\alpha_1 - 3) , \quad \mu > 0 , \tag{5.8.12}$$

where μ is the zero-stress shear modulus and $f(J_e)$ is a function that satisfies the conditions

$$f(1) = 0\,, \quad \frac{df}{dJ_e}(1) = 0\,, \quad \frac{d^2 f}{dJ_e^2}(1) > 0\,. \tag{5.8.13}$$

It then follows from (4.1.16)

$$J_e = \frac{\rho_z}{\rho}\,, \tag{5.8.14}$$

and (5.8.10) that the pressure p and deviatoric stress $\mathbf{T}''$ for this material are given by

$$p = -\frac{df}{dJ_e}\,, \quad \mathbf{T}'' = J_e^{-1}\mu\mathbf{B}_e''\,. \tag{5.8.15}$$

A Compressible Mooney–Rivlin Material
For a compressible Mooney–Rivlin material the strain energy function is specified by

$$\rho_z\Sigma = f(J_e) + \frac{1}{2}\mu[(1-4C)(\alpha_1 - 3) + C(\alpha_2 - 3)]\,, \tag{5.8.16}$$

where $f(J_e)$ satisfies the conditions (5.8.13), μ is the positive zero-stress shear modulus and C is a material constant. Then, using (5.8.14) the associated constitutive equations for p and $\mathbf{T}''$ are given by

$$p = -\frac{df}{dJ_e}\,, \quad \mathbf{T}'' = J_e^{-1}\mu\left[(1-4C)\mathbf{B}_e'' + 2C\left(\mathbf{B}_e'^2 - \frac{1}{3}\alpha_2\mathbf{I}\right)\right]. \tag{5.8.17}$$

A Specific Function for Dilatation
As a special case, consider a polyconvex function $f(J_e)$ for the strain energy of dilatation given by (e.g., [45])

$$f(J_e) = \frac{1}{2}k\left[\frac{1}{2}(J_e^2 - 1) - \ln(J_e)\right], \tag{5.8.18}$$

with the positive constant k being the zero-stress bulk modulus. It then follows from (5.8.15) that the pressure is given by

$$p = \frac{1}{2}k\left(\frac{1}{J_e} - J_e\right). \tag{5.8.19}$$

This function has the property that the pressure becomes infinite as J_e approaches zero and it approaches negative infinity as J_e approaches infinity.

Incompressible Neo-Hookean and Mooney–Rivlin Materials
Most often, Neo-Hookean and Mooney–Rivlin materials are considered to be incompressible. Specifically, using the constraint (5.7.1), the separation (5.7.7), the constraint response (5.7.17) and the constitutive equations (5.8.15), (5.8.17) and (5.8.19),

it follows that the stress for an incompressible Neo-Hookean material is given by

$$\mathbf{T} = -\bar{p}\,\mathbf{I} + \mu \mathbf{B}_e'' , \tag{5.8.20}$$

and the stress for an incompressible Mooney–Rivlin material is given by

$$\mathbf{T} = -\bar{p}\,\mathbf{I} + \left[(1-4C)\mathbf{B}_e'' + 2C\left(\mathbf{B}_e'^2 - \frac{1}{3}\alpha_2 \mathbf{I} \right) \right] , \tag{5.8.21}$$

where $\bar{p}$ is an arbitrary function of $\mathbf{x}$ and t determined by the equations of motion and boundary conditions.

An Elastic Material with a Quadratic Strain Energy Function
For an elastic material with a quadratic strain energy function, use is made of the elastic strains e_{ij} defined in (3.11.33)

$$e_{ij} = \frac{1}{2}(m_{ij} - \delta_{ij}) , \tag{5.8.22}$$

relative to zero-stress material states defined in (3.11.9)

$$m_{ij} = \delta_{ij} \quad \text{for any zero-stress material state} , \tag{5.8.23}$$

to express Σ in the form

$$\rho_z \Sigma = \frac{1}{2} K_{ijkl} e_{ij} e_{kl} , \tag{5.8.24}$$

where K_{ijkl} are constant components of a fourth-order stiffness tensor having the symmetries

$$K_{jikl} = K_{ijlk} = K_{klij} = K_{ijkl} . \tag{5.8.25}$$

It then follows from (5.3.4) that the Cauchy stress for this material is given by

$$\mathbf{T} = J_e^{-1} K_{ijkl} e_{kl}\, \mathbf{m}_i \otimes \mathbf{m}_j . \tag{5.8.26}$$

To analyze the material symmetry of the strain energy function (5.8.24), use is made of the condition (5.6.6) to deduce that

$$[K_{ijkl} - H_{im} H_{jn} H_{kr} H_{ls} K_{mnrs}] e_{ij} e_{kl} = 0 , \tag{5.8.27}$$

for all strains e_{ij} which requires K_{ijkl} to satisfy the condition that

$$K_{ijkl} = H_{im} H_{jn} H_{kr} H_{ls} K_{mnrs} , \tag{5.8.28}$$

where H_{ij} is an orthogonal tensor which characterizes the symmetry of the material defined in its zero-stress material state with microstructural vectors $\mathbf{m}_i$ forming a right-handed orthonormal triad.

The following considers four cases of materials:

Case I: General Anisotropic

If the material posseses no symmetry then the symmetry group of H_{ij} consists only of $H_{ij} = \delta_{ij}$ and the $3^4 = 81$ constants K_{ijkl} are restricted only by the symmetries (5.8.25) which reduce the number of independent constants to the 21 constants given by

$$K_{ijkl} = \begin{pmatrix} K_{1111} & K_{1112} & K_{1113} & K_{1122} & K_{1123} & K_{1133} & K_{1212} \\ K_{1213} & K_{1222} & K_{1223} & K_{1233} & K_{1313} & K_{1322} & K_{1323} \\ K_{1333} & K_{2222} & K_{2223} & K_{2233} & K_{2323} & K_{2333} & K_{3333} \end{pmatrix} . \tag{5.8.29}$$

Case II: Symmetry About One Plane

If the material possesses symmetry about the plane normal to $\mathbf{m}_3$ in a zero-stress material state then the restrictions (5.8.28) must hold for the group H_{ij} that includes

$$H_{ij} = \begin{pmatrix} 1 & 0 & 0 \\ 0 & 1 & 0 \\ 0 & 0 & -1 \end{pmatrix} , \tag{5.8.30}$$

so that from (5.8.29) and (5.8.30), it follows that any component in which the index 3 appears an odd number of times must vanish

$$K_{1113} = K_{1123} = K_{1213} = K_{1223} = K_{1322} = K_{1333} = K_{2223} = K_{2333} = 0 . \tag{5.8.31}$$

Thus, the remaining 13 independent constants are given by

$$K_{ijkl} = \begin{pmatrix} K_{1111} & K_{1112} & K_{1122} & K_{1133} & K_{1212} & K_{1222} & K_{1233} \\ K_{1313} & K_{1323} & K_{2222} & K_{2233} & K_{2323} & K_{3333} & \end{pmatrix} . \tag{5.8.32}$$

Case III: Symmetry About Two Orthogonal Planes

If the material possesses symmetry about both planes with normals the $\mathbf{m}_3$ and $\mathbf{m}_2$ in a zero-stress material state, then the restrictions (5.8.28) must hold for the group H_{ij} that includes (5.8.30) and

$$H_{ij} = \begin{pmatrix} 1 & 0 & 0 \\ 0 & -1 & 0 \\ 0 & 0 & 1 \end{pmatrix} , \tag{5.8.33}$$

so that from (5.8.32) and (5.8.33), it follows that any component in which the index 2 appears an odd number of times must vanish

$$K_{1112} = K_{1222} = K_{1233} = K_{1323} = 0 . \tag{5.8.34}$$

Thus, the remaining 9 independent constants are given by

$$K_{ijkl} = \begin{pmatrix} K_{1111} & K_{1122} & K_{1133} & K_{1212} & K_{1313} & K_{2222} & K_{2233} \\ K_{2323} & K_{3333} \end{pmatrix} . \tag{5.8.35}$$

Notice from (5.8.35) that the index 1 only appears an even number of times so that the material also possesses symmetry about the plane normal to $\mathbf{m}_1$ in a zero-stress material state. This stiffness characterizes an *orthotropic elastic material.*

Case IV: Isotropic Elastic Material

If the material possesses symmetry with respect to the full orthogonal group then the material is called *isotropic* with a center of symmetry. Using the results in Appendix E, it follows that the material is characterized by only two independent constants λ and μ, called *Lame's constants*, such that

$$\begin{gathered} K_{1111} = K_{2222} = K_{3333} = \lambda + 2\mu\,, \quad K_{1122} = K_{1133} = K_{2233} = \lambda\,, \\ K_{1212} = K_{1313} = K_{2323} = \mu\,, \end{gathered} \tag{5.8.36}$$

and the fourth-order tensor K_{ijkl} can be expressed in the form

$$K_{ijkl} = \lambda\delta_{ij}\delta_{kl} + \mu(\delta_{ik}\delta_{jl} + \delta_{il}\delta_{jk}) . \tag{5.8.37}$$

It also follows that the strain energy (5.8.24) and the stress (5.8.26) can be written in the forms

$$\begin{gathered} \rho_z\Sigma = \frac{1}{2}\lambda\, e_{ii}e_{jj} + \mu\, e_{ij}e_{ij}\,, \\ \mathbf{T} = J_e^{-1}(\lambda e_{mm}\delta_{ij} + 2\mu e_{ij})(\mathbf{m}_i \otimes \mathbf{m}_j) . \end{gathered} \tag{5.8.38}$$

Notice that this strain energy is a function of the invariants of e_{ij}, as it should be for an isotropic material.

Linearized Constitutive Equations

To obtain the fully linearized constitutive equation, it is convenient to consider the initial state of the material to be at zero stress with the vectors $\mathbf{m}_i$ specified by the orthonormal triad $\mathbf{M}_i$, such that

$$\mathbf{m}_i(0) = \mathbf{M}_i\,, \quad \mathbf{M}_i \cdot \mathbf{M}_j = \delta_{ij}\,, \quad \mathbf{M}_1 \times \mathbf{M}_2 \cdot \mathbf{M}_3 = 1\,. \tag{5.8.39}$$

Recalling that the displacement $\mathbf{u}$ relative to this initial state is given by

$$\mathbf{u} = \mathbf{x} - \mathbf{X}\,, \quad \mathbf{X} = \mathbf{x}(0)\,, \tag{5.8.40}$$

and taking $t_n = 0$ in (3.13.3), the relative deformation gradient $\mathbf{F}_r$ is given by

$$\mathbf{F}_r = \mathbf{I} + \partial\mathbf{u}/\partial\mathbf{X}\,. \tag{5.8.41}$$

Then, neglecting quadratic terms in the displacement **u** and its derivatives, it follows from (3.13.9) that $\mathbf{m}_i$ can be approximated by

$$\mathbf{m}_i = (\mathbf{I} + \partial\mathbf{u}/\partial\mathbf{X})\,\mathbf{M}_i\,. \tag{5.8.42}$$

Next, separating the displacement gradient into its symmetric and skew-symmetric parts like in (4.10.7) yields

$$\begin{aligned}
\partial\mathbf{u}/\partial\mathbf{X} &= (\varepsilon_{ij} + \omega_{ij})\,\mathbf{M}_i \otimes \mathbf{M}_j\,,\\
\varepsilon_{ij} &= \frac{1}{2}[\partial\mathbf{u}/\partial\mathbf{X} + (\partial\mathbf{u}/\partial\mathbf{X})^T]\cdot\mathbf{M}_i \otimes \mathbf{M}_j\,,\\
\omega_{ij} &= \frac{1}{2}[\partial\mathbf{u}/\partial\mathbf{X} - (\partial\mathbf{u}/\partial\mathbf{X})^T]\cdot\mathbf{M}_i \otimes \mathbf{M}_j\,,
\end{aligned} \tag{5.8.43}$$

so the vectors $\mathbf{m}_i$ can be approximated by

$$\mathbf{m}_i = (\delta_{ij} + \varepsilon_{ij} + \omega_{ij})\,\mathbf{M}_i\,. \tag{5.8.44}$$

It then follows that the metric m_{ij} and the strains e_{ij} are approximated by

$$m_{ij} = \delta_{ij} + 2\varepsilon_{ij}\,, \quad e_{ij} = \varepsilon_{ij}\,, \tag{5.8.45}$$

and the stress is approximated by

$$\mathbf{T} = (\lambda\varepsilon_{mm}\delta_{ij} + 2\mu\varepsilon_{ij})(\mathbf{M}_i \otimes \mathbf{M}_j)\,. \tag{5.8.46}$$

Restrictions on the Material Constants
From physical considerations it is expected that any strain from a zero-stress material state should cause an increase in strain energy. Mathematically this means that the strain energy function is positive-definite

$$\Sigma > 0 \text{ for any } e_{ij} \neq 0\,. \tag{5.8.47}$$

Recalling that the strain e_{ij} can be separated into its spherical and deviatoric parts

$$e_{ij} = \frac{1}{3}e_{mm}\delta_{ij} + e''_{ij}\,, \quad e''_{mm} = 0\,, \tag{5.8.48}$$

the isotropic strain energy function (5.8.38) can be rewritten in the form

$$\rho_z\Sigma = \frac{1}{2}\left(\frac{3\lambda + 2\mu}{3}\right)(e_{ii}e_{jj}) + \mu e''_{ij}e''_{ij}\,. \tag{5.8.49}$$

Since the terms e_{ii} and $e''_{ij}e''_{ij}$ are independent quantities, this strain energy will be positive-definite whenever

Table 5.1 Relationships between the material constants for an isotropic linear elastic material

	λ	μ	E	ν	k
$\lambda\,,\mu$			$\frac{\mu(3\lambda+2\mu)}{\lambda+\mu}$	$\frac{\lambda}{2(\lambda+\mu)}$	$\frac{3\lambda+2\mu}{3}$
$\lambda\,,\nu$		$\frac{\lambda(1-2\nu)}{2\nu}$	$\frac{\lambda(1+\nu)(1-2\nu)}{\nu}$		$\frac{\lambda(1+\nu)}{3\nu}$
$\lambda\,,k$		$\frac{3(k-\lambda)}{2}$	$\frac{9k(k-\lambda)}{3k-\lambda}$	$\frac{\lambda}{3k-\lambda}$	
$\mu\,,E$	$\frac{\mu(2\mu-E)}{E-3\mu}$			$\frac{E-2\mu}{2\mu}$	$\frac{\mu E}{3(3\mu-E)}$
$\mu\,,\nu$	$\frac{2\mu\nu}{1-2\nu}$		$2\mu(1+\nu)$		$\frac{2\mu(1+\nu)}{3(1-2\nu)}$
$\mu\,,k$	$\frac{3k-2\mu}{3}$		$\frac{9k\mu}{3k+\mu}$	$\frac{3k-2\mu}{2(3k+\mu)}$	
$E\,,\nu$	$\frac{E\nu}{(1+\nu)(1-2\nu)}$	$\frac{E}{2(1+\nu)}$			$\frac{E}{3(1-2\nu)}$
$E\,,k$	$\frac{3k(3k-E)}{9k-E}$	$\frac{3Ek}{9k-E}$		$\frac{3k-E}{6k}$	
$\nu\,,k$	$\frac{3k\nu}{1+\nu}$	$\frac{3k(1-2\nu)}{2(1+\nu)}$	$3k(1-2\nu)$		

$\mu = \frac{(E-3\lambda)+\sqrt{(E-3\lambda)^2+8\lambda E}}{4}\,, \quad \nu = \frac{-(E+\lambda)+\sqrt{(E+\lambda)^2+8\lambda^2}}{4\lambda}$

$k = \frac{(3\lambda+E)+\sqrt{(3\lambda+E)^2-4\lambda E}}{6}$

$$\frac{3\lambda + 2\mu}{3} > 0\,, \quad \mu > 0\,. \tag{5.8.50}$$

For the linearized theory, e_{ii} characterizes dilatational deformations and e''_{ij} characterizes distortional deformations.

Moreover, it is noted that this isotropic elastic material with a quadratic strain energy function can be characterized by any two of the following material constants: λ (Lame's constant); μ (shear modulus); E (Young's modulus); ν (Poisson's ratio); or k (bulk modulus), which are interrelated by the expressions in Table 5.1. Using these expressions it can be shown that the restrictions (5.8.50) also require

$$k > 0\,, \quad E > 0\,, \quad -1 < \nu < \frac{1}{2} > 0\,. \tag{5.8.51}$$

Limitations of a Quadratic Strain Energy Function

The anisotropic elastic material characterized by (5.8.24) and (5.8.26), and the isotropic elastic material characterized by (5.8.38) both have a strain energy function that is quadratic in the strains e_{ij}, with the Cauchy stress $\mathbf{T}$ depending nonlinearly on J_e and e_{ij} since the vectors $\mathbf{m}_i$ also depend on the strains e_{ij}. These constitutive equations are valid for large rotations and moderate strains e_{ij}.

To see that these quadratic strain energy functions are limited to moderate strains consider the simple case of an isotropic elastic material (5.8.38) experiencing uniaxial stress in the $\mathbf{m}_1$ direction for which

$$\begin{gathered} \mathbf{m}_1 = a\mathbf{e}_1\,, \quad \mathbf{m}_2 = b\mathbf{e}_2\,, \quad \mathbf{m}_3 = b\mathbf{e}_3\,, \quad J_e = ab^2\,, \\ e_{11} = \frac{1}{2}(a^2 - 1)\,, \quad e_{22} = \frac{1}{2}(b^2 - 1)\,, \end{gathered} \tag{5.8.52}$$

where $\mathbf{e}_i$ are fixed rectangular Cartesian base vectors. In these expressions, a is the axial stretch and b is the lateral stretch, both measured from a zero-stress material state. For uniaxial stress

$$\mathbf{T} = T_{11}\mathbf{e}_1 \otimes \mathbf{e}_1 \,, \tag{5.8.53}$$

and the constitutive equations yield the restrictions

$$\begin{aligned}
&\mathbf{T} = T_{11}\mathbf{e}_1 \otimes \mathbf{e}_1 \,, \\
&T_{11} = \mathbf{T}\cdot\mathbf{e}_1 \otimes \mathbf{e}_1 = \frac{a}{b^2}[(\lambda + 2\mu)e_{11} + 2\lambda e_{22}] \\
&\qquad = \frac{2\mu}{(1-2\nu)}\left(\frac{a}{b^2}\right)[(1-\nu)e_{11} + 2\nu e_{22}] \,, \\
&\mathbf{T}\cdot\mathbf{e}_2 \otimes \mathbf{e}_2 = \mathbf{T}\cdot\mathbf{e}_3 \otimes \mathbf{e}_3 = \frac{1}{a}[\lambda e_{11} + 2(\lambda + \mu)e_{22}] \\
&\qquad = \frac{2\mu}{(1-2\nu)}\left(\frac{1}{a}\right)(\nu e_{11} + e_{22}) = 0 \,,
\end{aligned} \tag{5.8.54}$$

where use has been made of Table 5.1 to write λ in terms of the zero-stress shear modulus μ and Poisson's ratio ν. Then, the solution of these equations is given by

$$e_{22} = -\nu e_{11} \,, \quad T_{11} = 2\mu(1+\nu)\left(\frac{a}{b^2}\right) e_{11} \,, \tag{5.8.55}$$

and the restrictions on the strains can be solved to obtain

$$b = \sqrt{1 + \nu(1 - a^2)} \,. \tag{5.8.56}$$

For Poisson's ratio in the range

$$0 < \nu \le \frac{1}{2} \,, \tag{5.8.57}$$

it can be seen that the maximum axial stretch a_{max} occurs when b vanishes and that the maximum lateral stretch b_{max} occurs when a vanishes, for which

$$\begin{aligned}
a_{max} &= \sqrt{\tfrac{1+\nu}{\nu}} \,, \quad b = 0 \,, \quad J_e = 0 \,, \quad T_{11} = \infty \,, \\
b_{max} &= \sqrt{1+\nu} \,, \quad a = 0 \,, \quad J_e = 0 \,, \quad T_{11} = 0 \,.
\end{aligned} \tag{5.8.58}$$

These results are unphysical because they indicate that infinite tension causes a finite axial stress with zero volume and that the material can be compressed to zero length with a finite cross section, zero volume and zero stress.

In contrast, the stress $\mathbf{T}$ for the compressible Neo-Hookean material characterized by (5.8.1), (5.8.12), (5.8.15), (5.8.18) and (5.8.19) is given by

$$\mathbf{T} = -\frac{1}{2}k\left(\frac{1}{J_e} - J_e\right)\mathbf{I} + J_e^{-1}\mu\mathbf{B}_e'' \,, \quad k = \frac{2\mu(1+\nu)}{3(1-2\nu)} \,, \tag{5.8.59}$$

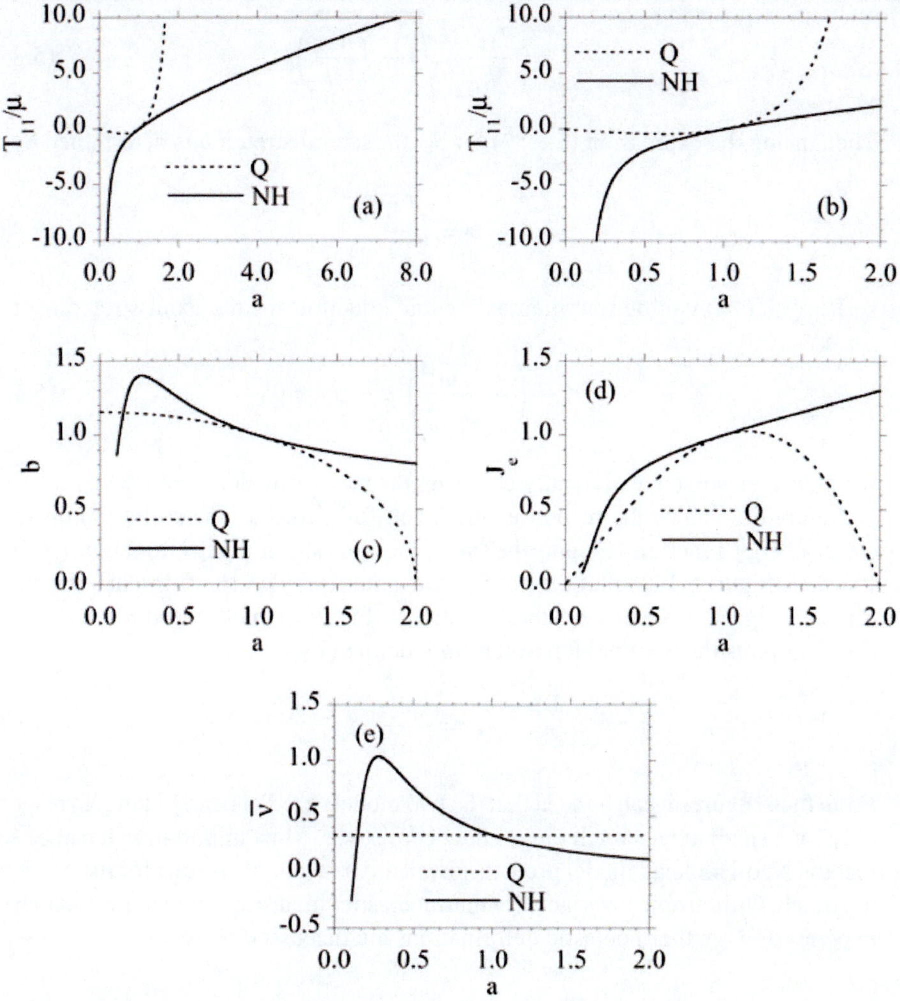

Fig. 5.1 Uniaxial tension: comparison of the responses predicted by the quadratic strain energy function (Q) and the Neo-Hookean strain energy function (NH) for $\nu = 1/3$

which for uniaxial stress (5.8.53) yields the restrictions

$$T_{11} = \mathbf{T} \cdot (\mathbf{e}_1 \otimes \mathbf{e}_1 - \mathbf{e}_2 \otimes \mathbf{e}_2) = J_e^{-5/3} \mu (a^2 - b^2)\,, \tag{5.8.60a}$$

$$p = -\frac{T_{11}}{3} = \frac{k}{2}\left(J_e - \frac{1}{J_e}\right)\,. \tag{5.8.60b}$$

The solution of these equations can be parameterized by the axial stress T_{11}. Specifically, (5.8.60b) can be solved for J_e to obtain

$$J_e = \frac{T_{11}}{3k} + \sqrt{1 + \left(\frac{T_{11}}{3k}\right)^2} \,. \tag{5.8.61}$$

Then, using the expression (5.8.52) for J_e, the lateral stretch b is determined by

$$b = \sqrt{\frac{J_e}{a}} \,, \tag{5.8.62}$$

so Eq. (5.8.60a) can be rewritten as a cubic equation for the axial stretch a of the form

$$a^3 - \left(\frac{J_e^{5/3} T_{11}}{\mu}\right) a - J_e = 0 \,, \tag{5.8.63}$$

which can be solved analytically choosing the root for which $a = 1$ when $T_{11} = 0$.

Figure 5.1 shows the responses predicted for uniaxial tension by the quadratic strain energy function (Q) and the Neo-Hookean strain energy function (NH) for $\nu = 1/3$. Figure 5.1a, b plot the normalized axial stress T_{11} for different axial stretch regions, Fig. 5.1c plots the lateral stretch b, Fig. 5.1d plots the dilatation J_e and Fig. 5.1e plots the nominal Poisson ratio $\bar{\nu}$ defined by

$$\bar{\nu} = -\frac{e_{22}}{e_{11}} \,. \tag{5.8.64}$$

From these figures it can be seen that the two models predict nearly identical response only for a small axial stretch range about zero stress. Most importantly it can be seen that the Neo-Hookean model predicts physically reasonable results for the full range of stretch. Orthotropic invariants for thermoelastic–inelastic soft materials which can experience large thermoelastic deformations are discussed in Sect. 6.6.

5.9 Viscous and Inviscid Fluids

This section discusses purely mechanical constitutive equations for compressible viscous and inviscid fluids. From a physical point of view it is clear that the stress **T** in a compressible fluid must depend on the elastic dilatation J_e, which is a measure of the fluid's density. Moreover, experience with stirring honey indicates that it is harder to stir the honey faster. This suggests that **T** will also depend on the velocity gradient **L**. In addition, the pressure required to pump a viscous fluid through a pipe depends on the flow rate. Therefore, **T** might also depend on the velocity **v**. Based on these observations, as a first attempt to propose a constitutive equation for fluids, it is assumed that the stress can be expressed in the form

$$\mathbf{T} = \tilde{\mathbf{T}}(J_e, \mathbf{v}, \mathbf{D}, \mathbf{W}) \,, \tag{5.9.1}$$

where for convenience $\mathbf{L}$ has been separated into its symmetric part $\mathbf{D}$ and its skew-symmetric part $\mathbf{W}$.

In the following, use will be made of invariance under SRBM to develop restrictions on the functional form (5.9.1). Since (5.9.1) must hold for all motions it must also hold for SRBM so that

$$\mathbf{T}^{+} = \tilde{\mathbf{T}}(J_e^{+}, \mathbf{v}^{+}, \mathbf{D}^{+}, \mathbf{W}^{+}) . \tag{5.9.2}$$

However, under SRBM the Cauchy stress $\mathbf{T}$ transforms by

$$\mathbf{T}^{+} = \mathbf{Q}\mathbf{T}\mathbf{Q}^{T} , \tag{5.9.3}$$

where $\mathbf{Q}$ is a proper orthogonal tensor function of time only. Thus, the functional form (5.9.1) must satisfy the restrictions

$$\tilde{\mathbf{T}}(J_e^{+}, \mathbf{v}^{+}, \mathbf{D}^{+}, \mathbf{W}^{+}) = \mathbf{Q}\tilde{\mathbf{T}}(J_e, \mathbf{v}, \mathbf{D}, \mathbf{W})\mathbf{Q}^{T} . \tag{5.9.4}$$

Recalling that under SRBM

$$\begin{aligned}
\dot{\mathbf{Q}} &= \boldsymbol{\Omega}\mathbf{Q} , & \boldsymbol{\Omega}^{T} &= -\boldsymbol{\Omega} , \\
J_e^{+} &= J_e , & \mathbf{v}^{+} &= \dot{\mathbf{c}} + \boldsymbol{\Omega}\mathbf{Q}\mathbf{x} + \mathbf{Q}\mathbf{v} , \\
\mathbf{D}^{+} &= \mathbf{Q}\mathbf{D}\mathbf{Q}^{T} , & \mathbf{W}^{+} &= \mathbf{Q}\mathbf{W}\mathbf{Q}^{T} + \boldsymbol{\Omega} ,
\end{aligned} \tag{5.9.5}$$

equation (5.9.4) requires

$$\tilde{\mathbf{T}}(J_e, \dot{\mathbf{c}} + \boldsymbol{\Omega}\mathbf{Q}\mathbf{x} + \mathbf{Q}\mathbf{v}, \mathbf{Q}\mathbf{D}\mathbf{Q}^{T}, \mathbf{Q}\mathbf{W}\mathbf{Q}^{T} + \boldsymbol{\Omega}) = \mathbf{Q}\tilde{\mathbf{T}}(J_e, \mathbf{v}, \mathbf{D}, \mathbf{W})\mathbf{Q}^{T} . \tag{5.9.6}$$

Since (5.9.6) must hold for all motions and all $SRMB$, necessary restrictions on the functional form $\tilde{\mathbf{T}}$ can be obtained by considering special SRBMs.

Superposed Translational Velocity
This case considers superposed translational velocity with

$$\dot{\mathbf{c}} \neq 0 , \quad \mathbf{Q} = \mathbf{I} , \quad \dot{\mathbf{Q}} = 0 . \tag{5.9.7}$$

Substituting (5.9.7) into (5.9.6) yields

$$\tilde{\mathbf{T}}(J_e, \dot{\mathbf{c}} + \mathbf{v}, \mathbf{D}, \mathbf{W}) = \tilde{\mathbf{T}}(J_e, \mathbf{v}, \mathbf{D}, \mathbf{W}) . \tag{5.9.8}$$

Since this equation must hold for arbitrary values of $\dot{\mathbf{c}}$ and the right-hand side is independent of $\dot{\mathbf{c}}$, it follows that the Cauchy stress cannot depend on the velocity $\mathbf{v}$. Thus, $\mathbf{T}$ must be expressed as another function $\bar{\mathbf{T}}$ of J_e, $\mathbf{D}$ and $\mathbf{W}$ only

$$\mathbf{T} = \bar{\mathbf{T}}(J_e, \mathbf{D}, \mathbf{W}) , \tag{5.9.9}$$

and the restriction (5.9.6) requires

$$\bar{\mathbf{T}}(J_e, \mathbf{QDQ}^T, \mathbf{QWQ}^T + \boldsymbol{\Omega}) = \mathbf{Q}\bar{\mathbf{T}}(J_e, \mathbf{D}, \mathbf{W})\mathbf{Q}^T . \tag{5.9.10}$$

Superposed Rate of Rotation
This case considers superposed rate of rotation. Since (5.9.10) must hold for all skew-symmetric tensors $\boldsymbol{\Omega}$ and the right-hand side of this equation is independent of $\boldsymbol{\Omega}$, it follows that the Cauchy stress $\mathbf{T}$ cannot depend on the spin tensor $\mathbf{W}$. Thus, the most general viscous fluid is characterized by the constitutive equation

$$\mathbf{T} = \hat{\mathbf{T}}(J_e) + \overset{v}{\mathbf{T}}(J_e, \mathbf{D}) , \quad \overset{v}{\mathbf{T}}(J_e, 0) = 0 , \tag{5.9.11}$$

where $\hat{\mathbf{T}}(J_e)$ characterizes the elastic response due to dilatation and $\overset{v}{\mathbf{T}}(J_e, \mathbf{D})$ characterizes the viscous response. Also, these constitutive equations must satisfy the restrictions that under SRBM

$$\hat{\mathbf{T}}(J_e) = \mathbf{Q}\hat{\mathbf{T}}(J_e)\mathbf{Q}^T , \quad \overset{v}{\mathbf{T}}(J_e, \mathbf{QDQ}^T) = \mathbf{Q}\overset{v}{\mathbf{T}}(J_e, \mathbf{D})\,\mathbf{Q}^T , \tag{5.9.12}$$

which require $\hat{\mathbf{T}}$ to be an isotropic tensor and $\overset{v}{\mathbf{T}}$ to be an isotropic tensor function of $\mathbf{D}$.

Reiner-Rivlin Fluid
Since the restrictions (5.9.12) must hold for all proper orthogonal $\mathbf{Q}$ the function $\overset{v}{\mathbf{T}}$ is called an isotropic tensor function of its argument $\mathbf{D}$. This notion of an isotropic tensor function should not be confused with the notion of an isotropic tensor as discussed in Appendix E. Furthermore, since the restriction (5.9.12) is unaltered by the interchange of $\mathbf{Q}$ with $-\mathbf{Q}$, it follows that $\overset{v}{\mathbf{T}}$ is a hemotropic function of $\mathbf{D}$ (isotropic with a center of symmetry). Now, using a result from the theory of invariants, it follows that the most general form of $\hat{\mathbf{T}}$ and $\overset{v}{\mathbf{T}}$ can be expressed as

$$\hat{\mathbf{T}}(J_e) = -\hat{p}(J_e)\,\mathbf{I} , \quad \overset{v}{\mathbf{T}} = d_0\,\mathbf{I} + d_1\,\mathbf{D} + d_2\,\mathbf{D}^2 , \tag{5.9.13}$$

where $\hat{p}(J_e)$ is a function of J_e only, d_0, d_1 and d_2 are scalar functions of J_e and the three independent invariants of $\mathbf{D}$. Alternatively, using the separation of deformation rate into dilatational and distortional deformation rates, $\overset{v}{\mathbf{T}}$ can be written in the form

$$\overset{v}{\mathbf{T}} = \bar{d}_0\,(\mathbf{D}\cdot\mathbf{I})\,\mathbf{I} + \bar{d}_1\,\mathbf{D}'' + \bar{d}_2\,\mathrm{Sign}(\mathbf{D}''^3\cdot\mathbf{I})\left[\mathbf{D}''^2 - \frac{1}{3}(\mathbf{D}''\cdot\mathbf{I})\,\mathbf{I}\right] , \tag{5.9.14}$$

where $\mathbf{D}''$ is the deviatoric part of $\mathbf{D}$, $\bar{d}_0$, $\bar{d}_1$ and $\bar{d}_2$ are scalar functions of J_e, $\mathbf{D}\cdot\mathbf{I}$, the two independent invariants of $\mathbf{D}''$ and the function $\mathrm{Sign}(x)$ is defined by

$$\mathrm{Sig}(x) = 1 \quad \text{for} \quad x \geq 0\,, \quad \mathrm{Sig}(x) = -1 \quad \text{for} \quad x < 0\,. \tag{5.9.15}$$

The constitutive equation characterized by (5.9.11) with the expressions (5.9.13) for the stress $\mathbf{T}$ described a Reiner-Rivlin fluid. An alternative form of this Reiner-Rivlin fluid is characterized by (5.9.11) for the total stress $\mathbf{T}$, (5.9.13) for the elastic stress $\hat{\mathbf{T}}$ and (5.9.14) for the viscous stress $\overset{v}{\mathbf{T}}$. Moreover, the strain energy is taken to be a function of the dilatation

$$\Sigma = \hat{\Sigma}(J_e)\,, \tag{5.9.16}$$

so the rate material dissipation (4.5.7) requires

$$\mathcal{D} = \left[-\hat{p}(J_e) - \rho_z \frac{\partial \hat{\Sigma}}{\partial J_e} \right] \mathbf{D} \cdot \mathbf{I} + \overset{v}{\mathbf{T}} \cdot \mathbf{D} \geq 0\,, \tag{5.9.17}$$

where use has been made of the expression (5.8.14) for the elastic dilatation J_e.

Inviscid Fluid

For an inviscid fluid the Cauchy stress is independent of the rate of deformation $\mathbf{D}$ so that $\overset{v}{\mathbf{T}}$ vanishes in (5.9.13) and (5.9.14) and (5.9.17) requires

$$\mathcal{D} = \left[-\hat{p}(J_e) - \rho_z \frac{\partial \hat{\Sigma}}{\partial J_e} \right] \mathbf{D} \cdot \mathbf{I} \geq 0\,. \tag{5.9.18}$$

Since the coefficient of $\mathbf{D} \cdot \mathbf{I}$ is independent of rate, it can be shown that for an inviscid fluid

$$\mathbf{T} = \hat{\mathbf{T}} = -\hat{p}(J_e)\,\mathbf{I}\,, \quad \hat{p}(J_e) = -\rho_z \frac{\partial \hat{\Sigma}}{\partial J_e}\,. \tag{5.9.19}$$

This means for an inviscid fluid the traction vector $\mathbf{t}$ always acts normal to the surface on which it is applied

$$\mathbf{t} = \mathbf{T}\mathbf{n} = -\hat{p}\,\mathbf{n}\,, \tag{5.9.20}$$

and the pressure $\hat{p}$ is a function of the elastic dilatation J_e only.

Restrictions on a Reiner-Rivlin Fluid

Without specifying the functional form of $\overset{v}{\mathbf{T}}$ it is not possible to obtain further restrictions using the dissipation equation (5.9.17). However, it is reasonable to assume that the elastic part of the stress is the same as that for an inviscid fluid which is given by (5.9.19) so the stress and the rate of material dissipation (5.9.17) associated with the viscous stress (5.9.14) become

$$\begin{gathered} \mathbf{T} = -\hat{p}(J_e)\,\mathbf{I} + \overset{v}{\mathbf{T}}(J_e, \mathbf{D})\,, \quad \hat{p}(J_e) = -\rho_z \frac{\partial \hat{\Sigma}}{\partial J_e}\,, \\ \mathcal{D} = \overset{v}{\mathbf{T}}(J_e, \mathbf{D}) \cdot \mathbf{D} = \bar{d}_0\,(\mathbf{D} \cdot \mathbf{I})^2 + \bar{d}_1\,\mathbf{D}'' \cdot \mathbf{D}'' + \bar{d}_2\,|\mathbf{D}''^3 \cdot \mathbf{I}| \geq 0\,, \end{gathered} \tag{5.9.21}$$

with the rate of material dissipation restricting the functional form for the viscous stress $\overset{v}{\mathbf{T}}$. Sufficient but not necessary conditions for $\mathcal{D} \geq 0$ are given by

$$\bar{d}_0 \geq 0 \,, \quad \bar{d}_1 \geq 0 \,, \quad \bar{d}_2 \geq 0 \,. \tag{5.9.22}$$

Newtonian Viscous Fluid
A Newtonian viscous fluid is a special case of a Reiner-Rivlin fluid in which the viscous stress $\overset{v}{\mathbf{T}}$ is a linear function of the rate of deformation $\mathbf{D}$. For this case, $\overset{v}{\mathbf{T}}$ reduces to

$$\overset{v}{\mathbf{T}} = \lambda(\mathbf{D} \cdot \mathbf{I})\,\mathbf{I} + 2\mu\mathbf{D}'' \,, \tag{5.9.23}$$

where λ and μ are scalar functions of J_e only. Moreover, it follows that $\overset{v}{\mathbf{T}}$ can be rewritten in the alternative form

$$\begin{aligned} &\mathbf{T} = -\hat{p}(J_e)\,\mathbf{I} + \overset{v}{\mathbf{T}} \,, \qquad \overset{v}{\mathbf{T}} = -\overset{v}{p}\,\mathbf{I} + 2\mu\mathbf{D}'' \,, \\ &p = -\tfrac{1}{3}\mathbf{T} \cdot \mathbf{I} = \hat{p} + \overset{v}{p} \,, \;\; \overset{v}{p} = -\tfrac{1}{3}\overset{v}{\mathbf{T}} \cdot \mathbf{I} = -\lambda\,\mathbf{D} \cdot \mathbf{I} \,, \end{aligned} \tag{5.9.24}$$

which shows that the total pressure p has an elastic part $\hat{p}$ and a viscous part $\overset{v}{p}$ that depends on the rate of volume expansion $\mathbf{D} \cdot \mathbf{I}$ with λ being the dilatational viscosity coefficient. Also, the rate of material dissipation (4.5.7) is satisfied provided that

$$\begin{aligned} &\mathcal{D} = \overset{v}{\mathbf{T}} \cdot \mathbf{D} = \lambda\,(\mathbf{D} \cdot \mathbf{I})^2 + 2\mu\mathbf{D}'' \cdot \mathbf{D}'' \geq 0 \,, \\ &\lambda \geq 0 \,, \quad \mu \geq 0 \,. \end{aligned} \tag{5.9.25}$$

5.10 Viscous Dissipation

A simple generalized nonlinear Kelvin–Voigt model (see Fig. 5.2) for viscous dissipation can be proposed by adding the response of the viscous part of a Newtonian viscous fluid to that of a general elastic material. Specifically, for this model the Cauchy stress $\mathbf{T}$ is proposed in the form

$$\mathbf{T} = \hat{\mathbf{T}} + \overset{v}{\mathbf{T}} \,, \quad \overset{v}{\mathbf{T}} = \lambda\,(\mathbf{D} \cdot \mathbf{I})\,\mathbf{I} + 2\mu\mathbf{D}'' \,, \quad \lambda \geq 0 \quad \mu \geq 0 \,, \tag{5.10.1}$$

where $\hat{\mathbf{T}}$ is the response of a general nonlinear elastic material with strain energy Σ that satisfies equation

$$\hat{\mathbf{T}} \cdot \mathbf{D} = \rho\dot{\Sigma} \,, \tag{5.10.2}$$

for all motions and λ, μ are non-negative functions of J_e that control the viscosity to dilatational deformation rate and to distortional deformation rate, respectively. Also, for this material the rate of material dissipation (4.5.7) requires

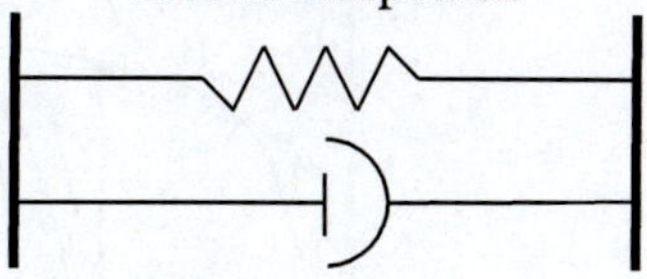

Fig. 5.2 Sketch of a nonlinear Kelvin–Voigt model with an elastic component in parallel with a viscous component

$$\mathcal{D} = \overset{\nu}{\mathbf{T}} \cdot \mathbf{D} = \lambda\,(\mathbf{D} \cdot \mathbf{I})^2 + 2\mu\,(\mathbf{D}'' \cdot \mathbf{D}'') \geq 0\,, \tag{5.10.3}$$

which is automatically satisfied. Moreover, it follows that when λ and μ are both positive, dissipation continues until the rate of deformation vanishes $\mathbf{D} = 0$ with $\mathbf{T} = \hat{\mathbf{T}}$.

If the elastic part of the response is isotropic then the strain energy is given by (5.8.5) and the stress $\hat{\mathbf{T}}$ is given by (5.8.10). Alternatively, if the elastic part of the response is anisotropic then the strain energy is given by (5.3.9) and the stress $\hat{\mathbf{T}}$ is given by (5.3.11). For either case, this model proposes isotropic viscous dissipation.

5.11 Elastic–Inelastic Materials

Figure 5.3a shows a sketch of the stress–strain response of a typical metal to uniaxial stress loading. The quantity T_{11} is the total axial component of the Cauchy stress $\mathbf{T}$ and the quantity E_1 is the total axial extension. The material is loaded in tension along the path $OABCD$, unloaded along DE, reloaded along $EFGH$, unloaded along HI and reloaded in compression along $IJKL$. Inspection of the points C, E and L in Fig. 5.3a reveals that the stress in an elastic–plastic material can have significantly different values for the same value of axial extension E_1. This means that the response of an elastic–plastic material depends on the past history of deformation (i.e., the responses to the deformation histories $OABC$, $OAB - E$ and $OAB - L$ are different).

The points A, F, J in Fig. 5.3a represent points on the loading paths beyond which the stress–strain relationship becomes nonlinear. Although the curve $OABCD$ is nonlinear it is not possible to determine whether the response is elastic or elastic–inelastic until unloading is considered. Since the response shown in Fig. 5.3a does not unload along the same loading path, it is clear that the response is not elastic, but rather is elastic–inelastic. Moreover, B, G and K represent the points on the loading paths beyond which some detectable value of strain relative to the peak strain (normally taken to be 0.2%) remains when the material is unloaded to zero stress. These points are called the yield points and deformation beyond them causes permanent changes in the response of the material. It is also important to mention that the paths BCD, GH and KL represent strain hardening paths where the magnitude of the stress increases with increasing effective inelastic deformation.

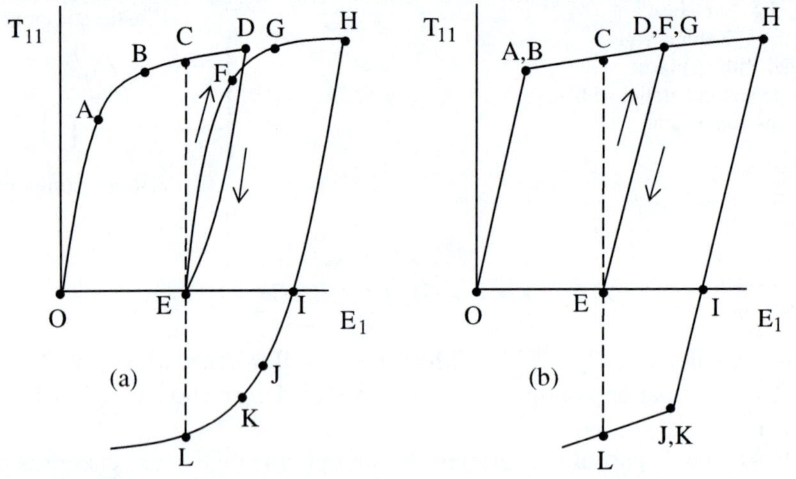

Fig. 5.3 **a** A sketch of the stress–strain response of a typical metal to uniaxial stress; **b** idealization of the stress–strain response of a metal to uniaxial stress

To model the material response shown in Fig. 5.3a it is common to separate the response into two parts: elastic response which is reversible and inelastic response which is irreversible. Also, the material response is idealized as shown in Fig. 5.3b by making the following assumptions:

(a) There are distinct yield points (A, B), (D, F, G) and (J, K) that form the boundary between elastic and inelastic responses.
(b) Unloading along DE and reloading along EF follow the same path.

Lagrangian Formulations
Lagrangian formulations of plasticity (inelasticity) enrich the theory of hyperelastic solids with a plastic deformation measure that captures observed effects of history and rate dependence of material response. A summary of the small deformation theory within the context of thermodynamics can be found in the classical paper by Naghdi [28]. Unfortunately, the large deformation theory of plasticity still is plagued with controversies, some of which have been discussed in the critical review [29]. This section discusses three prominent formulations of large deformation theory: one by Green and Naghdi [16], another attributed to Bilby et al. [6], Kröner [21] and Lee [23], and another attributed to Besseling [4].

Green–Naghdi Formulation
Green and Naghdi [16] developed a large deformation thermomechanical theory of plasticity. Confining attention to the purely mechanical response and using the notation in this book, this theory introduces the total deformation gradient $\mathbf{F}$ and the right Cauchy–Green deformation tensor $\mathbf{C}$, which satisfy equations

$$\dot{\mathbf{F}} = \mathbf{L}\mathbf{F}, \quad \mathbf{C} = \mathbf{F}^T\mathbf{F}, \quad \dot{\mathbf{C}} = 2\mathbf{F}^T\mathbf{D}\mathbf{F}. \tag{5.11.1}$$

The theory of hyperelasticity is enriched by introducing a symmetric plastic deformation tensor $\mathbf{C}_p$ (similar to $\mathbf{C}$) and a scalar measure of isotropic hardening κ by the evolution equations

$$\dot{\mathbf{C}}_p = \mathbf{A}_p\,, \quad \mathbf{A}_p = \Gamma \bar{\mathbf{A}}_p\,, \quad \dot{\kappa} = \Gamma H\,, \quad \Gamma \geq 0\,, \tag{5.11.2}$$

where $\bar{\mathbf{A}}_p$ controls the direction of plastic deformation rate, Γ is a non-negative function that controls the magnitude of plastic deformation rate $\mathbf{A}_p$ and H controls the rate of hardening. For metals, plastic deformation rate is isochoric so $\mathbf{C}_p$ remains unimodular, which requires

$$\det(\mathbf{C}_p) = 1\,, \quad \bar{\mathbf{A}}_p \cdot \mathbf{C}_p^{-1} = 0\,. \tag{5.11.3}$$

Under SRBM the total deformation tensor $\mathbf{F}$, the right Cauchy–Green tensor $\mathbf{C}$, the plastic deformation $\mathbf{C}_p$ and the hardening variable κ transform to $\mathbf{F}^+$, $\mathbf{C}^+$, $\mathbf{C}_p^+$ and κ^+, such that

$$\mathbf{F}^+ = \mathbf{QF}\,, \quad \mathbf{C}^+ = \mathbf{C}\,, \quad \mathbf{C}_p^+ = \mathbf{C}_p\,, \quad \kappa^+ = \kappa\,, \tag{5.11.4}$$

which place restrictions on the functional forms of Γ, $\bar{\mathbf{A}}_p$, H.

In this theory, the strain energy Σ is assumed to be a function of $\mathbf{F}$, $\mathbf{C}_p$ and κ but since Σ is uninfluenced by SRBM, it must depend on $\mathbf{F}$ only through the deformation tensor $\mathbf{C}$ so that

$$\Sigma = \Sigma(\mathbf{C}, \mathbf{C}_p, \kappa)\,. \tag{5.11.5}$$

For both rate-independent and rate-dependent material response, the constitutive equation for stress is taken in the form

$$\mathbf{T} = 2\rho \mathbf{F} \frac{\partial \Sigma}{\partial \mathbf{C}} \mathbf{F}^T\,. \tag{5.11.6}$$

Moreover, the rate of material dissipation (4.5.7) requires

$$\mathcal{D} = -\Gamma \rho \left(\frac{\partial \Sigma}{\partial \mathbf{C}_p} \cdot \bar{\mathbf{A}}_p + \frac{\partial \Sigma}{\partial \kappa} H \right) \geq 0\,, \tag{5.11.7}$$

which places restrictions on the functional forms of Σ, $\bar{\mathbf{A}}_p$ and H.

In addition to solving the balance of linear momentum (4.4.5), this theory requires solution of the evolution equations (5.11.1) and (5.11.2) with initial conditions

$$\mathbf{F}(0)\,, \mathbf{C}_p(0)\,, \kappa(0)\,. \tag{5.11.8}$$

Bilby, Kröner, Lee Formulation

Bilby et al. [6], Kröner [21] and Lee [23] introduced a formulation that depends on a second-order non-symmetric plastic deformation tensor $\mathbf{F}_p$ (similar to $\mathbf{F}$) determined by an evolution equation of the form

$$\dot{\mathbf{F}}_p = \mathbf{\Lambda}_p \mathbf{F}_p\,, \quad \mathbf{\Lambda}_p = \Gamma \bar{\mathbf{\Lambda}}_p\,, \tag{5.11.9}$$

where $\bar{\mathbf{\Lambda}}_p$ controls the direction of plastic deformation rate and Γ is a non-negative function that controls the magnitude of plastic deformation rate $\dot{\mathbf{F}}_p$. Again, for metal plasticity the plastic deformation rate is isochoric so $\mathbf{F}_p$ is unimodular and $\bar{\mathbf{\Lambda}}_p$ is restricted, such that

$$\det(\mathbf{F}_p) = 1\,, \quad \bar{\mathbf{\Lambda}}_p \cdot \mathbf{I} = 0\,. \tag{5.11.10}$$

Moreover, an elastic deformation tensor $\mathbf{F}_e$ is defined by the multiplicative form

$$\mathbf{F}_e \equiv \mathbf{F}\mathbf{F}_p^{-1}\,, \tag{5.11.11}$$

and a hardening variable κ is introduced which satisfies the evolution equation in (5.11.2). Usually this equation is written in the form $\mathbf{F} = \mathbf{F}_e\mathbf{F}_p$, which suggests that $\mathbf{F}_p$ transforms the reference configuration into an intermediate zero-stress configuration and $\mathbf{F}_e$ transforms an intermediate configuration into the current configuration. For general inhomogeneous deformations, $\mathbf{F}$ describes a compatible field with the position $\mathbf{x}$ of a material point in the current configuration being a differentiable function of the position $\mathbf{X}$ of the same material point in the reference configuration. However, in general, both $\mathbf{F}_p$ and $\mathbf{F}_e$ are incompatible tensors which are not determined by differentiation of deformation fields so unloading the material yields a configuration which has residual stresses. In other words, in general, it is not possible to unload the material to a zero-stress intermediate configuration.

The constitutive equations are restricted so that under SRBM, Γ, $\bar{\mathbf{\Lambda}}_p$ and $\mathbf{F}_p$ transform to Γ^+, $\bar{\mathbf{\Lambda}}_p^+$ and $\mathbf{F}_p^+$, such that

$$\Gamma^+ = \Gamma\,, \quad \bar{\mathbf{\Lambda}}_p^+ = \bar{\mathbf{\Lambda}}_p\,, \quad \mathbf{F}_p^+ = \mathbf{F}_p\,. \tag{5.11.12}$$

It then follows from (5.11.4), (5.11.11) and (5.11.12) that under SRBM the elastic deformation tensors $\mathbf{F}_e$ and $\mathbf{C}_e$ transform to $\mathbf{F}_e^+$ and $\mathbf{C}_e^+$, such that

$$\mathbf{F}_e^+ = \mathbf{Q}\mathbf{F}_e\,, \quad \mathbf{C}_e = \mathbf{F}_e^T\mathbf{F}_e\,, \quad \mathbf{C}_e^+ = \mathbf{C}_e\,. \tag{5.11.13}$$

Using the fact that

$$\dot{\overline{\mathbf{F}_p^{-1}}} = -\Gamma \mathbf{F}_p^{-1}\bar{\mathbf{\Lambda}}_p\,, \tag{5.11.14}$$

it follows that $\mathbf{F}_e$ and $\mathbf{C}_e$ satisfy the evolution equations

$$\begin{aligned} \dot{\mathbf{F}}_e &= (\mathbf{L} - \Gamma \mathbf{F}_e\bar{\mathbf{\Lambda}}_p\mathbf{F}_e^{-1})\mathbf{F}_e\,, \\ \dot{\mathbf{C}}_e &= \mathbf{F}_e^T[2\mathbf{D} - \Gamma(\mathbf{F}_e^{-T}\bar{\mathbf{\Lambda}}_p^T\mathbf{F}_e^T + \mathbf{F}_e\bar{\mathbf{\Lambda}}_p\mathbf{F}_e^{-1})]\mathbf{F}_e\,. \end{aligned} \tag{5.11.15}$$

In this theory, the strain energy Σ is assumed to be a function of $\mathbf{F}_e$ and κ but since Σ is uninfluenced by SRBM, it must depend on $\mathbf{F}_e$ only through the deformation tensor $\mathbf{C}_e$ so that

$$\Sigma = \Sigma(\mathbf{C}_e, \kappa) . \tag{5.11.16}$$

For both rate-independent and rate-dependent material response the stress is specified by

$$\mathbf{T} = 2\rho \mathbf{F}_e \frac{\partial \Sigma}{\partial \mathbf{C}_e} \mathbf{F}_e^T , \tag{5.11.17}$$

and the rate of material dissipation (4.5.7) requires

$$\mathcal{D} = \Gamma \rho \left(2\mathbf{C}_e \frac{\partial \Sigma}{\partial \mathbf{C}_e} \cdot \bar{\mathbf{\Lambda}}_p - \frac{\partial \Sigma}{\partial \kappa} H \right) \geq 0 , \tag{5.11.18}$$

which places restrictions on the functional forms of Σ, $\bar{\mathbf{\Lambda}}_p$ and H.

In addition to solving the balance of linear momentum (4.4.5), this theory requires solution of the evolution equations (5.11.1) for $\mathbf{F}$, (5.11.2) for κ and (5.11.9) for $\mathbf{F}_p$ with initial conditions

$$\mathbf{F}(0), \mathbf{F}_p(0), \kappa(0) . \tag{5.11.19}$$

Besseling Formulation

The formulation discussed by Besseling [4] (see also Besseling and van der Giessen [5]) was motivated by the work of Eckart [12] and Mandel [26] and can be interpreted as proposing an evolution equation for a second-order non-symmetric tensor $\mathbf{F}_e$ with positive determinant directly by the evolution equation

$$\dot{\mathbf{F}}_e = \mathbf{L}_e \mathbf{F}_e , \quad \mathbf{L}_e = \mathbf{L} - \mathbf{L}_p , \quad \mathbf{L}_p = \Gamma \bar{\mathbf{L}}_p , \tag{5.11.20}$$

where $\mathbf{L}_e$ is the elastic deformation rate, $\bar{\mathbf{L}}_p$ controls the direction of inelastic rate $\mathbf{L}_p$, Γ is a non-negative function that controls the magnitude of inelastic rate and $\mathbf{F}_e$ measures elastic deformations from a zero-stress intermediate configuration.

Moreover, the evolution equation (5.11.20) will be identical to the evolution equation for $\mathbf{F}_e$ in (5.11.15) if $\bar{\mathbf{L}}_p$ is specified by

$$\bar{\mathbf{L}}_p = \mathbf{F}_e \bar{\mathbf{\Lambda}}_p \mathbf{F}_e^{-1} , \tag{5.11.21}$$

which under SRBM satisfies the transformation relation

$$\bar{\mathbf{L}}_p^+ = \mathbf{Q} \bar{\mathbf{L}}_p \mathbf{Q}^T . \tag{5.11.22}$$

For this theory, the strain energy function Σ is specified by (5.11.16), the stress $\mathbf{T}$ is specified by (5.11.17) and the rate of material dissipation $\mathcal{D}$ requires (5.11.18). In addition to solving the balance of linear momentum (4.4.5), this theory requires solution of the evolution equations (5.11.20) for $\mathbf{F}_e$ and (5.11.2) for κ with initial conditions

$$\mathbf{F}_e(0), \kappa(0) . \tag{5.11.23}$$

Unphysical Arbitrariness of the Lagrangian Formulation
Unphysical arbitrariness of the Lagrangian formulation has been discussed in a series of papers [36–38]. Specifically, for a fixed value of elastic deformation $\mathbf{F}_e$ and an arbitrary nonsingular tensor $\mathbf{A}$ with $\det\mathbf{A} > 0$, it follows from (5.11.11) that

$$\mathbf{F}_e = (\mathbf{F}\mathbf{A})(\mathbf{F}_p\mathbf{A})^{-1} . \tag{5.11.24}$$

This means that the reference configuration associated with $\mathbf{F}$ and $\mathbf{F}_p$ is arbitrary. In particular, $\mathbf{A}$ can be used to set the initial value of $\mathbf{F} = \mathbf{I}$ or to set the initial value of $\mathbf{F}_p = \mathbf{I}$ so the choice of total deformation measure $\mathbf{F}$ or the plastic deformation measure $\mathbf{F}_p$ is arbitrary.

In addition, the elastic deformation tensor $\mathbf{F}_e$ and the plastic deformation tensor $\mathbf{F}_p$ in (5.11.11) are usually presented as a separation of the total deformation gradient $\mathbf{F}$ into elastic and plastic parts

$$\mathbf{F} = \mathbf{F}_e\mathbf{F}_p . \tag{5.11.25}$$

Using the expression (3.3.1a) which shows that $\mathbf{F}$ transforms a material line element $d\mathbf{X}$ in the reference configuration to its deformed line element $d\mathbf{x}$ in the current configuration, the separation (5.11.25) is often interpreted as $\mathbf{F}_p$ transforming the line element $d\mathbf{X}$ to $d\mathbf{y}$ in an intermediate configuration and $\mathbf{F}_e$ transforming $d\mathbf{y}$ to $d\mathbf{x}$

$$d\mathbf{x} = \mathbf{F}d\mathbf{X} , \quad d\mathbf{y} = \mathbf{F}_p d\mathbf{X} , \quad d\mathbf{x} = \mathbf{F}_e d\mathbf{y} . \tag{5.11.26}$$

Letting $\mathbf{O}$ be a proper orthogonal tensor

$$\mathbf{O}\mathbf{O}^T = \mathbf{I} , \quad \det(\mathbf{O}) = +1 , \tag{5.11.27}$$

the separation (5.11.25) can be rewritten in the form

$$\mathbf{F} = (\mathbf{F}_e\mathbf{O}^T)(\mathbf{O}\mathbf{F}_p) , \tag{5.11.28}$$

which shows that both the plastic deformation tensor $\mathbf{F}_p$ and the elastic deformation tensor $\mathbf{F}_e$ contain arbitrariness to rotations of the intermediate configuration.

Eulerian Formulation of Elastically Anisotropic Elastic–Inelastic Materials
The Eulerian formulation for nonlinear elastic solids in Sect. 5.3 can be generalized for elastically anisotropic elastic–inelastic materials by modifying the evolution equation for the microstructural vectors $\mathbf{m}_i$ to include a second-order tensor $\mathbf{L}_p$ that characterizes the inelastic rate. Specifically, an Eulerian formulation for elastically anisotropic inelastic material response, which was motivated by the work of Eckart [12] and Leonov [24], was developed in [35]. The main idea is to model the following physical features of inelastic flow in metals:

- elastic deformations of the atomic lattice cause stress.
- elastic deformations of the atomic lattice remain small after dislocations have moved through the lattice.

- the atoms in a specific lattice change with time as dislocations move through the lattice.
- edges of the parallelepiped formed by the atomic lattice do not rotate as material line elements.

In this model, the elastic deformations and orientation of the atomic lattice are modeled by the parallelepiped formed by the triad $\mathbf{m}_i$ $(i = 1, 2, 3)$ of linearly independent microstructural vectors

$$J_e = \mathbf{m}_1 \times \mathbf{m}_2 \cdot \mathbf{m}_3 \geq 0\,, \tag{5.11.29}$$

where the elastic dilatation J_e is an internal state variable that can be determined by the current state of stress in the material. These microstructural vectors are determined by the evolution equations

$$\dot{\mathbf{m}}_i = (\mathbf{L} - \mathbf{L}_p)\,\mathbf{m}_i\,, \quad \mathbf{L}_p = \Gamma \bar{\mathbf{L}}_p\,, \quad \Gamma \geq 0\,, \tag{5.11.30}$$

where Γ controls the magnitude and $\bar{\mathbf{L}}_p$ controls the direction of the inelastic rate tensor $\mathbf{L}_p$, both of which require a constitutive equation. If $\mathbf{L}_p$ vanishes, then the solution of (5.11.30) causes $\mathbf{m}_i$ to evolve as material line elements so these equations characterize an Eulerian formulation of a general anisotropic hyperelastic solid. Otherwise, $\mathbf{m}_i$ characterize elastic deformations and the orientation of the atomic lattice, which is not directly connected to material line elements.

In addition, an isotropic hardening variable κ is determined by the evolution equation

$$\dot{\kappa} = \Gamma H\,, \tag{5.11.31}$$

where H is a function that controls the rate of hardening. More general directional hardening can be modeled by introducing directional hardening variables $\beta_{ij} = \beta_{ji}$ which satisfy the evolution equations

$$\dot{\beta}_{ij} = \Gamma H_{ij}\,, \tag{5.11.32}$$

where $H_{ij} = H_{ji}$ are functions that control the relative magnitudes of β_{ij}. These functions should not be confused with the components H_{ij} of the proper orthogonal matrix used to discuss material symmetry in Sect. 5.6.

Under SRBM the microstructural vectors $\mathbf{m}_i$, the inelastic deformation rate Γ, its direction $\bar{\mathbf{L}}_p$, the hardening variables κ and β_{ij} and the hardening functions H and H_{ij} transform to $\mathbf{m}_i^+$, Γ^+, $\bar{\mathbf{L}}_p^+$, κ^+, β_{ij}^+, H^+ and H_{ij}^+, such that

$$\begin{aligned} &\mathbf{m}_i^+ = \mathbf{Q}\mathbf{m}_i\,, \quad \Gamma^+ = \Gamma\,, \quad \bar{\mathbf{L}}_p^+ = \mathbf{Q}\bar{\mathbf{L}}_p\mathbf{Q}^T\,, \\ &\kappa^+ = \kappa\,, \qquad \beta_{ij}^+ = \beta_{ij}\,, \quad H^+ = H\,, \qquad H_{ij}^+ = H_{ij}\,. \end{aligned} \tag{5.11.33}$$

The strain energy Σ is assumed to be a function of $\mathbf{m}_i$, κ and β_{ij}, but since Σ must be unaffected by SRBM it can depend on $\mathbf{m}_i$ only through the metric m_{ij} of elastic deformation, which satisfies equations

$$m_{ij} = \mathbf{m}_i \cdot \mathbf{m}_j\,, \quad m^+_{ij} = m_{ij}\,. \tag{5.11.34}$$

Moreover, using (5.11.30) and (5.11.34), it follows that the elastic metric satisfies the evolution equation

$$\dot{m}_{ij} = 2(\mathbf{D} - \mathbf{D}_p) \cdot (\mathbf{m}_i \otimes \mathbf{m}_j)\,, \tag{5.11.35}$$

where the inelastic deformation rate $\mathbf{D}_p$ is defined by

$$\mathbf{D}_p = \frac{1}{2}(\mathbf{L}_p + \mathbf{L}_p^T) = \Gamma\bar{\mathbf{D}}_p\,, \quad \bar{\mathbf{D}}_p = \frac{1}{2}(\bar{\mathbf{L}}_p + \bar{\mathbf{L}}_p^T)\,. \tag{5.11.36}$$

For this model the strain energy function and the stress are proposed in the forms

$$\Sigma = \Sigma(m_{ij}, \kappa, \beta_{ij})\,, \quad \mathbf{T} = \mathbf{T}(\mathbf{m}_i, \kappa, \beta_{ij})\,. \tag{5.11.37}$$

It then follows that the rate of material dissipation (4.5.7) requires

$$\begin{aligned} \mathcal{D} = &\left[\mathbf{T} - 2\rho\frac{\partial\Sigma}{\partial m_{ij}}(\mathbf{m}_i \otimes \mathbf{m}_j)\right] \cdot \mathbf{D} \\ &+ \Gamma\left[2\rho\frac{\partial\Sigma}{\partial m_{ij}}(\mathbf{m}_i \otimes \mathbf{m}_j) \cdot \bar{\mathbf{D}}_p - \rho\frac{\partial\Sigma}{\partial\kappa}H - \rho\frac{\partial\Sigma}{\partial\beta_{ij}}H_{ij}\right] \geq 0\,. \end{aligned} \tag{5.11.38}$$

Without specifying details of inelastic deformation rate and the hardening functions $\Gamma, \bar{\mathbf{D}}_p, H$ and H_{ij} it is not possible to obtain necessary restrictions on the constitutive equation for stress. However, motivated by the constitutive equation (5.3.4) for a hyperelastic material and by the requirement that the constitutive equation for elastic–inelastic response contain that for a hyperelastic material as a special case, the constitutive equation for stress in an elastic–inelastic material is specified by

$$\mathbf{T} = 2\rho\frac{\partial\Sigma}{\partial m_{ij}}(\mathbf{m}_i \otimes \mathbf{m}_j)\,. \tag{5.11.39}$$

Then, the rate of material dissipation (4.5.7) requires the total dissipation due to the inelastic rate and the rate of hardening to be non-negative

$$\mathcal{D} = \Gamma\left(\mathbf{T} \cdot \bar{\mathbf{D}}_p - \rho\frac{\partial\Sigma}{\partial\kappa}H - \rho\frac{\partial\Sigma}{\partial\beta_{ij}}H_{ij}\right) \geq 0\,. \tag{5.11.40}$$

Also, the constitutive equation for stress is assumed to be restricted so that a zero-stress material state is characterized by

$$\mathbf{T} = 0\,, \quad \frac{\partial\Sigma}{\partial m_{ij}} = 0 \quad \text{for } m_{ij} = \delta_{ij}\,, \tag{5.11.41}$$

where δ_{ij} is the Kronecker delta. This means that the triad $\mathbf{m}_i$ has been defined so that $\mathbf{m}_i$ are orthonormal vectors in a zero-stress material state. In particular, it is noted that the vectors $\mathbf{m}_i$ in any zero-stress material state are usually not parallel to the lattice vectors in that state.

Since $\mathbf{m}_i$ are linearly independent and not necessarily orthonormal in the current configuration, it is convenient to introduce their reciprocal vectors $\mathbf{m}^i$ by

$$\mathbf{m}^1 = J_e^{-1}(\mathbf{m}_2 \times \mathbf{m}_3)\,, \quad \mathbf{m}^2 = J_e^{-1}(\mathbf{m}_3 \times \mathbf{m}_1)\,, \quad \mathbf{m}^3 = J_e^{-1}(\mathbf{m}_1 \times \mathbf{m}_2)\,, \tag{5.11.42}$$

so that

$$\dot{J}_e = J_e(\mathbf{D} - \mathbf{D}_p) \cdot \mathbf{I}\,. \tag{5.11.43}$$

Moreover, the evolution equations (5.11.30) for $\mathbf{m}_i$, (5.11.31) for κ and (5.11.32) for β_{ij} require initial conditions

$$\mathbf{m}_i(0),\, \kappa(0),\, \beta_{ij}(0)\,. \tag{5.11.44}$$

Separation of Elastic Dilatation and Distortional Deformations

To introduce separate control over the response of the material to dilatation and distortional rates of deformation, it is convenient to use the elastic dilatation J_e, the distortional deformation vectors $\mathbf{m}'_i$ and the elastic distortional deformation metric m'_{ij}, which satisfy the Eqs. (5.11.43), (3.11.14), (5.11.30) and (3.11.16),

$$\begin{aligned}
&J_e = \mathbf{m}_1 \times \mathbf{m}_2 \cdot \mathbf{m}_3 > 0\,, && \dot{J}_e = J_e(\mathbf{D} \cdot \mathbf{I} - \Gamma \bar{\mathbf{D}}_p)\,,\\
&\mathbf{m}'_i = J_e^{-1/3}\,\mathbf{m}_i\,, && \dot{\mathbf{m}}'_i = (\mathbf{L}'' - \Gamma \bar{\mathbf{L}}''_p)\,\mathbf{m}'_i\,,\\
&m'_{ij} = \mathbf{m}'_i \cdot \mathbf{m}'_j = m'_{ji}\,, && \dot{m}'_{ij} = 2\left(\mathbf{m}'_i \otimes \mathbf{m}'_j - \tfrac{1}{3} m'_{ij}\mathbf{I}\right) \cdot (\mathbf{D} - \Gamma \bar{\mathbf{D}}_p)\,,
\end{aligned} \tag{5.11.45}$$

where $\mathbf{L}''$ is the deviatoric part of the velocity gradient $\mathbf{L}$ and $\bar{\mathbf{L}}''_p$ is the deviatoric part of $\bar{\mathbf{L}}_p$. Then, the strain energy function and stress are proposed in the forms

$$\Sigma = \tilde{\Sigma}(J_e, m'_{ij}, \kappa, \beta_{ij})\,, \quad \mathbf{T} = \tilde{\mathbf{T}}(J_e, \mathbf{m}'_i, \kappa, \beta_{ij})\,, \tag{5.11.46}$$

and the rate of material dissipation (4.5.7) requires

$$\begin{aligned}
\mathcal{D} = &\left[\mathbf{T} - \rho J_e \frac{\partial \tilde{\Sigma}}{\partial J_e}\mathbf{I} - 2\rho \frac{\partial \tilde{\Sigma}}{\partial m'_{ij}}\left(\mathbf{m}'_i \otimes \mathbf{m}'_j - \frac{1}{3} m'_{ij}\mathbf{I}\right)\right] \cdot \mathbf{D}\\
&+ \Gamma \left[\rho J_e \frac{\partial \tilde{\Sigma}}{\partial J_e}\bar{\mathbf{D}}_p \cdot \mathbf{I} + 2\rho \frac{\partial \tilde{\Sigma}}{\partial m'_{ij}}\left(\mathbf{m}'_i \otimes \mathbf{m}'_j - \frac{1}{3} m'_{ij}\mathbf{I}\right) \cdot \bar{\mathbf{D}}_p\right.\\
&\left. -\rho \frac{\partial \tilde{\Sigma}}{\partial \kappa} H - \rho \frac{\partial \tilde{\Sigma}}{\partial \beta_{ij}} H_{ij}\right] \geq 0\,.
\end{aligned} \tag{5.11.47}$$

Again, without specifying details of the inelastic rate and the hardening functions Γ, $\bar{\mathbf{D}}_p$, H and H_{ij} it is not possible to obtain necessary restrictions on the constitutive equation for stress. However, motivated by the constitutive equation (5.3.11) for a hyperelastic material and by the requirement that the constitutive equation for elastic–inelastic response contain that for a hyperelastic material as a special case, the constitutive equation for stress in an elastic–inelastic material is specified by

$$\begin{aligned} \mathbf{T} &= -p\,\mathbf{I} + \mathbf{T}'' = \tilde{\mathbf{T}}\,, \quad p = \tilde{p} = -\rho J_e \frac{\partial \tilde{\Sigma}}{\partial J_e}\,, \\ \mathbf{T}'' &= \tilde{\mathbf{T}}'' = 2\rho \frac{\partial \tilde{\Sigma}}{\partial m'_{ij}} \left(\mathbf{m}'_i \otimes \mathbf{m}'_j - \frac{1}{3} m'_{ij} \mathbf{I} \right). \end{aligned} \tag{5.11.48}$$

Then, the rate of material dissipation requires the total dissipation due to the inelastic rate and the rate of hardening to be non-negative

$$\mathcal{D} = \Gamma \left[-\tilde{p}\,(\bar{\mathbf{D}}_p \cdot \mathbf{I}) + \tilde{\mathbf{T}}'' \cdot \bar{\mathbf{D}}_p - \rho \frac{\partial \tilde{\Sigma}}{\partial \kappa} H - \rho \frac{\partial \tilde{\Sigma}}{\partial \beta_{ij}} H_{ij} \right] \geq 0\,. \tag{5.11.49}$$

Also, the constitutive equation for stress is assumed to be restricted so that a zero-stress material state is characterized by

$$\mathbf{T} = 0\,, \quad \frac{\partial \tilde{\Sigma}}{\partial J_e} = 0\,, \quad \frac{\partial \tilde{\Sigma}}{\partial m'_{ij}} = \frac{1}{3} \frac{\partial \tilde{\Sigma}}{\partial m'_{nn}} \delta_{ij} \quad \text{for } J_e = 1 \text{ and } m'_{ij} = \delta_{ij}\,. \tag{5.11.50}$$

This means that the triad $\mathbf{m}'_i$ has been defined so that $\mathbf{m}'_i$ are orthonormal vectors in a zero-stress material state.

Moreover, the evolution equations (5.11.43) for J_e, (5.11.45) for $\mathbf{m}'_i$, (5.11.31) for κ and (5.11.32) for β_{ij} require initial conditions

$$J_e(0),\, \mathbf{m}'_i(0),\, \kappa(0),\, \beta_{ij}(0)\,. \tag{5.11.51}$$

Eulerian Formulation of Elastically Isotropic Elastic–Inelastic Materials
For elastically isotropic elastic–inelastic materials experiments on identical samples of the material in its current state cannot distinguish between the microstructural vectors $\mathbf{m}'_1$, $\mathbf{m}'_2$ and $\mathbf{m}'_3$ so the material response functions must be insensitive to this arbitrariness of $\mathbf{m}'_i$. Consequently, J_e in (5.11.29) characterizes the elastic dilatation and satisfies the evolution equation (5.11.43). Also, the symmetric, positive-definite, unimodular tensor $\mathbf{B}'_e$ defined in (5.8.1)

$$\mathbf{B}'_e = \mathbf{m}'_i \otimes \mathbf{m}'_i\,, \tag{5.11.52}$$

characterizes elastic distortional deformations. Using (5.11.45), it can be shown that $\mathbf{B}'_e$ satisfies the evolution equation

$$\dot{\mathbf{B}}'_e = \mathbf{L}''\mathbf{B}'_e + \mathbf{B}'_e\mathbf{L}''^T - \Gamma\bar{\mathbf{A}}_p\,, \quad \bar{\mathbf{A}}_p = \bar{\mathbf{L}}''_p\mathbf{B}'_e + \mathbf{B}'_e\bar{\mathbf{L}}''^T_p\,, \tag{5.11.53}$$

where $\bar{\mathbf{L}}''_p$ is the deviatoric part of $\bar{\mathbf{L}}_p$. This evolution equation automatically satisfies the condition (3.11.26) that $\mathbf{B}'_e$ remains unimodular [det $\mathbf{B}'_e = 1$] since

$$\dot{\mathbf{B}}'_e \cdot \mathbf{B}'^{-1}_e = 0\,, \quad \bar{\mathbf{A}}_p \cdot \mathbf{B}'^{-1}_e = 0\,. \tag{5.11.54}$$

Following the work of Eckart [12] and Leonov [24] for elastically isotropic elastic–inelastic materials, an evolution equation for the elastic distortional deformation tensor $\mathbf{B}'_e$ can be proposed directly and independently of the microstructural vectors $\mathbf{m}'_i$. This means that instead of specifying a constitutive equation for $\bar{\mathbf{L}}_p$, it is possible to propose an evolution equation for $\mathbf{B}'_e$ directly in the form

$$\dot{\mathbf{B}}'_e = \mathbf{L}''\mathbf{B}'_e + \mathbf{B}'_e\mathbf{L}''^T - \Gamma\mathbf{A}_p\,, \tag{5.11.55}$$

where $\mathbf{A}_p$ is a symmetric tensor that controls the direction of inelastic distortional deformation rate. This tensor must satisfy the restriction

$$\mathbf{A}_p \cdot \mathbf{B}'^{-1}_e = 0\,, \tag{5.11.56}$$

which ensures that $\mathbf{B}'_e$ remains unimodular.

In this model, the strain energy function for elastically isotropic response is taken to be a function of the elastic dilatation J_e, the elastic distortional deformation $\mathbf{B}'_e$ and the hardening κ. However, under SRBM J_e, $\mathbf{B}'_e$ and $\mathbf{A}_p$ transform to J^+_e, $\mathbf{B}'^+_e$ and $\mathbf{A}^+_p$, such that

$$J^+_e = J_e\,, \quad \mathbf{B}'^+_e = \mathbf{Q}\mathbf{B}'_e\mathbf{Q}^T\,, \quad \mathbf{A}^+_p = \mathbf{Q}\mathbf{A}_p\mathbf{Q}^T\,, \tag{5.11.57}$$

so the strain energy function can depend on $\mathbf{B}'_e$ only through its two independent invariants α_1 and α_2, defined by

$$\alpha_1 = \mathbf{B}'_e \cdot \mathbf{I}\,, \quad \alpha_1 = \mathbf{B}'_e \cdot \mathbf{B}'_e\,, \tag{5.11.58}$$

which satisfy the evolution equations

$$\begin{aligned} \dot{\alpha}_1 &= 2\mathbf{B}''_e \cdot \mathbf{D} - \Gamma\mathbf{A}_p \cdot \mathbf{I}\,, \\ \dot{\alpha}_2 &= 4\left(\mathbf{B}'^2_e - \frac{1}{3}\alpha_2\mathbf{I}\right) \cdot \mathbf{D} - 2\Gamma\mathbf{A}_p \cdot \mathbf{B}'_e\,. \end{aligned} \tag{5.11.59}$$

Thus, the strain energy function Σ and the stress are proposed in the forms

$$\Sigma = \Sigma(J_e, \alpha_1, \alpha_2, \kappa)\,, \quad \mathbf{T} = \mathbf{T}(J_e, \mathbf{B}'_e, \kappa)\,, \tag{5.11.60}$$

and for both rate-independent and rate-dependent response the stress is specified by

$$\begin{aligned} \mathbf{T} &= -p\mathbf{I} + \mathbf{T}'' , \quad p = -\rho J_e \frac{\partial \Sigma}{\partial J_e} , \\ \mathbf{T}'' &= 2\rho \left[\frac{\partial \Sigma}{\partial \alpha_1} \mathbf{B}_e'' + 2 \frac{\partial \Sigma}{\partial \alpha_2} \left(\mathbf{B}_e'^2 - \frac{1}{3} \alpha_2 \mathbf{I} \right) \right] , \end{aligned} \tag{5.11.61}$$

where $\mathbf{T}''$ is the deviatoric part of $\mathbf{T}$. Also, the rate of material dissipation $\mathcal{D}$ in (4.5.7) requires

$$\mathcal{D} = \Gamma \left[-p\,(\bar{\mathbf{D}}_p \cdot \mathbf{I}) + \rho \left(\frac{\partial \Sigma}{\partial \alpha_1} \mathbf{A}_p \cdot \mathbf{I} + 2 \frac{\partial \Sigma}{\partial \alpha_2} \mathbf{A}_p \cdot \mathbf{B}_e' - \frac{\partial \Sigma}{\partial \kappa} H \right) \right] \geq 0 . \tag{5.11.62}$$

In addition, the constitutive equation for stress is assumed to be restricted so that a zero-stress material state is characterized by

$$\mathbf{T} = 0 , \quad \frac{\partial \Sigma}{\partial J_e} = 0 \quad \text{for } J_e = 1 \text{ and } \mathbf{B}_e' = \mathbf{I} . \tag{5.11.63}$$

The evolution equations (5.11.43) for J_e, (5.11.55) for $\mathbf{B}_e'$ and (5.11.31) for κ require initial conditions

$$J_e(0), \mathbf{B}_e'(0), \kappa(0) . \tag{5.11.64}$$

In this regard, it is assumed that the constitutive equation (5.11.61) for stress is invertible and that experiments can be performed to determine the values of hardening variable κ at any state of the material. In particular, the values of J_e and $\mathbf{B}_e'$ in any zero-stress material state are given by (5.11.63). Also, when Γ vanishes, the theory represents an Eulerian formulation of a general elastically isotropic hyperelastic material.

Since the inelastic deformation rate causes a tendency for the deviatoric stress $\mathbf{T}''$ to approach zero, Rubin and Attia [42] proposed $\mathbf{A}_p$ in the form

$$\mathbf{A}_p = \mathbf{B}_e' - \left(\frac{3}{\mathbf{B}_e'^{-1} \cdot \mathbf{I}} \right) \mathbf{I} , \tag{5.11.65}$$

so the evolution equation (5.11.55) is given by

$$\dot{\mathbf{B}}_e' = \mathbf{L}'' \mathbf{B}_e' + \mathbf{B}_e' \mathbf{L}''^T - \Gamma \mathbf{A}_p , \quad \mathbf{A}_p = \mathbf{B}_e' - \left(\frac{3}{\mathbf{B}_e'^{-1} \cdot \mathbf{I}} \right) \mathbf{I} . \tag{5.11.66}$$

As discussed in [42], since $\mathbf{B}_e'$ is a unimodular positive-definite tensor, the spectral form of $\mathbf{B}_e'$ can be used to show that

$$\mathbf{A}_p \cdot \mathbf{I} \geq 0 , \quad \mathbf{A}_p \cdot \mathbf{B}_e' \geq 0 . \tag{5.11.67}$$

As a special case, the strain energy function is given by a compressible Neo-Hookean form

$$\rho_z \Sigma = f(J_e) + \frac{1}{2}\mu(\alpha_1 - 3)\,, \tag{5.11.68}$$

where ρ_z is a constant that is not necessarily the zero-stress density, $f(J_e)$ controls the response to elastic dilatation and μ is the positive zero-stress shear modulus. Moreover, from (5.11.61) the stress is specified by

$$\mathbf{T} = -p\,\mathbf{I} + \mathbf{T}''\,, \quad p = -\left(\frac{\rho J_e}{\rho_z}\right)\frac{df}{dJ_e}\,, \quad \mathbf{T}'' = \left(\frac{\rho}{\rho_z}\right)\mu\,\mathbf{B}_e''\,, \tag{5.11.69}$$

with the function $f(J_e)$ satisfying the restrictions

$$f(1) = 0\,, \quad \frac{df}{dJ_e}(1) = 0\,, \quad \frac{d^2 f}{dJ_e^2}(1) > 0\,, \tag{5.11.70}$$

imposed by the condition (5.11.63) for a zero-stress material state and the condition that the bulk modulus is positive. Also, the rate of material dissipation (5.11.62) requires

$$\mathcal{D} = \Gamma\left[-p\,(\bar{\mathbf{D}}_p \cdot \mathbf{I}) + \frac{1}{2}\left(\frac{\rho}{\rho_z}\right)\mu \mathbf{A}_p \cdot \mathbf{I}\right] \geq 0\,. \tag{5.11.71}$$

In Sect. 6.8 the volumetric inelastic rate $\bar{\mathbf{D}}_p \cdot \mathbf{I}$ will be related to the rate of change of porosity in a porous material. However, for nonporous metals plastic deformation rate is considered to be isochoric, which requires

$$\bar{\mathbf{D}}_p \cdot \mathbf{I} = 0\,, \tag{5.11.72}$$

and the rate of material dissipation (5.11.71) requires

$$\mathcal{D} = \frac{1}{2}\left(\frac{\rho}{\rho_z}\right)\mu\Gamma(\mathbf{A}_p \cdot \mathbf{I}) \geq 0\,, \tag{5.11.73}$$

which in view of (5.11.67), is automatically satisfied. In this expression, use has been made of the evolution equation (5.11.45) for J_e, the expressions (4.1.16) and (5.11.69)–(5.11.73) to deduce that

$$\begin{gathered} J_e = \frac{\rho}{\rho_z}\,, \quad \mathbf{T} = -p\,\mathbf{I} + \mathbf{T}''\,, \quad p = -\frac{df}{dJ_e}\,, \\ \mathbf{T}'' = \mu\,\mathbf{B}_e''\,, \quad \mathcal{D} = \frac{1}{2}\mu\Gamma(\mathbf{A}_p \cdot \mathbf{I}) \geq 0\,, \end{gathered} \tag{5.11.74}$$

where ρ_z is the mass density in any zero-stress state.

Additional Comments on Arbitrariness

From the perspective of the definition of internal state variables by Onat [31], the total deformation tensor $\mathbf{F}$, the plastic deformation tensors $\mathbf{C}_p$ and $\mathbf{F}_p$ and the elastic deformation tensors $\mathbf{F}_e$ and $\mathbf{C}_e$ are not internal state variables since they cannot be measured, in principle, by experiments on identical samples of the material in its current state. In particular, they are affected by arbitrariness of the choices of: the reference configuration; an intermediate configuration; a total deformation measure and a plastic deformation measure, which have been discussed in [36–38].

In [38] it was proved that when this arbitrariness is removed from the Lagrangian multiplicative formulation associated with (5.11.11), that formulation must reduce to the Eulerian formulation based on the microstructural vectors $\mathbf{m}_i$. Moreover in [38] it was shown that $\mathbf{m}_i$ are internal state variables in the sense of Onat [31] because their initial values can be measured, in principle, by experiments on identical samples of the material in its current state.

Elastic anisotropy of a material with the strain energy function specified by (5.11.37) is characterized by the dependence of the strain energy on the vectors $\mathbf{m}_i$. It is important to emphasize that the index (i) in $\mathbf{m}_i$ refers to distinct directions of the atomic lattice. If any of these directions cannot be distinguished by experiments, then the strain energy function must satisfy symmetry conditions which ensure that the material response is also insensitive to these indistinguishable directions.

Comparison of the evolution equation (5.11.20) for $\mathbf{F}_e$ and (5.11.30) for $\mathbf{m}_i$ suggests that these formulations may be identical. The discussion in Sect. 5.4, which describes the difference between $\mathbf{F}_e$ and $\mathbf{m}_i$ for an elastic material, is similar for an elastic–inelastic material. Specifically, consider an *arbitrary* right-handed orthonormal set of constant base vectors $\mathbf{M}_i$ and define the elastic deformation tensor $\mathbf{F}_e$ by

$$\mathbf{F}_e = \mathbf{m}_i \otimes \mathbf{M}_i \,, \tag{5.11.75}$$

which satisfies the evolution equation and initial condition

$$\dot{\mathbf{F}}_e = (\mathbf{L} - \mathbf{L}_p)\mathbf{F}_e \,, \quad \mathbf{F}_e(0) = \mathbf{m}_i(0) \otimes \mathbf{M}_i \,. \tag{5.11.76}$$

However, in [38] it was shown that the elastic response of the material depends on $\mathbf{m}_i$ through the evolution equation (5.11.30) and on their initial values $\mathbf{m}_i(0)$. Although $\mathbf{m}_i(0)$ are measurable, the tensor $\mathbf{F}_e$ contains unphysical arbitrariness of the orientation of $\mathbf{M}_i$ which can be removed by considering the Eulerian formulation based on $\mathbf{m}_i$.

Rate-Independent Inelasticity with a Yield Function

For rate-independent inelasticity a yield function g is introduced which characterizes elastic response for $g < 0$ and the elastic–inelastic boundary for $g = 0$. For states at the elastic–inelastic boundary, it is necessary to specify unloading, neutral loading and loading conditions which have zero inelastic rate for unloading and neutral loading, and nonzero inelastic rate for loading. Differences in the loading conditions for stress-space and strain-space formulations have been discussed in [30]. In particular, the strain-space formulation can model strain softening with decrease in stress that occurs due to damage mechanisms. Moss [27] pointed out that the numerical algorithm developed by Wilkins [48] is consistent with the loading conditions in the strain-space formulation developed by Naghdi and Trapp [30].

Here, use is made of the strain-space loading conditions and the yield function for elastically anisotropic response is specified by

$$g = g(m_{ij}, \kappa, \beta_{ij}) \le 0 \,. \tag{5.11.77}$$

With the help of the evolution equations (5.11.35) for m_{ij}, (5.11.31) for κ and (5.11.32) for β_{ij}, it follows that

$$\begin{aligned} \dot{g} &= \hat{g} - \Gamma \bar{g} \,, \\ \hat{g} &= 2\left(\frac{\partial g}{\partial m_{ij}}\right)(\mathbf{m}_i \otimes \mathbf{m}_j) \cdot \mathbf{D} \,, \\ \bar{g} &= 2\left(\frac{\partial g}{\partial m_{ij}}\right)(\mathbf{m}_i \otimes \mathbf{m}_j) \cdot \bar{\mathbf{D}}_p - \left(\frac{\partial g}{\partial \kappa}\right) H - \left(\frac{\partial g}{\partial \beta_{ij}}\right) H_{ij} > 0 \,, \end{aligned} \tag{5.11.78}$$

where the functional form of g has been restricted so that $\bar{g}$ remains positive. Then, the values of Γ for elastic response, unloading from the elastic–inelastic boundary, neutral loading on the elastic–inelastic boundary and inelastic loading on the elastic–inelastic boundary are specified by

$$\Gamma = \begin{cases} 0 & \text{for elastic response} \quad g < 0 \,, \\ 0 & \text{for unloading} \quad g = 0 \text{ and } \hat{g} < 0 \,, \\ 0 & \text{for neutral loading} \quad g = 0 \text{ and } \hat{g} = 0 \,, \\ \dfrac{\hat{g}}{\bar{g}} > 0 & \text{for inelastic loading} \quad g = 0 \text{ and } \hat{g} > 0 \,, \end{cases} \tag{5.11.79}$$

where the value of Γ for loading has been determined by the consistency condition which ensures that g remains zero ($\dot{g} = 0$) during inelastic loading. From these conditions it can be seen that for elastic response with $\Gamma = 0$, the rate of change of the yield function $\dot{g} = \hat{g}$ so that $\hat{g} > 0$ on the elastic–inelastic boundary requires nonzero inelastic deformation rate ($\Gamma > 0$) to satisfy the consistency condition.

Also, since during loading Γ is linear in the rate $\mathbf{D}$, it follows that the evolution equations (5.11.31), (5.11.32) and (5.11.35) are homogeneous of order one in time when H and H_{ij} are independent of $\mathbf{D}$, so the material response is rate independent.

Rate-Dependent Response

For the rate-independent theory, the rate of inelastic deformation Γ is a homogeneous function of order one in the total rate of deformation $\mathbf{D}$. In contrast, if Γ is not is a homogeneous function of order one in $\mathbf{D}$, then the material response is rate dependent. Examples of rate-dependent response can be found in [7–11, 25, 33, 34].

A model exhibiting a smooth elastic–inelastic transition for both rate-independent and rate-dependent response can be found in [18, 19]. In this model the function Γ in

(5.11.30) for elastically anisotropic response or in (5.11.55) for elastically isotropic response, which controls the magnitude of inelastic deformation rate, is specified in the form

$$\begin{aligned} &\Gamma = \Gamma_0 + \Gamma_1 \langle g \rangle\,, \quad \Gamma_0 = a_0 + b_0 \dot{\varepsilon}\,, \quad \Gamma_1 = a_1 + b_1 \dot{\varepsilon}\,, \\ &a_0 \ge 0\,, \quad b_0 \ge 0\,, \quad a_1 \ge 0\,, \quad b_1 \ge 0\,, \\ &\dot{\varepsilon} = \sqrt{\frac{2}{3} \mathbf{D}'' \cdot \mathbf{D}''}\,, \end{aligned} \tag{5.11.80}$$

where $\dot{\varepsilon}$ is the effective total distortional deformation rate, g is a yield function and the Macaulay brackets $\langle g \rangle$ are defined by

$$\langle g \rangle = \max(g, 0)\,. \tag{5.11.81}$$

When $a_0 = b_0 = b_1 = 0$ this form yields a rate-dependent overstress model like that developed in [25, 33]. Also, when $a_0 = b_0 = a_1 = 0$ the model yields a rate-independent overstress model, which approximates a standard rate-independent yield function when b_1 is large enough to ensure that g remains a small positive value during inelastic loading. In addition, the constants a_0 and b_0 control the inelastic rate that is active for all nonzero values of $\bar{\mathbf{L}}_p$ in (5.11.30) or $\mathbf{A}_p$ in (5.11.55), which can model the response observed in soils. It is also noted that this smooth-transition model has been generalized and numerical algorithms have been developed in [20].

Strongly Objective, Robust Numerical Integration Algorithms

Elastically Isotropic Response
Strongly objective, robust numerical algorithms for integrating the evolution equations for elastic–inelastic response have been discussed in [18, 19, 32, 40, 43, 44]. In this section, attention is limited to elastically isotropic elastic–inelastic material response of metals for which the elastic dilatation J_e and the symmetric, positive-definite, unimodular elastic distortional deformation tensor $\mathbf{B}'_e$ satisfy the evolution equation (3.11.30) for J_e and (5.11.66) for $\mathbf{B}'_e$

$$\begin{aligned} &\dot{J}_e = J_e \mathbf{D} \cdot \mathbf{I}\,, \\ &\dot{\mathbf{B}}'_e = \mathbf{L}'' \mathbf{B}'_e + \mathbf{B}'_e \mathbf{L}''^T - \Gamma \left[\mathbf{B}'_e - \left(\frac{3}{\mathbf{B}'^{-1}_e \cdot \mathbf{I}} \right) \mathbf{I} \right]. \end{aligned} \tag{5.11.82}$$

Moreover, the deviatoric part of the evolution equation for $\mathbf{B}'_e$ can be written in the form

$$\dot{\mathbf{B}}''_e = \mathbf{L}'' \mathbf{B}'_e + \mathbf{B}'_e \mathbf{L}''^T - \frac{2}{3} (\mathbf{B}''_e \cdot \mathbf{D}'') \mathbf{I} - \Gamma \mathbf{B}''_e\,, \tag{5.11.83}$$

where $\mathbf{L}''$ and $\mathbf{D}''$ are the deviatoric parts of $\mathbf{L}$ and $\mathbf{D}$, respectively, and $\mathbf{B}''_e$ is the deviatoric part of $\mathbf{B}'_e$ (3.11.38).

Following the discussion in Sect. 3.13 and the work in [14, 46] and considering a typical time step which begins at $t = t_n$, ends at $t = t_{n+1}$, with time increment $\Delta t = t_{n+1} - t_n$, the relative dilatation J_r and unimodular part $\mathbf{F}'_r$ of the relative deformation gradient during the time step satisfy the evolution equations and initial conditions (3.13.5) and (3.13.7)

$$\begin{aligned} \dot{J}_r &= J_r \mathbf{D} \cdot \mathbf{I}\,, \quad J_r(t_n) = 1\,, \\ \dot{\mathbf{F}}'_r &= \mathbf{L}''\,\mathbf{F}'_r\,, \quad \mathbf{F}'_r(t_n) = \mathbf{I}\,. \end{aligned} \tag{5.11.84}$$

Then, the exact solution of the evolution equation for J_e is given by

$$J_e(t_{n+1}) = J_r(t_{n+1})J_e(t_n)\,. \tag{5.11.85}$$

Also, the elastic trial value $\mathbf{B}''^{*}_e(t)$ defined by (3.13.9) and (3.13.11)

$$\mathbf{B}''^{*}_e = \mathbf{B}'^{*}_e - \frac{1}{3}(\mathbf{B}'^{*}_e \cdot \mathbf{I})\,\mathbf{I}\,, \quad \mathbf{B}'^{*}_e(t) = \mathbf{F}'_r(t)\mathbf{B}'_e(t_n)\mathbf{F}'^{T}_r(t) \tag{5.11.86}$$

satisfies the evolution equation and initial condition

$$\dot{\mathbf{B}}''^{*}_e = \mathbf{L}''\mathbf{B}'^{*}_e + \mathbf{B}'^{*}_e\mathbf{L}''^{T} - \frac{2}{3}(\mathbf{B}''^{*}_e \cdot \mathbf{D}'')\,\mathbf{I}\,, \quad \mathbf{B}''^{*}_e(t_n) = \mathbf{B}''_e(t_n)\,. \tag{5.11.87}$$

Consequently, $\mathbf{B}''^{*}_e(t_{n+1})$ is the exact solution of (5.11.83) when inelastic deformation rate vanishes (i.e., $\Gamma = 0$).

Next, the evolution equation (5.11.83) is approximated by

$$\dot{\mathbf{B}}''_e = \dot{\mathbf{B}}''^{*}_e - \Gamma\mathbf{B}''_e\,, \tag{5.11.88}$$

which with the help of a backward Euler approximation of the derivative can be solved to obtain

$$\mathbf{B}''_e(t_{n+1}) = \left(\frac{1}{1 + \Delta\Gamma}\right) \mathbf{B}''^{*}_e(t_{n+1})\,, \tag{5.11.89a}$$

$$\Delta\Gamma = \Delta t\,\Gamma(t_{n+1})\,, \tag{5.11.89b}$$

where $\Gamma(t_{n+1})$ is an approximation of Γ at the end of the time step that is uninfluenced by SRBM. This expression is similar to the radial-return numerical algorithm developed by Wilkins [48] which scales the trial deviatoric stress to obtain the solution at the end of the time step.

For a general functional form of Γ it is necessary to iterate on the guess for $\Delta\Gamma$ and integrate the other evolution equations for the values of the history-dependent variables at the end of the time step. This procedure continues until (5.11.89b) is consistent with the guess for $\Delta\Gamma$ and the functional form for Γ evaluated using the predicted values of the history-dependent variables at the end of the time step. It is important to emphasize that each iteration step must start with the initial values of the history-dependent variables equal to their accepted values at the beginning of the step to not accumulate history dependence of inaccurate trial solutions.

As described in [42], once the value of $\mathbf{B}_e''$ has been determined at the end of the time step, the value of $\mathbf{B}_e'$ at the end of the time step is determined by solving the cubic equation

$$\det \mathbf{B}_e' = \det\left(\frac{1}{3}\alpha_1 + \mathbf{B}_e''\right) = 1\,, \tag{5.11.90}$$

for the associated value of the invariant α_1. In this regard, it was noted in [18] that the solution (49a) in [42] is more accurate than the solution (54) there.

As a simple example, the strain energy function is specified by (5.11.68) and the von Mises effective stress σ_e is determined by

$$\begin{aligned}
\sigma_e &= \sqrt{\frac{3}{2}\mathbf{T}''\cdot\mathbf{T}''} = J_e^{-1}\mu\sqrt{\frac{3}{2}\mathbf{B}_e''\cdot\mathbf{B}_e''} = 2J_e^{-1}\mu\gamma_e\,,\\
\gamma_e &= \sqrt{\frac{3}{2}\mathbf{g}_e''\cdot\mathbf{g}_e''} = \frac{1}{2}\sqrt{\frac{3}{2}\mathbf{B}_e''\cdot\mathbf{B}_e''}\,,\quad \mathbf{g}_e'' = \frac{1}{2}\mathbf{B}_e''\,,
\end{aligned} \tag{5.11.91}$$

where $\mathbf{g}_e''$ is the elastic distortional strain tensor defined in (3.11.37) and γ_e is a scalar measure of elastic distortional strain. Motivated by these expressions a simple form for the yield function g is specified by

$$g = 1 - \frac{\kappa}{\gamma_e}\,, \tag{5.11.92}$$

which indicates that the onset of yield occurs when $\gamma_e = \kappa$.

Next, the elastic trial value $\gamma_e^*(t_{n+1})$ and the value $\gamma_e(t_{n+1})$ at the end of the time step are defined by

$$\begin{aligned}
\gamma_e^*(t_{n+1}) &= \frac{1}{2}\sqrt{\frac{3}{2}\mathbf{B}_e''^*(t_{n+1})\cdot\mathbf{B}_e''^*(t_{n+1})}\,,\\
\gamma_e(t_{n+1}) &= \frac{1}{2}\sqrt{\frac{3}{2}\mathbf{B}_e''(t_{n+1})\cdot\mathbf{B}_e''(t_{n+1})}\,.
\end{aligned} \tag{5.11.93}$$

It then follows from (5.11.89a) that

$$\gamma_e(t_{n+1}) = \left(\frac{1}{1+\Delta\Gamma}\right)\gamma_e^*(t_{n+1})\,. \tag{5.11.94}$$

Moreover, the elastic trial value $g^*(t_{n+1})$ of the yield function (5.11.92) at the end of the time step is given by

$$g^*(t_{n+1}) = 1 - \frac{\kappa(t_n)}{\gamma_e^*(t_{n+1})}\,. \tag{5.11.95}$$

If $g^*(t_{n+1}) \leq 0$, then the response during the time step is elastic with

$$\Delta\Gamma = 0\,, \quad \kappa(t_{n+1}) = \kappa(t_n) \quad \text{for } g^*(t_{n+1}) \leq 0\,. \tag{5.11.96}$$

On the other hand, if $g^*(t_{n+1}) > 0$, then the response during the time step is inelastic and the value of $\Delta\Gamma$ is determined by requiring the yield function at the end of the time step to vanish

$$g(t_{n+1}) = 1 - \frac{\kappa(t_{n+1})}{\gamma_e(t_{n+1})} = 1 - \frac{(1+\Delta\Gamma)\kappa(t_{n+1})}{\gamma_e^*(t_{n+1})} = 0\,, \tag{5.11.97}$$

where $\kappa(t_{n+1})$ is an estimate of the value of κ at the end of the time step that must satisfy the restriction

$$\kappa(t_{n+1}) < \gamma_e^*(t_{n+1})\,. \tag{5.11.98}$$

Then, for inelastic response the solution of (5.11.97) yields

$$\Delta\Gamma = \frac{\gamma_e^*(t_{n+1})}{\kappa(t_{n+1})} - 1 > 0 \quad \text{for } g^*(t_{n+1}) > 0\,. \tag{5.11.99}$$

Although the Eulerian formulations does not introduce a measure of inelastic strain, many evolution equations for hardening are formulated in terms of an effective inelastic strain rate $\dot{\varepsilon}_p$. To help translate these evolution equations into an Eulerian formulation, with the help of (3.11.37) and (5.11.91), the evolution equation (5.11.83) suggests that the effective inelastic strain rate $\dot{\varepsilon}_p$ be defined by

$$\dot{\varepsilon}_p = \Gamma\sqrt{\frac{2}{3}\mathbf{g}_e'' \cdot \mathbf{g}_e''} = \frac{2}{3}\Gamma\,\gamma_e\,, \tag{5.11.100}$$

which can be integrated by the expression

$$\varepsilon_p(t_{n+1}) = \varepsilon_p(t_n) + \frac{2}{3}\,\Delta\Gamma\,\gamma_e(t_{n+1}) = \varepsilon_p(t_n) + \frac{2}{3}\left(\frac{\Delta\Gamma}{1+\Delta\Gamma}\right)\gamma_e^*(t_{n+1})\,, \tag{5.11.101}$$

where use has been made of (5.11.94).

Elastically Anisotropic Response

Recently Kroon and Rubin [22] developed a strongly objective, robust numerical algorithm for integrating the evolution equations (5.11.45) for the elastic dilatation J_e and for the elastic distortional deformation vectors $\mathbf{m}_i'$ as well as evolution equations for isotropic κ in (5.11.31) and directional hardening β_{ij} in (5.11.32).

$$\begin{aligned} \dot{J}_e &= J_e(\mathbf{D}\cdot\mathbf{I} - \Gamma\bar{\mathbf{D}}_p)\,, & \dot{\mathbf{m}}_i' &= (\mathbf{L}'' - \Gamma\bar{\mathbf{L}}_p'')\,\mathbf{m}_i'\,, \\ \dot{\kappa} &= \Gamma H\,, & \dot{\beta}_{ij} &= \Gamma H_{ij}\,. \end{aligned} \tag{5.11.102}$$

To present the main idea of this algorithm, consider a fully anisotropic elastic–inelastic material with a strain energy function Σ of the form

$$\Sigma = \Sigma(J_e, m'_{ij}, \kappa, \beta_{ij}), \tag{5.11.103}$$

for which the Cauchy stress is given by (5.11.48)

$$\mathbf{T} = J_e\rho\frac{\partial\Sigma}{\partial J_e}\mathbf{I} + 2\rho\frac{\partial\Sigma}{\partial m'_{ij}}\left(\mathbf{m}'_i \otimes \mathbf{m}'_j - \frac{1}{3}m'_{ij}\mathbf{I}\right). \tag{5.11.104}$$

Also, for definiteness the function Γ in the evolution equations (5.11.45) is specified by the form (5.11.80) proposed for the model with a smooth elastic–inelastic transition. Specifically, Γ is specified by

$$\Gamma = \Gamma(J_e, m'_{ij}, \kappa, \beta_{ij}, \dot{\varepsilon}), \tag{5.11.105}$$

where the effective total distortional deformation rate $\dot{\varepsilon}$ is defined in (5.11.80).

For the numerical algorithm, $\Delta\Gamma_I$ represents the Ith estimate of $\Delta t\Gamma(t_{n+1})$ evaluated at the end of the time step. The evolution equations (5.11.102) are solved for the values $J_e(t_{n+1})$ and $\mathbf{m}'_i(t_{n+1})$ at the end of the time step, which together with estimates of the hardening variables $\kappa(t_{n+1})$ and $\beta_{ij}(t_{n+1})$ are used to obtain the value $\Gamma(t_{n+1})$ of Γ at the end of the time step. Convergence of the algorithm is obtained by iterating on the value $\Delta\Gamma_I$ until the function

$$f(\Delta\Gamma_I) = \Delta\Gamma_I - \Delta t\Gamma(t_{n+1}), \quad \Delta\Gamma_I \geq 0 \tag{5.11.106}$$

is sufficiently small.

Using the relative dilatation J_r in (5.11.84), the elastic trial J_e^* of the elastic dilatation J_e satisfies equations

$$J_e^*(t) = J_r(t)J_e(t_n), \quad \dot{J}_e^* = J_e^*\mathbf{D}\cdot\mathbf{I}, \quad J_e^*(t_n) = J_e(t_n), \tag{5.11.107}$$

so the evolution equation (5.11.102) for J_e can be rewritten in the form

$$\frac{d}{dt}\left(\frac{J_e}{J_e^*}\right) = -\Gamma\bar{\mathbf{D}}_p\cdot\mathbf{I}, \tag{5.11.108}$$

which can be integrated approximately to obtain

$$J_e(t_{n+1}) = J_e^*(t_{n+1})\exp[-\Delta\Gamma_I\bar{\mathbf{D}}_p(t_{n+1})\cdot\mathbf{I}], \quad \Delta\Gamma_I = \Delta t\Gamma(t_{n+1}). \tag{5.11.109}$$

In this equation, $\bar{\mathbf{D}}_p(t_{n+1})$ is an estimate of the value of $\bar{\mathbf{D}}_p$ at the end of the time step and $\Delta\Gamma_I$ is the Ith estimate of $\Delta t\Gamma(t_{n+1})$ evaluated at the end of the time step.

Next, using (5.11.84) the elastic trial values $\mathbf{m}_i^{\prime *}$ of the elastic distortional deformation vectors $\mathbf{m}_i^{\prime}$ satisfy equations

$$\mathbf{m}_i^{\prime *} = \mathbf{F}_r^{\prime} \, \mathbf{m}_i^{\prime}(t_n) \,, \quad \dot{\mathbf{m}}_i^{\prime *} = \mathbf{L}^{\prime\prime} \, \mathbf{m}_i^{\prime *} \,, \quad \mathbf{m}_i^{\prime *}(t_n) = \mathbf{m}_i^{\prime}(t_n) \,. \tag{5.11.110}$$

Then, the evolution equation (5.11.102) for $\mathbf{m}_i^{\prime}$ is approximated by

$$\dot{\mathbf{m}}_i^{\prime} = \dot{\mathbf{m}}_i^{\prime *} - \Gamma \bar{\mathbf{L}}_p^{\prime\prime} \, \mathbf{m}_i^{\prime} \,. \tag{5.11.111}$$

Using a backward Euler approximation of the derivative, this equation integrates to obtain

$$\mathbf{m}_i^{\prime}(t_{n+1}, I) = \mathbf{A}^{*-1}(I) \mathbf{m}_i^{\prime *}(t_{n+1}) \,, \quad \mathbf{A}^*(I) = \frac{\mathbf{I} + \Delta \Gamma_I \bar{\mathbf{L}}_p^{\prime\prime *}}{[\det(\mathbf{I} + \Delta \Gamma_I \bar{\mathbf{L}}_p^{\prime\prime *})]^{1/3}} \,, \tag{5.11.112}$$

where $\mathbf{A}^*$ has been normalized to be unimodular [i.e., $\det \mathbf{A}^* = 1$] which ensures that the vectors $\mathbf{m}_i^{\prime}(t_{n+1}, I)$ satisfy the condition

$$\mathbf{m}_1^{\prime}(t_{n+1}, I) \times \mathbf{m}_2^{\prime}(t_{n+1}, I) \cdot \mathbf{m}_3^{\prime}(t_{n+1}, I) = 1 \,. \tag{5.11.113}$$

The tensor $\bar{\mathbf{L}}_p^{\prime\prime *}$ in (5.11.112) is an estimate of $\bar{\mathbf{L}}_p^{\prime\prime}$ defined by

$$\bar{\mathbf{L}}_p^{\prime\prime *} = \mathbf{\Lambda}_p - \frac{1}{3}(\mathbf{\Lambda}_p \cdot \mathbf{I}) \, \mathbf{I} \,. \tag{5.11.114}$$

For an arbitrary time step $t = t_n$ with $n > 1$, $\mathbf{\Lambda}_p$ is specified by

$$\begin{aligned} \mathbf{\Lambda}_p &= \bar{L}_p^{\prime\prime ij} \, [\mathbf{m}_i^{\prime *}(t_{n+1}) \otimes \mathbf{m}_j^{\prime *}(t_{n+1})] \,, \\ \bar{L}_p^{\prime\prime ij} &= [\bar{\mathbf{L}}_p^{\prime\prime}(t_n) \cdot \mathbf{m}^{i\prime}(t_n) \otimes \mathbf{m}^{j\prime}(t_n)] \quad \text{for} \quad n > 1 \,, \end{aligned} \tag{5.11.115}$$

where $\bar{\mathbf{L}}_p^{\prime\prime}(t_n)$ and $\mathbf{m}^{i\prime}(t_n)$ are the converged values of $\bar{\mathbf{L}}_p^{\prime\prime}$ and $\mathbf{m}^{i\prime}$ from the previous time step with the reciprocal vectors $\mathbf{m}^{i\prime}$ defined in (3.11.18). The value of $\mathbf{\Lambda}_p$ at the beginning of the integration process $t = t_1$ is specified to be a fraction of its elastic trial value

$$\mathbf{\Lambda}_p = \alpha \bar{\mathbf{L}}_p^{\prime\prime *}(t_{n+1}) \,, \quad 0 < \alpha < 1 \,, \quad \text{for} \quad n = 1 \,, \tag{5.11.116}$$

where $\bar{\mathbf{L}}_p^{\prime\prime *}(t_{n+1})$ is the value of $\bar{\mathbf{L}}_p^{\prime\prime}$ evaluated using the elastic trial values $J_e^*(t_{n+1})$ and $\mathbf{m}_i^*(t_{n+1})$ and estimates of the hardening variables $\kappa(t_{n+1})$ and $\beta_{ij}(t_{n+1})$ at the end of the time step. Also, the strongly objective average total distortional deformation rate $\tilde{\mathbf{D}}^{\prime\prime}$ developed in [41] and recorded in (3.13.20) can be used for a strongly objective approximation of $\dot{\varepsilon}$ at the end of the time step

$$\dot{\varepsilon} = \sqrt{\frac{2}{3}\tilde{\mathbf{D}}'' \cdot \tilde{\mathbf{D}}''}\,, \quad \tilde{\mathbf{D}}'' = \frac{1}{2\Delta t}\left[\mathbf{I} - \left\{\frac{3}{\mathbf{B}_r'^{-1}(t_{n+1}) \cdot \mathbf{I}}\right\} \mathbf{B}_r'^{-1}(t_{n+1})\right], \quad (5.11.117)$$

where the relative deformation $\mathbf{B}_r'$ is defined by

$$\mathbf{B}_r' = \mathbf{F}_r'\mathbf{F}_r'^T\,. \quad (5.11.118)$$

Since under SRBM the quantities J_e, $\mathbf{m}_i'$, J_r and $\mathbf{F}_r'$ transform to J_e^+, $\mathbf{m}_i'^+$, J_r^+ and $\mathbf{F}_r'^+$ according to the transformation relations (3.11.31) and (3.13.8)

$$J_e^+ = J_e\,, \quad \mathbf{m}^{i\prime+} = \mathbf{Q}\mathbf{m}^{i\prime}\,, \quad J_r^+ = J_r\,, \quad \mathbf{F}_r'^+ = \mathbf{Q}\mathbf{F}_r'\,, \quad (5.11.119)$$

it follows that the numerical estimates $J_e(t_{n+1})$ and $\mathbf{m}_i'(t_{n+1})$ transform to $J_e^+(t_{n+1})$ and $\mathbf{m}_i'^+(t_{n+1})$ under SRBM, such that

$$J_e^+(t_{n+1}) = J_e(t_{n+1})\,, \quad \mathbf{m}^{i\prime+}(t_{n+1}) = \mathbf{Q}\mathbf{m}^{i\prime}(t_{n+1})\,, \quad (5.11.120)$$

when the estimates $\kappa(t_{n+1})$ and $\beta_{ij}(t_{n+1})$ are insensitive to SRBM. This means that these numerical estimates are strongly objective since the vector and tensor estimates satisfy the same invariance transformation relations under SRBM as the exact values.

Robustness of the numerical algorithm developed in [22] was tested by taking large time steps which in one time step load the material from zero stress to a point in the inelastic range. It was found that the algorithm worked well for the constant α in (5.11.116) specified by

$$\alpha = 0.18\,. \quad (5.11.121)$$

It is emphasized that if the first time step causes elastic response, then there is no influence of the parameter α since $\Delta\Gamma_1 = 0$.

Elastically Isotropic Response to Simple Shear

With reference to fixed rectangular Cartesian base vectors $\mathbf{e}_i$, the velocity gradient $\mathbf{L}$ for simple shear can be specified by

$$\mathbf{L} = L_{12}\mathbf{e}_1 \otimes \mathbf{e}_2\,. \quad (5.11.122)$$

Using the zero-stress initial conditions (5.11.63), the solution of the evolution equation (5.11.43) for J_e requires for a metal with isochoric inelasticity (5.11.72) that

$$J_e = 1\,, \quad (5.11.123)$$

and the evolution equation (5.11.66) admits a solution for the elastic distortional deformation $\mathbf{B}_e'$ of the form

$$\mathbf{B}'_e = a\mathbf{e}_1 \otimes \mathbf{e}_1 + b\mathbf{e}_2 \otimes \mathbf{e}_2 + c\mathbf{e}_3 \otimes \mathbf{e}_3 + d(\mathbf{e}_1 \otimes \mathbf{e}_2 + \mathbf{e}_2 \otimes \mathbf{e}_1)\,,$$
$$c = \frac{1}{\sqrt{ab - d^2}}\,, \tag{5.11.124}$$

where a, b and d are functions of time determined by solving the three nontrivial scalar evolution equations associated with (5.11.66). Moreover, it was shown in [44] that for monotonic loading with

$$L_{12} = \gamma_s \Gamma > 0\,, \tag{5.11.125}$$

and constant γ_s, that these evolution equations admit a steady-state solution for which

$$a = \frac{1 + 2\gamma_s^2}{(1 + \gamma_s^2)^{1/3}}\,, \quad b = c = \frac{1}{(1 + \gamma_s^2)^{1/3}}\,, \quad d = \frac{\gamma_s}{(1 + \gamma_s^2)^{1/3}}\,. \tag{5.11.126}$$

Also, the steady-state values of $\dot{\varepsilon}$ in (5.11.80) and γ_e in (5.11.91) are given by

$$\dot{\varepsilon} = \frac{\gamma_s \Gamma}{\sqrt{3}}\,, \quad \gamma_e = \frac{\gamma_s \sqrt{3 + 4\gamma_s^2}}{2(1 + \gamma_s^2)^{1/3}}\,. \tag{5.11.127}$$

For simplicity, consider the case when the yield function is specified by (5.11.92)

$$g = 1 - \frac{\kappa}{\gamma_e}\,, \tag{5.11.128}$$

with the hardening variable κ being constant. It then follows that for standard rate-independent inelasticity, the loading conditions (5.11.79) require $g = 0$ during inelastic loading, which determines the steady-state value of γ_s by the solution of equation

$$\kappa = \frac{\gamma_s \sqrt{3 + 4\gamma_s^2}}{2(1 + \gamma_s^2)^{1/3}}\,. \tag{5.11.129}$$

Alternatively, for the simple rate-independent smooth elastic–inelastic transition model (5.11.80) characterized by

$$\Gamma = b_1 \dot{\varepsilon} \langle g \rangle\,, \tag{5.11.130}$$

the steady-state value of γ_s is determined by equation

$$\kappa = \frac{(b_1 \gamma_s - \sqrt{3})\sqrt{3 + 4\gamma_s^2}}{2b_1(1 + \gamma_s^2)^{1/3}} \quad \text{for} \quad b_1 \gamma_s > \sqrt{3}\,. \tag{5.11.131}$$

Using the Neo-Hookean model, the pressure vanishes and deviatoric stress is given by (5.11.69). To examine the influence of the constant b_1 on the solution of

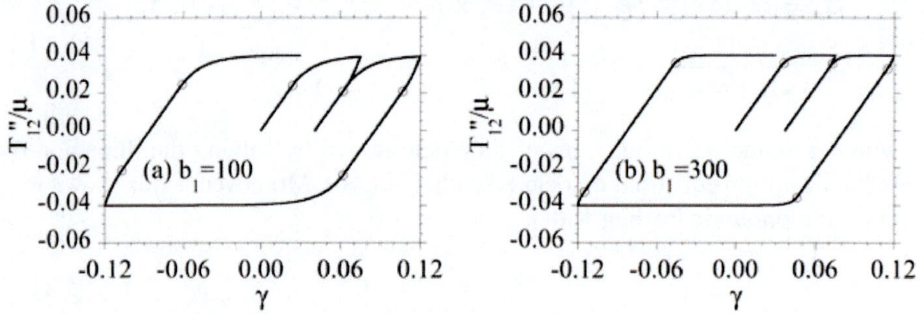

Fig. 5.4 Cyclic loading of the smooth-transition model in simple shear. Plots of the shear stress T''_{12} versus the total shear strain γ for different material constants: **a** $b_1 = 100$, $\kappa = 0.019670$ and **b** $b_1 = 300$, $\kappa = 0.029675$, which produce the same steady-state value of shear stress. The symbols indicate the locations of the elastic–inelastic transitions

the smooth model, it is convenient to determine values of the pair of constants b_1, κ which yield the same steady-state value of the shear stress. Specifically, as a special case, the steady-state value of γ_s is determined by solving equation

$$\frac{T''_{12}}{\mu} = \frac{\gamma_s}{(1+\gamma_s^2)^{1/3}} = 0.04\,, \tag{5.11.132}$$

and (5.11.131) is used to determine the values

$$\begin{aligned} \kappa &= 0.019670 \quad \text{for} \quad b_1 = 100\,, \\ \kappa &= 0.029675 \quad \text{for} \quad b_1 = 300\,. \end{aligned} \tag{5.11.133}$$

Figure 5.4 shows the transient solution of shear stress T''_{12} versus total shear strain γ determined by integrating the evolution equation

$$\dot{\gamma} = L_{12}\,, \tag{5.11.134}$$

subject to the initial condition $\gamma(0) = 0$. The symbols in Fig. 5.4a, b indicate the locations of the elastic–inelastic transitions for cyclic simple shear loading. Figure 5.4a shows that for $b_1 = 100$, the response exhibits significant overstress with inelasticity continuing to occur during the onset of unloading. Even though the hardening parameter κ is constant, the model exhibits effective hardening due to the overstress. Figure 5.4b shows that for $b_1 = 300$ the effects of the overstress are significantly reduced. In this regard, it is noted that in the limit that $b_1 \to \infty$, the smooth model yields standard rate-independent response with the yield function $g = 0$ during inelastic loading.

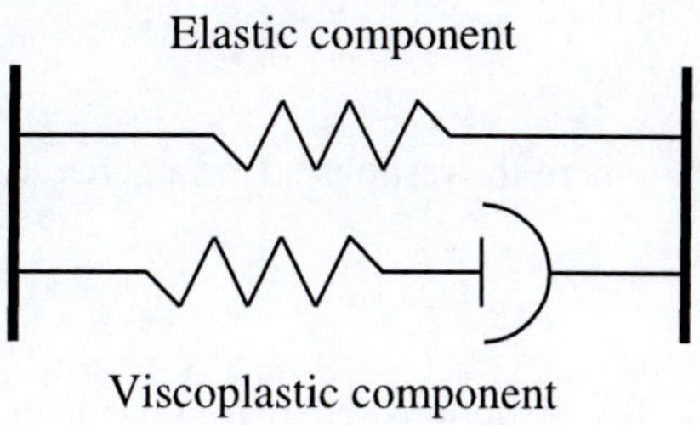

Fig. 5.5 Sketch of a nonlinear Maxwell model with an elastic component in parallel with a viscoplastic component

5.12 Viscoelastic Response

A simple generalized nonlinear Maxwell model (see Fig. 5.5) for viscoelastic response can be proposed by adding the response of a viscoplastic material to that of a general elastic material. To model dilatational dissipation of viscoplastic component, it is necessary to enhance the model described in Sect. 5.11. To this end, the viscoplastic component is modeled by the elastic dilatation $J_v > 0$ and the unimodular elastic distortional deformation tensor $\mathbf{B}'_e$ which satisfy the evolution equations

$$\frac{\dot{J}_v}{J_v} = \mathbf{D} \cdot \mathbf{I} - \Gamma_v \ln(J_v) \,,$$
$$\dot{\mathbf{B}}'_e = \mathbf{L}\mathbf{B}'_e + \mathbf{B}'_e\mathbf{L}^T - \frac{2}{3}(\mathbf{D} \cdot \mathbf{I})\mathbf{B}'_e - \Gamma \mathbf{A}_p \,, \quad \mathbf{A}_p = \mathbf{B}'_e - \left(\frac{3}{\mathbf{B}'^{-1}_e \cdot \mathbf{I}}\right)\mathbf{I} \,,$$
$$\Gamma_v > 0 \,, \quad \Gamma > 0 \,, \tag{5.12.1}$$

where Γ_v and Γ are positive constants that, respectively, control the time-dependent relaxation of J_v toward unity and $\mathbf{B}'_e$ toward the unity tensor $\mathbf{I}$. The functional form of the evolution equation for J_v is motivated by the work in [39] which introduced a modified evolution equation for a cardiac muscle that simplified the numerical integration algorithm. Also, the first invariant of $\mathbf{B}'_e$ satisfies equations

$$\alpha_1 = \mathbf{B}'_e \cdot \mathbf{I} \,, \quad \dot{\alpha}_1 = 2\mathbf{B}''_e \cdot \mathbf{D} - \Gamma \mathbf{A}_p \,, \tag{5.12.2}$$

where $\mathbf{B}''_e$ is the deviatoric part of $\mathbf{B}'_e$.

Now, the strain energy function of the viscoelastic material is specified in the form

$$\Sigma = \hat{\Sigma} + \overset{v}{\Sigma}(J_v, \alpha_1) \,, \quad \rho_z \overset{v}{\Sigma} = K_v[J_v - 1 - \ln(J_v)] + \frac{1}{2}\mu_v(\alpha_1 - 3) \,, \tag{5.12.3}$$

where ρ_z is a constant density not necessarily equal to a zero-stress density, $\hat{\Sigma}$ characterizes the strain energy of a general nonlinear elastic material and $\overset{v}{\Sigma}$ characterizes the strain energy of the viscoplastic component, with K_v being the positive elastic bulk modulus and μ_v being the positive shear modulus of the viscoplastic component.

For this model the Cauchy stress $\mathbf{T}$ is proposed in the form

$$\mathbf{T} = \hat{\mathbf{T}} + \overset{v}{\mathbf{T}}, \tag{5.12.4}$$

where the response $\hat{\mathbf{T}}$ of a general nonlinear elastic material satisfies equation

$$\hat{\mathbf{T}} \cdot \mathbf{D} = \rho \dot{\hat{\Sigma}}, \tag{5.12.5}$$

for all motions. Then, for this material, the rate of material dissipation (4.5.7) requires

$$\mathcal{D} = \overset{v}{\mathbf{T}} \cdot \mathbf{D} - \rho \dot{\overset{v}{\Sigma}} \geq 0. \tag{5.12.6}$$

Specifying $\overset{v}{\mathbf{T}}$ by

$$\begin{aligned} &\overset{v}{\mathbf{T}} = -\overset{v}{p}\mathbf{I} + \overset{v}{\mathbf{T}}{}'', \quad \overset{v}{p} = -\rho J_v \frac{\partial \overset{v}{\Sigma}}{\partial J_v} = \left(\frac{\rho}{\rho_z}\right) K_v (1 - J_v), \\ &\overset{v}{\mathbf{T}}{}'' = 2\rho \frac{\partial \overset{v}{\Sigma}}{\partial \alpha_1} \mathbf{B}_e'' = \left(\frac{\rho}{\rho_z}\right) \mu_v \mathbf{B}_e'', \end{aligned} \tag{5.12.7}$$

the rate of material dissipation requires

$$\mathcal{D} = -\Gamma_v \overset{v}{p} \ln(J_v) + \frac{1}{2}\Gamma \left(\frac{\rho}{\rho_z}\right) \mu_v \mathbf{A}_p \cdot \mathbf{I} \geq 0, \tag{5.12.8}$$

which in view of the constitutive equation (5.12.7) for the pressure $\overset{v}{p}$ and (5.11.67) is automatically satisfied. Since Γ_v and Γ are both positive, dissipation continues until $J_v = 1$ and $\mathbf{B}_e' = \mathbf{I}$ with $\mathbf{T} = \hat{\mathbf{T}}$.

If the elastic part of the response is isotropic, then the strain energy is given by (5.8.5) and the stress $\hat{\mathbf{T}}$ is given by (5.8.10). Alternatively, if the elastic part of the response is anisotropic, then the strain energy is given by (5.3.9) and the stress $\hat{\mathbf{T}}$ is given by (5.3.11). For either case, this model proposes elastically isotropic viscoplastic dissipation.

A robust, strongly objective numerical integration algorithm for the evolution equation for $\mathbf{B}_e'$ was discussed in Sect. 5.11. To develop a robust, strongly objective numerical integration algorithm for the evolution equation (5.12.1) for the elastic dilatation J_v, consider the time interval $t_n \leq t \leq t_{n+1}$ with time increment $\Delta t = t_{n+1} - t_n$ and recall that the relative dilatation J_r satisfies the evolution equation (5.11.84) and initial condition

$$\dot{J}_r = J_r \mathbf{D} \cdot \mathbf{I}, \quad J_r(t_n) = 1. \tag{5.12.9}$$

Thus, $(5.12.1)_1$ can be rewritten in the form

$$\frac{d}{dt}\ln\left[\frac{J_v}{J_r J_v(t_n)}\right] = -\Gamma_v \ln(J_v)\,. \qquad (5.12.10)$$

Next, using a backward Euler approximation of the derivative yields equation

$$\ln\left[\frac{J_v}{J_r J_v(t_n)}\right] = -\Delta t \Gamma \ln(J_v)\,, \qquad (5.12.11)$$

which can be solved to obtain

$$J_v(t_{n+1}) = [J_r(t_{n+1}) J_v(t_n)]^{1/(1+\Delta t\,\Gamma_v)}\,. \qquad (5.12.12)$$

5.13 Crystal Plasticity

Crystal plasticity models (e.g., [1, 17]) identify a finite number N of slip planes in the crystal with unit normals ${}_I\mathbf{n}$ and unit slip directions ${}_I\mathbf{s}$ in the slip planes. In addition, a constitutive equation for the inelastic rate $\mathbf{L}_p$ is proposed in the form

$$\mathbf{L}_p = \sum_{I=1}^{N} {}_I\Gamma\; {}_I\mathbf{s} \otimes {}_I\,\mathbf{n}\,, \quad {}_I\mathbf{s} \cdot {}_I\,\mathbf{n} = 0\,, \qquad (5.13.1)$$

where ${}_I\Gamma$ characterizes the inelastic rate on the Ith slip plane, which typically is a function of history-dependent variables. This form for $\mathbf{L}_p$ includes all slip rates on all of the slip planes and is applicable to metal plasticity with no inelastic dilatation rate

$$\mathbf{L}_p \cdot \mathbf{I} = \mathbf{D}_p \cdot \mathbf{I} = 0\,. \qquad (5.13.2)$$

Within the context of the Eulerian formulation with evolution equations (5.11.30) for the microstructural vectors

$$\dot{\mathbf{m}}_i = (\mathbf{L} - \mathbf{L}_p)\mathbf{m}_i\,, \qquad (5.13.3)$$

the microstructural vectors can be used to characterize the deformation and orientation of the average crystal lattice. Moreover, the elastic distortional deformation microstructural vectors $\mathbf{m}_i'$ satisfy the evolution equations (5.11.45)

$$\dot{\mathbf{m}}_i' = (\mathbf{L}'' - \mathbf{L}_p'')\mathbf{m}_i'\,, \qquad (5.13.4)$$

where $\mathbf{L}''$ is the deviatoric part of the velocity gradient and $\mathbf{L}_p''$ is the deviatoric part of the inelastic rate $\mathbf{L}_p$. Since the elastic distortional microstructural vectors $\mathbf{m}_i'$ can be used to model the crystal, the values of ${}_In_i$ of the unit normals ${}_I\mathbf{n}$ to slip systems and ${}_Is^i$ of the unit vectors ${}_I\mathbf{s}$ in the slip directions in a zero-stress state (with $\mathbf{m}_i' = \mathbf{m}^{i\prime}$

being orthonormal vectors) can be constants for general stress states with

$$
{}_I\mathbf{n} = \frac{{}_I n_i \mathbf{m}^{i'}}{|{}_I n_j \mathbf{m}^{j'}|}, \quad {}_I\mathbf{s} = \frac{{}_I s^i \mathbf{m}'_i}{|{}_I s^j \mathbf{m}'_j|}, \qquad (5.13.5)
$$
$$
{}_I n_i {}_I n_i = 1, \quad {}_I s^i {}_I s^i = 1, \quad {}_I n_i {}_I s^i = 0,
$$

where there is no sum on the repeated capital index I.

This formulation will be properly invariant under Superposed Rigid Body Motions SRBM if ${}_I\Gamma$ are uninfluenced by SRBM

$$
{}_I\Gamma^+ = {}_I\Gamma. \qquad (5.13.6)
$$

If ${}_I\Gamma$ are determined by consistency conditions for standard rate-independent yield functions, then the active slip systems may not be determined uniquely. However, if ${}_I\Gamma$ are determined by functions similar to those (5.11.80) of the smooth elastic–inelastic transition model developed in [18, 19], then loading and unloading conditions are not needed and all slip systems are simultaneously active even for rate-independent response. Examples for standard small strain formulations of crystal plasticity which has been modified to use the smooth elastic–inelastic transition model can be found in [13, 15].

References

1. Asaro RJ (1983) Micromechanics of crystals and polycrystals. Adv Appl Mech 23:1–115
2. Barenblatt GI, Joseph DD (2013) Collected papers of RS Rivlin: volume I and II. Springer Science & Business Media, New York
3. Bernstein B (1960) Hypo-elasticity and elasticity. Arch Rational Mech Anal 6:89–104
4. Besseling JF (1968) A thermodynamic approach to rheology. Irreversible aspects of continuum mechanics and transfer of physical characteristics in moving fluids, pp 16–53
5. Besseling JF, Van Der Giessen E (1994) Mathematical modeling of inelastic deformation. CRC Press, Boca Raton
6. Bilby BA, Gardner LRT, Stroh AN (1957) Continuous distributions of dislocations and the theory of plasticity. In: Proceedings of the 9th international congress of applied mechanics, vol 9. University de Brussels, pp 35–44
7. Bodner SR (1968) Constitutive equations for dynamic material behavior. Mechanical behavior of materials under dynamic loads. Springer, Berlin, pp 176–190
8. Bodner SR (1987) Review of a unified elastic-viscoplastic theory. Unified constitutive equations for creep and plasticity, pp 273–301
9. Bodner SR (2002) Unified plasticity for engineering applications. Mathematical concepts and methods in science and engineering, vol 47. Kluwer, New York
10. Bodner SR, Partom Y (1972) A large deformation elastic-viscoplastic analysis of a thick-walled spherical shell. J Appl Mech 39:751–757
11. Bodner SR, Partom Y (1975) Constitutive equations for elastic-viscoplastic strain-hardening materials. J Appl Mech 42:385–389
12. Eckart C (1948) The thermodynamics of irreversible processes. IV. The theory of elasticity and anelasticity. Phys Rev 73:373–382

13. Farooq H, Cailletaud G, Forest S, Ryckelynck D (2020) Crystal plasticity modeling of the cyclic behavior of polycrystalline aggregates under non-symmetric uniaxial loading: global and local analyses. Int J Plast 126:102619
14. Flory PJ (1961) Thermodynamic relations for high elastic materials. Trans Faraday Soc 57:829–838
15. Forest S, Rubin MB (2016) A rate-independent crystal plasticity model with a smooth elastic-plastic transition and no slip indeterminacy. Eur J Mech-A/Solids 55:278–288
16. Green AE, Naghdi PM (1965) A general theory of an elastic-plastic continuum. Arch Rational Mech Anal 18:251–281
17. Hill R (1966) Generalized constitutive relations for incremental deformation of metal crystals by multislip. Arch Rational Mech Anal 14:95–102
18. Hollenstein M, Jabareen M, Rubin MB (2013) Modeling a smooth elastic-inelastic transition with a strongly objective numerical integrator needing no iteration. Comput Mech 52:649–667
19. Hollenstein M, Jabareen M, Rubin MB (2015) Erratum to: modeling a smooth elastic-inelastic transition with a strongly objective numerical integrator needing no iteration. Comput Mech 55:453–453
20. Jabareen M (2015) Strongly objective numerical implementation and generalization of a unified large inelastic deformation model with a smooth elastic-inelastic transition. Int J Eng Sci 96:46–67
21. Kröner E (1959) General continuum theory of dislocations and intrinsic stresses. Arch Rational Mech Anal 4:273–334
22. Kroon M, Rubin MB (2020) A strongly objective, robust integration algorithm for Eulerian evolution equations modeling general anisotropic elastic-inelastic material response. Finite Elem Anal Des 177:103422
23. Lee EH (1969) Elastic-plastic deformation at finite strains. J Appl Mech 36:1–6
24. Leonov AI (1976) Nonequilibrium thermodynamics and rheology of viscoelastic polymer media. Rheol Acta 15:85–98
25. Malvern LE (1951) The propagation of longitudinal waves of plastic deformation in a bar of material exhibiting a strain-rate effect. J Appl Mech 18:203–208
26. Mandel J (1973) Equations constitutives et directeurs dans les milieux plastiques et viscoplastiques. Int J Solids Struct 9:725–740
27. Moss WC (1984) On the computational significance of the strain space formulation of plasticity theory. Int J Numer Methods Eng 20:1703–1709
28. Naghdi PM (1960) Stress-strain relations in plasticity and thermoplasticity. In: Plasticity: proceedings of the second symposium on naval structural mechanics, pp 121–169
29. Naghdi PM (1990) A critical review of the state of finite plasticity. Zeitschrift für angewandte Mathematik und Physik ZAMP 41:315–394
30. Naghdi PM, Trapp JA (1975) The significance of formulating plasticity theory with reference to loading surfaces in strain space. Int J Eng Sci 13:785–797
31. Onat ET (1968) The notion of state and its implications in thermodynamics of inelastic solids. Irreversible aspects of continuum mechanics and transfer of physical characteristics in moving fluids, pp 292–314
32. Papes O (2013) Nonlinear continuum mechanics in modern engineering applications. PhD dissertation DISS ETH NO 19956
33. Perzyna P (1963) The constitutive equations for rate sensitive plastic materials. Q Appl Math 20:321–332
34. Rubin MB (1987) An elastic-viscoplastic model exhibiting continuity of solid and fluid states. Int J Eng Sci 25:1175–1191
35. Rubin MB (1994) Plasticity theory formulated in terms of physically based microstructural variables - Part I. Theory. Int J Solids Struct 31:2615–2634
36. Rubin MB (1996) On the treatment of elastic deformation in finite elastic-viscoplastic theory. Int J Plast 12:951–965
37. Rubin MB (2001) Physical reasons for abandoning plastic deformation measures in plasticity and viscoplasticity theory. Arch Mech 53:519–539

38. Rubin MB (2012) Removal of unphysical arbitrariness in constitutive equations for elastically anisotropic nonlinear elastic-viscoplastic solids. Int J Eng Sci 53:38–45
39. Rubin MB (2016) A viscoplastic model for the active component in cardiac muscle. Biomech Model Mechanobiol 15:965–982
40. Rubin MB (2019) An Eulerian formulation of inelasticity - from metal plasticity to growth of biological tissues. Trans R Soc A 377:20180071
41. Rubin MB (2020) A strongly objective expression for the average deformation rate with application to numerical integration algorithms. Finite Elem Anal Des 175:103409
42. Rubin MB, Attia AV (1996) Calculation of hyperelastic response of finitely deformed elastic-viscoplastic materials. Int J Numer Methods Eng 39:309–320
43. Rubin MB, Cardiff P (2017) Advantages of formulating an evolution equation directly for elastic distortional deformation in finite deformation plasticity. Comput Mech 60:703–707
44. Rubin MB, Papes O (2011) Advantages of formulating evolution equations for elastic-viscoplastic materials in terms of the velocity gradient instead of the spin tensor. J Mech Mater Struct 6:529–543
45. Schröder J, Neff P (2003) Invariant formulation of hyperelastic transverse isotropy based on polyconvex free energy functions. Int J Solids Struct 40:401–445
46. Simo JC (1988) A framework for finite strain elastoplasticity based on maximum plastic dissipation and the multiplicative decomposition. Part II: computational aspects. Comput Methods Appl Mech Eng 68:1–31
47. Truesdell C, Noll W (2004) The non-linear field theories of mechanics. Springer, Berlin, pp 1–579
48. Wilkins ML (1963) Calculation of elastic-plastic flow. Technical report, UCRL7322, California University Livermore Radiation Lab

Chapter 6
Thermomechanical Theory

Abstract The objective of this chapter is to present the balance laws for the thermomechanical theory. Specifically, the balances of entropy and energy are presented and different forms of second law of thermodynamics are discussed. Invariance under Superposed Rigid Body Motions (SRBM) is considered for the new thermal quantities and thermal constraints on material response are discussed. In addition, specific nonlinear constitutive equations are presented for a number of materials modeling: thermoelastic, thermoelastic–inelastic and porous responses. Also, constitutive equations for growth of thermoelastic–inelastic biological tissues are presented.

6.1 Thermomechanical Processes

A thermomechanical process is characterized by its velocity field $\mathbf{v}$ and its absolute temperature field θ

$$\mathbf{v} = \mathbf{v}(\mathbf{x}, t)\,, \quad \theta = \theta(\mathbf{x}, t)\,, \tag{6.1.1}$$

the position of a material point $\mathbf{x}$ is determined by integrating the equation

$$\dot{\mathbf{x}} = \mathbf{v}(\mathbf{x}, t)\,, \tag{6.1.2}$$

and the velocity gradient $\mathbf{L}$, rate of deformation tensor $\mathbf{D}$ and temperature gradient $\mathbf{g}$ are defined by

$$\mathbf{L} = \partial\mathbf{v}/\partial\mathbf{x}\,, \quad \mathbf{D} = \frac{1}{2}(\mathbf{L} + \mathbf{L}^T)\,, \quad \mathbf{g} = \frac{\partial\theta}{\partial\mathbf{x}}\,. \tag{6.1.3}$$

These quantities are defined at every material point in the material region P and on its closed boundary ∂P.

M. B. Rubin, *Continuum Mechanics with Eulerian Formulations of Constitutive Equations*, Solid Mechanics and Its Applications 265,
https://doi.org/10.1007/978-3-030-57776-6_6

Within the context of the thermomechanical theory proposed by Green and Naghdi [7, 8], in addition to the current mass density ρ, the specific (per unit mass) body force $\mathbf{b}$, the Cauchy stress $\mathbf{T}$, the unit outward normal $\mathbf{n}$ to ∂P and the traction vector $\mathbf{t} = \mathbf{Tn}$ per unit current area on ∂P, it is necessary to introduce the specific entropy η, the specific external rate of entropy supply s, the specific internal rate of entropy production ξ, the specific internal energy ε, the specific external rate of energy supply r on P and the entropy flux $\mathbf{p}$ and energy flux $\mathbf{q}$ vectors, both per unit present area on ∂P.

The external fields

$$\mathbf{b}\,,\; s\,, \tag{6.1.4}$$

need to be specified and constitutive equations must be provided for the response functions

$$\mathbf{T}\,,\; \eta\,,\; \xi\,,\; \mathbf{p}\,,\; \varepsilon\,, \tag{6.1.5}$$

with r and $\mathbf{q}$ determined by

$$r = \theta s\,, \quad \mathbf{q} = \theta \mathbf{p}\,. \tag{6.1.6}$$

6.2 Balance Laws for the Thermomechanical Theory

Within the context of the thermomechanical theory proposed by Green and Naghdi [7, 8] the current mass density ρ, the current position $\mathbf{x}$ of a material point and the absolute temperature θ are determined by the global forms of the conservation of mass and the balances of linear momentum and entropy

$$\begin{aligned}
&\frac{d}{dt}\int_P \rho dv = 0 \\
&\frac{d}{dt}\int_P \rho \mathbf{v} dv = \int_P \rho \mathbf{b} dv + \int_{\partial P} \mathbf{t} da\,, \\
&\frac{d}{dt}\int_P \rho \eta dv = \int_P \rho (s + \xi) dv - \int_{\partial P} \mathbf{p}\cdot\mathbf{n} da\,.
\end{aligned} \tag{6.2.1}$$

The minus sign appears before the integral over the entropy flux because $\mathbf{p}\cdot\mathbf{n}$ is the rate of entropy expelled by the body through its surface. The global form of the balance of angular momentum is given by

$$\frac{d}{dt}\int_P (\mathbf{x}\times\rho\mathbf{v}) dv = \int_P (\mathbf{x}\times\rho\mathbf{b}) dv + \int_{\partial P} \mathbf{x}\times\mathbf{t} da\,, \tag{6.2.2}$$

and the balance of energy (i.e., the first law of thermodynamics) takes the form

$$\dot{\mathcal{E}} + \dot{\mathcal{K}} = \mathcal{W} + \mathcal{H}\,. \tag{6.2.3}$$

In words, the first law of thermodynamics states that the rate of change of the total internal energy $\mathcal{E}$ plus the rate of change of the total kinetic energy $\mathcal{K}$ is balanced by the total rate of work $\mathcal{W}$ done on the body and total rate of heat $\mathcal{H}$ supplied to the body, which indicates the equivalence of thermal and mechanical supplies of energy. Specifically, these quantities are defined by

$$\begin{aligned} &\mathcal{E} = \int_P \rho\varepsilon dv\,, && \mathcal{K} = \int_P \tfrac{1}{2}\rho\mathbf{v}\cdot\mathbf{v}dv\,, \\ &\mathcal{W} = \int_P \rho\mathbf{b}\cdot\mathbf{v}dv + \int_{\partial P}\mathbf{t}\cdot\mathbf{v}da\,, && \mathcal{H} = \int_P \rho\theta s dv - \int_{\partial P}\theta\mathbf{p}\cdot\mathbf{n}da\,. \end{aligned} \tag{6.2.4}$$

Using standard continuity conditions, the local forms of the conservation of mass and balances of linear momentum and entropy are given by

$$\dot{\rho} + \rho\mathbf{D}\cdot\mathbf{I} = 0\,, \quad \rho\dot{\mathbf{v}} = \rho\mathbf{b} + \mathrm{div}\mathbf{T}\,, \quad \rho\dot{\eta} = \rho(s+\xi) - \mathrm{div}\mathbf{p}\,. \tag{6.2.5}$$

Also, using these balance laws, the reduced local form of the balance of angular momentum requires the Cauchy stress $\mathbf{T}$ to be symmetric

$$\mathbf{T}^T = \mathbf{T}\,, \tag{6.2.6}$$

and the reduced local form of the balance of energy requires

$$\rho\dot{\varepsilon} = \rho\theta s - \mathrm{div}(\theta\mathbf{p}) + \mathbf{T}\cdot\mathbf{D}\,. \tag{6.2.7}$$

Next, multiplying the balance of entropy in (6.2.5) by θ and using the expressions (6.1.6) it can be shown that

$$\rho\theta s - \mathrm{div}(\theta\mathbf{p}) = \rho\theta\dot{\eta} - \rho\theta\xi - \mathbf{p}\cdot\mathbf{g}\,. \tag{6.2.8}$$

Also, the internal rate of entropy production is separated into two parts [17]: a thermal part $-\mathbf{p}\cdot\mathbf{g}$ due to heat conduction and another part $\rho\theta\xi'$ due to the rate of material dissipation

$$\rho\theta\xi = -\mathbf{p}\cdot\mathbf{g} + \rho\theta\xi'\,, \tag{6.2.9}$$

so that the external rate of energy supply can be written in the form

$$\rho\theta s - \mathrm{div}(\theta\mathbf{p}) = -\rho\theta\xi' + \rho\theta\dot{\eta}\,. \tag{6.2.10}$$

In addition, the specific Helmholtz free energy ψ is defined by

$$\psi = \varepsilon - \theta\eta\,, \tag{6.2.11}$$

and the balance of energy (6.2.7) yields a constitutive equation for the rate of material dissipation $\rho\theta\xi'$

$$\rho\theta\xi' = \mathbf{T}\cdot\mathbf{D} - \rho(\dot{\psi} + \eta\dot{\theta})\,, \tag{6.2.12}$$

where use has been made of (6.2.10).

In this formulation of thermomechanics, ρ, $\mathbf{x}$ and θ are determined by the conservation of mass and the balances of linear momentum and entropy (6.2.5) and the balances of angular momentum and energy place restrictions on the constitutive equations which ensure that they are identically satisfied for all thermomechanical processes. Specifically, the reduced form of the balance of angular momentum (6.2.6) requires the Cauchy stress $\mathbf{T}$ to be symmetric and the reduced form of the balance of energy (6.2.12) determines a constitutive equation for the rate of material dissipation $\rho\theta\xi'$.

6.3 Second Laws of Thermomechanics

Observations indicate that thermomechanical processes progress is specific directions. For example, consider a body which is isolated with no rates of external work and heat supply

$$\begin{aligned} \mathbf{b} \cdot \mathbf{v} = 0\,, \quad r = 0 \text{ on } P\,, \\ \mathbf{t} \cdot \mathbf{v} = 0\,, \quad \mathbf{q} \cdot \mathbf{n} = 0 \text{ on } \partial P\,. \end{aligned} \tag{6.3.1}$$

Then, the global form (6.2.3) of the first law of thermodynamics indicates that an isolated body preserves total energy

$$\mathcal{E} + \mathcal{K} = \mathcal{E}(0) + \mathcal{K}(0) = \text{constant}\,. \tag{6.3.2}$$

Next, consider a body that is made from a homogeneous material which is in a zero-stress uniform material state at rest. In the absence of external forces and with no heat supply through its boundary, the body is heated by an external rate of energy supply to obtain an inhomogeneous temperature field in the body at rest. Then, in the absence of external forces and further external heat supply, the total energy would remain constant even if part of the body became hotter and another part of it became colder. However, observations indicate that this does not happen naturally. Instead, the body tends to reach a uniform temperature. Notions of entropy model the observed directions of thermomechanical processes.

Clausius–Duhem Inequality

In the classical approach to continuum thermomechanics proposed by Coleman and Noll [5], the conservation of mass and the balances of linear momentum, angular momentum and energy are supplimented by the Clausius–Duhem inequality

$$\frac{d}{dt}\int_P \rho\eta dv - \int_P \frac{\rho r}{\theta} dv + \int_{\partial P} \frac{\mathbf{q} \cdot \mathbf{n}}{\theta} da \geq 0\,, \tag{6.3.3}$$

which is a statement of the second law of thermodynamics that thermomechanical processes cause the internal rate of entropy production to have a tendency to increase.

Using standard continuity conditions, the local form of the conservation of mass and the definitions (6.1.3) and (6.1.6), the local form of (6.3.3) requires

$$\rho\theta\dot{\eta} - \rho\theta s + \theta \text{div}\mathbf{p} \geq 0 \,. \tag{6.3.4}$$

Moreover, with the help of the balance of energy (6.2.7) and the definition (6.2.11) of the Helmholtz free energy ψ, the Clausius–Duhem inequality requires

$$\mathbf{T}\cdot\mathbf{D} - \rho(\dot{\psi} + \eta\dot{\theta}) - \mathbf{p}\cdot\mathbf{g} \geq 0 \,, \tag{6.3.5}$$

which places restrictions on constitutive equations.

Green–Naghdi Formulation
In the classical approach to thermomechanics, the Clausius–Duhem inequalty (6.3.5) is a single statement of the second law of thermodynamics that places restrictions on the constitutive equations. In contrast, the Green and Naghdi formulation places restrictions on the constitutive equations by requiring the reduced forms of the balance of angular momentum (6.2.6) and the balance of energy (6.2.7) to be satisfied identically, without any statement of the second law of thermodynamics.

To compare the two approaches to thermomechanics, use is made of the separation (6.2.9) and the constitutive Eq. (6.2.12) to rewrite the Clausius–Duhem inequality (6.3.4) in the form

$$\rho\theta\xi = -\mathbf{p}\cdot\mathbf{g} + \rho\theta\xi' > 0 \,, \tag{6.3.6}$$

which requires the total internal rate of entropy production to be non-negative. However, the Green–Naghdi formulation allows for proposing different statements of the second law of thermodynmaics, as was discussed in [8].

Heat Flows From Hot to Cold Regions
One statement of the second law of thermodynamics is that heat flows from hot to cold regions

$$-\mathbf{p}\cdot\mathbf{g} > 0 \text{ for } \mathbf{g} \neq 0 \,. \tag{6.3.7}$$

This indicates that the thermal part of the internal rate of entropy production in the separation (6.2.9) is non-negative.

Rate of Material Dissipation
To motivate a second statement of the second law of thermodynamics, it is noted from (6.1.6) and (6.2.10) that the rate of heat expelled by the body is given by

$$-(\rho r - \text{div}\mathbf{q}) = -\big[\rho\theta s - \text{div}(\theta\mathbf{p})\big] = \rho\theta\xi' - \rho\theta\dot{\eta} \,. \tag{6.3.8}$$

For general thermomechanical processes heat can be supplied or expelled. However, the notions of friction and viscous effects in fluids indicate that the rate of material dissipation causes a tendency for heat to be expelled by the body. Noting that positive values of $\rho\theta\xi'$ cause a tendency for heat to be expelled by the body, this second statement of the second law of thermodynamics requires the rate of material dissipation to be non-negative

$$\rho\theta\xi' = \mathbf{T}\cdot\mathbf{D} - \rho(\dot{\psi} + \eta\dot{\theta}) \geq 0\,. \tag{6.3.9}$$

Although the two statements (6.3.7) and (6.3.9) combined are consistent with the Clausius–Duhem inequality (6.3.6), this latter single statement of the second law of thermodynamics does not demand that (6.3.7) and (6.3.9) be satisfied individually, as in the Green–Naghdi formulation.

6.4 Invariance Under Superposed Rigid Body Motions (SRBM)

Although temperature θ is not a kinematic variable, it is an independent variable like the position vector $\mathbf{x}$ which needs to be determined by the balance laws, boundary and initial conditions. Consequently, in addition to the kinematic conditions (3.8.13) and (3.8.16)

$$\mathbf{x}^+ = \mathbf{c}(t) + \mathbf{Q}(t)\mathbf{x}\,, \quad \mathbf{Q}\mathbf{Q}^T = \mathbf{I}\,, \quad \det\mathbf{Q} = +1\,, \tag{6.4.1}$$

it is proposed that θ remains unaltered under SRBM

$$\theta^+ = \theta\,. \tag{6.4.2}$$

This means that the temperature gradient $\mathbf{g}$ transforms to $\mathbf{g}^+$, such that

$$\mathbf{g} = \frac{\partial\theta}{\partial\mathbf{x}}\,, \quad \mathbf{g}^+ = \frac{\partial\theta^+}{\partial\mathbf{x}^+} = \mathbf{g}\,(\partial\mathbf{x}/\partial\mathbf{x}^+) = \mathbf{g}\mathbf{Q}^T = \mathbf{Q}\mathbf{g}\,. \tag{6.4.3}$$

Section 4.7 introduced the notion of invariance under SRBM which is based on the two restrictions

(R-1): The balance laws must be form-invariant under SRBM. (6.4.4a)

(R-2): The constitutive response of the material relative to its orientation is the same for all SRBM. (6.4.4b)

The first restriction (R-1) in (6.4.4a) requires the global forms of the balance laws to be form-invariant in the superposed configuration P^+ with all independent and kinetic quantities taking their superposed values in P^+. Using the transformation relations (4.7.21)

$$\rho^+ = \rho\,, \quad \mathbf{T}^+ = \mathbf{Q}\mathbf{T}\mathbf{Q}^T\,, \quad \mathbf{b}^+ = \dot{\mathbf{v}}^+ + \mathbf{Q}(\mathbf{b} - \dot{\mathbf{v}})\,, \tag{6.4.5}$$

the conservation of mass, the balance of linear momentum (6.2.1) and the balance of angular momentum (6.2.2) are already form-invariant under SRBM. Consequently, within the context of the thermomechanical theory, the physical restriction (R-1) in (6.4.4a) requires the balance of entropy in (6.2.1) and the balance of energy (6.2.3) to remain form-invariant under SRBM

$$\frac{d}{dt}\int_{P^+} \rho^+\eta^+ dv^+ = \int_{P^+} \rho^+(s^+ + \xi^+)dv^+ - \int_{\partial P^+} \mathbf{p}^+ \cdot \mathbf{n}^+ da^+ , \tag{6.4.6a}$$

$$\dot{\mathcal{E}}^+ + \dot{\mathcal{K}}^+ = \mathcal{W}^+ + \mathcal{H}^+ , \tag{6.4.6b}$$

with the specifications

$$\begin{aligned}
\mathcal{E}^+ &= \int_{P^+} \rho^+\varepsilon^+ dv^+ , \quad \mathcal{K}^+ = \int_{P^+} \frac{1}{2}\rho^+\mathbf{v}^+ \cdot \mathbf{v}^+ dv^+ , \\
\mathcal{W}^+ &= \int_{P^+} \rho^+\mathbf{b}^+ \cdot \mathbf{v}^+ dv^+ + \int_{\partial P^+} \mathbf{t}^+ \cdot \mathbf{v}^+ da^+ , \\
\mathcal{H}^+ &= \int_{P^+} \rho^+\theta^+ s^+ dv^+ - \int_{\partial P^+} \theta^+\mathbf{p}^+ \cdot \mathbf{n}^+ da^+ .
\end{aligned} \tag{6.4.7}$$

Using standard continuity arguments the local form the of the balance of entropy (6.4.6a) requires

$$(\dot{\rho}^+ + \rho^+\mathbf{D}^+ \cdot \mathbf{I})\, \eta^+ + \rho^+\dot{\eta}^+ = \rho^+(s^+ + \xi^+) - \mathrm{div}^+\mathbf{p}^+ , \tag{6.4.8}$$

and the local form of the balance of energy (6.4.6b) requires

$$\begin{aligned}
(\dot{\rho}^+ + \rho^+\mathbf{D}^+ \cdot \mathbf{I})\left(\varepsilon^+ + \frac{1}{2}\rho^+\mathbf{v}^+ \cdot \mathbf{v}^+\right) + (\rho^+\dot{\mathbf{v}}^+ - \rho^+\mathbf{b}^+ - \mathrm{div}^+\mathbf{T}^+) \cdot \dot{\mathbf{v}}^+ + \rho^+\dot{\varepsilon}^+ \\
= \mathbf{T}^+ \cdot \mathbf{L}^+ + \rho^+\theta^+ s^+ - \mathrm{div}^+(\theta^+\mathbf{p}^+) .
\end{aligned} \tag{6.4.9}$$

Then, using form-invariance of the local forms of the conservation of mass and the balances of linear and angular momentum, the local form of the balance of entropy requires

$$\rho^+\dot{\eta}^+ = \rho^+(s^+ + \xi^+) - \mathrm{div}^+\mathbf{p}^+ , \tag{6.4.10}$$

and the local form of the balance of energy requires

$$\rho^+\dot{\varepsilon}^+ = \mathbf{T}^+ \cdot \mathbf{D}^+ + \rho^+\theta^+ s^+ - \mathrm{div}^+(\theta^+\mathbf{p}^+) . \tag{6.4.11}$$

Now, with the help of (2.5.4) and (6.4.1) it can be shown that

$$\begin{aligned}
\mathrm{div}^+\mathbf{p}^+ &= (\partial\mathbf{p}^+/\partial\mathbf{x}^+) \cdot \mathbf{I} = (\partial\mathbf{p}^+/\partial\mathbf{x})(\partial\mathbf{x}/\partial\mathbf{x}^+) \cdot \mathbf{I} = (\partial\mathbf{p}^+/\partial\mathbf{x}) \cdot \mathbf{Q} , \\
\mathrm{div}^+\mathbf{p}^+ &= \mathrm{div}(\mathbf{Q}^T\mathbf{p}^+) .
\end{aligned} \tag{6.4.12}$$

Next, using the invariance of ρ in (6.4.5), the balance of entropy (6.4.10) can be solved for s^+ to obtain

$$s^+ = \dot{\eta}^+ - \xi^+ + \frac{1}{\rho}\mathrm{div}(\mathbf{Q}^T\mathbf{p}^+)\,. \tag{6.4.13}$$

Moreover, using the local balance of entropy in (6.2.5) it can be shown that

$$s^+ = s + (\dot{\eta}^+ - \dot{\eta}) - (\xi^+ - \xi) + \frac{1}{\rho}\mathrm{div}(\mathbf{Q}^T\mathbf{p}^+ - \mathbf{p})\,. \tag{6.4.14}$$

In this regard, it is noted that the restriction (R-1) tacitly assumes that the balance of entropy is valid for any specified external rate of entropy supply. Consequently, it is also valid for (6.4.14), which enforces SRBM.

To compete the restrictions for invariance under SRBM it is necessary to determine expressions for

$$\eta^+\,,\xi^+\,,\varepsilon^+\,,\mathbf{p}^+\,. \tag{6.4.15}$$

This requires use of the physical restriction (R-2) (6.4.4b) and recognition that in addition to the stress $\mathbf{T}$, the quantities η, ξ, ε and $\mathbf{p}$ characterize the material response for thermomechanical processes. This means that η, ξ and ε, which do not depend on the orientation of the material, must be uninfluenced by SRBM

$$\eta^+ = \eta\,, \quad \xi^+ = \xi\,, \quad \varepsilon^+ = \varepsilon\,. \tag{6.4.16}$$

Moreover, the response due to the entropy flux vector relative to the orientation of the material will be the same if the restriction

$$\mathbf{p}^+ \cdot \mathbf{n}^+ = \mathbf{p} \cdot \mathbf{n} \tag{6.4.17}$$

is valid for all material points, all unit normals $\mathbf{n}$ and all SRBM. Now, using the kinematic result (3.8.20) that $\mathbf{n}$ rotates under SRBM

$$\mathbf{n}^+ = \mathbf{Q}\mathbf{n}\,, \tag{6.4.18}$$

the expression (6.4.17) for the entropy flux vector can be rewritten in the form

$$(\mathbf{p}^+ - \mathbf{Q}\mathbf{p}) \cdot \mathbf{n}^+ = 0\,. \tag{6.4.19}$$

Then, since the coefficient of $\mathbf{n}^+$ in this equation is independent of $\mathbf{n}^+$, and $\mathbf{n}^+$ is an arbitrary unit vector, the entropy flux vector $\mathbf{p}$ and heat flux vector $\mathbf{q}$ defined in (6.1.6) must satisfy the transformation relations

$$\mathbf{p}^+ = \mathbf{Q}\mathbf{p}\,, \quad \mathbf{q}^+ = \mathbf{Q}\mathbf{q}\,. \tag{6.4.20}$$

This means that with the help of (6.4.16), the restriction (6.4.14) requires the external rate of entropy supply s and the external rate of heat supply r defined in (6.1.6) to be unaffected by SRBM

$$s^+ = s\,, \quad r^+ = r\,. \tag{6.4.21}$$

In summary, under superposed rigid body motions SRBM the thermomechanical quantities θ, $\mathbf{g}$, η, ε, ψ, s, r, ξ, ξ', $\mathbf{p}$ and $\mathbf{q}$ transform to θ^+, $\mathbf{g}^+$, η^+, ε^+, ψ^+, s^+, r^+, ξ^+, ξ'^+, $\mathbf{p}^+$ and $\mathbf{q}^+$, such that

$$\begin{aligned} &\theta^+ = \theta\,, \;\; \mathbf{g}^+ = \mathbf{Q}\mathbf{g}\,, \;\; \eta^+ = \eta \;\;\; \varepsilon^+ = \varepsilon\,, \;\;\; \psi^+ = \psi\,, \\ &s^+ = s\,, \;\; r^+ = r\,, \;\;\; \xi^+ = \xi\,, \;\; \xi'^+ = \xi'\,, \;\; \mathbf{p}^+ = \mathbf{Q}\mathbf{p}\,, \;\; \mathbf{q}^+ = \mathbf{Q}\mathbf{q}\,. \end{aligned} \tag{6.4.22}$$

6.5 Thermal Constraints

In general, it is possible to propose coupled thermomechanical constraints but such coupled constraints make it difficult to satisfy the forms (6.3.7) and (6.3.9) of the second law of thermodynamics individually. For this reason, this section considers thermal constraints which are independent of the kinematic constraints considered in Sect. 5.7. In this regard, it is noted that since the constraint response $\bar{\mathbf{T}}$ in (5.7.13) is workless (5.7.11)

$$\mathbf{T} \cdot \mathbf{D} = \hat{\mathbf{T}} \cdot \mathbf{D}\,, \tag{6.5.1}$$

so the constraint response makes no contribution to the rate of material dissipation in the second law of thermodynamics (6.3.9).

As a physical example of a thermal constraint, consider a material that has fibers in one direction that allow for very rapid heat conduction relative to the surrounding matrix material. For this case, the temperature gradient $\mathbf{g}$ in the direction of the fibers will be very small relative to the temperature gradient in directions perpendicular to the fibers due to slow conduction through the matrix material only. Motivated by this simple example, consider a thermal constraint which constrains the temperature gradient in the direction $\boldsymbol{\gamma}$ of the form

$$\boldsymbol{\gamma} \cdot \mathbf{g} = 0\,, \tag{6.5.2}$$

where $\boldsymbol{\gamma}$ is a vector that is independent of $\mathbf{g}$ and which under SRBM satisfies the transformation relation

$$\boldsymbol{\gamma}^+ = \mathbf{Q}\boldsymbol{\gamma}\,. \tag{6.5.3}$$

Moreover, consider a general unconstrained material that is characterized by a constitutive equation $\hat{\mathbf{p}}$ for the entropy flux $\mathbf{p}$. Next, consider a model of a constrained material for which $\mathbf{p}$ is additively separated into the constitutive part $\hat{\mathbf{p}}$ and a part $\bar{\mathbf{p}}$, called the constraint response, which enforces the thermal constraint (6.5.2)

$$\mathbf{p} = \hat{\mathbf{p}} + \bar{\mathbf{p}}\,. \tag{6.5.4}$$

Although $\hat{\mathbf{p}}$ characterizes the response to general temperature gradients, its value in (6.5.4) is determined by evaluating $\hat{\mathbf{p}}$ only for temperature gradients that satisfy the imposed thermal constraint. Moreover, $\hat{\mathbf{p}}$ automatically transforms under SRBM, such that

$$\hat{\mathbf{p}}^+ = \mathbf{Q}\hat{\mathbf{p}}\,. \tag{6.5.5}$$

Now, $\bar{\mathbf{p}}$ is assumed to satisfy the restriction

$$\bar{\mathbf{p}} \cdot \mathbf{g} = 0\,, \tag{6.5.6}$$

and to be independent of the rate $\mathbf{g}$.

Next, multiplying (6.5.2) by an arbitrary scalar γ and subtracting the result from (6.5.6) yields

$$(\bar{\mathbf{p}} - \gamma\,\boldsymbol{\gamma}) \cdot \mathbf{g} = 0\,. \tag{6.5.7}$$

Since $\boldsymbol{\gamma}$ is nonzero it is possible to specify γ by the equation

$$\gamma = \frac{\bar{\mathbf{p}} \cdot \boldsymbol{\gamma}}{|\boldsymbol{\gamma} \cdot \boldsymbol{\gamma}|}\,. \tag{6.5.8}$$

Then, the only nonzero components in (6.5.7) are perpendicular to $\boldsymbol{\gamma}$. Since this equation must hold for arbitrary temperature gradients $\mathbf{g}$ that satisfy the constraint (6.5.2) and the coefficient of $\mathbf{g}$ is independent of $\mathbf{g}$, it follows that the constraint response $\bar{\mathbf{p}}$ must be given by

$$\bar{\mathbf{p}} = \gamma\,\boldsymbol{\gamma}\,, \tag{6.5.9}$$

with γ being an arbitrary function of $\mathbf{x}$ and t that is determined by the balance laws and boundary conditions. Since the restriction (R-2) in (4.7.3b), which defines how the constitutive response of the material relative to its orientation is the same for all SRBM, requires $\mathbf{p}$ to satisfy the transformation relation (6.4.20) and since $\hat{\mathbf{p}}$ satisfies the transformaton relation (6.5.5), it follows from (6.5.4) that the constraint response $\bar{\mathbf{p}}$ satisfies the transformation relation

$$\bar{\mathbf{p}}^+ = \gamma^+\,\boldsymbol{\gamma}^+ = \mathbf{Q}\bar{\mathbf{p}} = \gamma\,\mathbf{Q}\boldsymbol{\gamma} \tag{6.5.10}$$

for all SRBM. Then, with the help of (6.5.3) it can be shown that the arbitrary function γ must be unaffected by SRBM

$$\gamma^+ = \gamma\,. \tag{6.5.11}$$

In addition, since the constraint response $\bar{\mathbf{p}}$ satisfies the restriction (6.5.6), it follows that

$$\mathbf{p} \cdot \mathbf{g} = \hat{\mathbf{p}} \cdot \mathbf{g}\,, \tag{6.5.12}$$

so the constraint response does not influence the restriction (6.3.7) of the second law of thermodynamics which requires heat to flow from hot to cold regions.

Furthermore, it is noted that up to two independent thermal constraints of the type (6.5.2) can be imposed simultaneously without causing $\mathbf{p}$ to be totally indeterminate.

6.6 Thermoelastic Materials

A thermoelastic solid is a special ideal material which is non-dissipative in the sense that the rate of material dissipation (6.2.12) vanishes

$$\rho\theta\xi' = \mathbf{T}\cdot\mathbf{D} - \rho(\dot{\psi} + \eta\dot{\theta}) = 0 \tag{6.6.1}$$

for all thermomechanical processes.

Within the context of the Eulerian formulation of constitutive equations, the microstructural vectors $\mathbf{m}_i$ and elastic metric m_{ij} described in Sect. 3.11 are determined by the evolution equations

$$\dot{\mathbf{m}}_i = \mathbf{L}\mathbf{m}_i\,, \quad m_{ij} = \mathbf{m}_i\cdot\mathbf{m}_j\,, \quad \dot{m}_{ij} = 2\mathbf{m}_i\otimes\mathbf{m}_j\cdot\mathbf{D}\,. \tag{6.6.2}$$

Moreover, for a thermoelastic material, the response functions ψ, η and $\mathbf{T}$ are specified in the forms

$$\psi = \hat{\psi}(m_{ij},\theta)\,, \quad \eta = \hat{\eta}(m_{ij},\theta)\,, \quad \mathbf{T} = \hat{\mathbf{T}}(\mathbf{m}_i,\theta)\,, \tag{6.6.3}$$

so the condition (6.6.1) requires

$$\left(\mathbf{T} - 2\rho\frac{\partial\hat{\psi}}{\partial m_{ij}}\mathbf{m}_i\otimes\mathbf{m}_j\right)\cdot\mathbf{D} - \rho(\frac{\partial\hat{\psi}}{\partial\theta} + \eta)\,\dot{\theta} = 0\,. \tag{6.6.4}$$

Since the coefficients of $\mathbf{D}$ and $\dot{\theta}$ are independent of these rates, and the coefficient of $\mathbf{D}$ is symmetric, $\mathbf{T}$ and η must be determined by the constitutive equations

$$\mathbf{T} = \hat{\mathbf{T}} = 2\rho\frac{\partial\hat{\psi}}{\partial m_{ij}}\mathbf{m}_i\otimes\mathbf{m}_j\,, \quad \eta = \hat{\eta} = -\frac{\partial\hat{\psi}}{\partial\theta}\,. \tag{6.6.5}$$

This expression for the Cauchy stress $\mathbf{T}$ automatically satisfies the restriction (6.2.6) due to angular momentum. Moreover, the entropy flux $\mathbf{p}$ takes the form

$$\mathbf{p} = p^i\mathbf{m}_i\,, \quad p^i = \hat{p}^i(m_{ij},\theta,g_i)\,, \quad g_i = \mathbf{g}\cdot\mathbf{m}_i\,, \tag{6.6.6}$$

which must satisfy the restriction (6.3.7) due to the second law of thermodynamics,

$$-\hat{\mathbf{p}} \cdot \mathbf{g} = -\hat{p}^i g_i > 0 \text{ for } \mathbf{g} \neq 0\,. \tag{6.6.7}$$

These functional forms are automatically properly invariant under SRBM. Also, using (6.2.11), it follows that the internal energy ε for a thermoelastic material is given by

$$\varepsilon = \hat{\varepsilon}(m_{ij}, \theta) = \hat{\psi}(m_{ij}, \theta) + \theta\hat{\eta}(m_{ij}, \theta)\,. \tag{6.6.8}$$

Furthermore, these constitutive equations are restricted so that the material is in a zero-stress material state whenever the elastic deformation metric $m_{ij} = \delta_{ij}$ and the temperature equals the reference zero-stress temperature $\theta = \theta_z$

$$\mathbf{T} = 0 \text{ whenever } m_{ij} = \delta_{ij} \text{ and } \theta = \theta_z\,. \tag{6.6.9}$$

Rate-Dependent Response

Although the evolution Eq. (6.6.2) for $\mathbf{m}_i$ are homogeneous equations of order one in time and thus predict rate-independent response, and the response functions (6.6.3) and (6.6.6) are explicitly independent of the rates $\mathbf{D}$ and $\dot{\theta}$, the response of a thermoelastic material is rate-dependent. This is because the balance of entropy in $(6.2.5)_3$ predicts time-dependent response of the temperature for transient processes.

Path-Independent Response

Since $\mathbf{m}_i$ arematerial line elements, it follows that the values of the response functions ψ, η, $\mathbf{T}$ and $\mathbf{p}$ at a specified state characterized by $\mathbf{m}_i$ and θ are independent of the path of the thermomechanical process that attains this state. This also means that for any thermomechanical process that starts at the state

$$\mathbf{m}_i(\mathbf{x}, t_1), \theta(\mathbf{x}, t_1), \mathbf{v}(\mathbf{x}, t_1)\,, \tag{6.6.10}$$

and ends at the state

$$\mathbf{m}_i(\mathbf{x}, t_2), \theta(\mathbf{x}, t_2), \mathbf{v}(\mathbf{x}, t_2)\,, \tag{6.6.11}$$

the changes in internal and kinetic energies

$$\Delta\mathcal{E} = \mathcal{E}(t_2) - \mathcal{E}(t_1)\,, \quad \Delta\mathcal{K} = \mathcal{K}(t_2) - \mathcal{K}(t_1) \tag{6.6.12}$$

are independent of the path of the thermodynamic process. Moreover, with the help of the first law of thermodynamics (6.2.3), it follows that the total work done on the body plus the total heat supplied to the body during this process is also independent of the path of the process

$$\int_{t=t_1}^{t=t_2} (\mathcal{W} + \mathcal{H})dt = \Delta\mathcal{E} + \Delta\mathcal{K}\,. \tag{6.6.13}$$

In addition, the total work done on the body plus the total heat supplied to the body vanishes for any cyclic process which starts and ends at the same state $\mathbf{m}_i$, θ and $\mathbf{v}$.

An Irreversible Process

Although a thermoelastic material is an ideal material with no material dissipation, it can experience an irreversible process. For the present discussion it is assumed that

$$\hat{\varepsilon} \to \infty \text{ whenever } \hat{\eta} \to \infty\,. \tag{6.6.14}$$

Now consider a cantilever beam made from a homogeneous thermoelastic material. The external body force and external rate of heat supply both vanish $\mathbf{b} = 0$ and $r = 0$. Also, the velocity field on the clamped boundary vanishes, all other boundaries are traction free $\mathbf{t} = 0$ and all boundaries are insulated from heat flow $\mathbf{q} \cdot \mathbf{n} = 0$. For this problem the total rate of work $\mathcal{W}$ done on the body vanishes and the total rate of heat supply $\mathcal{H}$ vanishes. Also, consider the case when the body is initially in a zero-stress material state at constant density ρ_z and constant zero-stress reference temperature θ_z, but it has an initial velocity field with a non-uniform rate of deformation so that

$$\begin{aligned} &m_{ij}(\mathbf{x},0) = \delta_{ij}\,, \quad \theta(\mathbf{x},0) = \theta_z\,, \quad \partial\mathbf{D}(\mathbf{x},0)/\partial\mathbf{x} \neq 0\,, \\ &\mathbf{T}(\mathbf{x},0) = 0\,, \qquad \eta(\mathbf{x},0) = 0\,, \quad \varepsilon(\mathbf{x},0) = 0\,. \end{aligned} \tag{6.6.15}$$

Consequently, from (6.2.3) the sum of the total internal and kinetic energies remains constant

$$\mathcal{E} + \mathcal{K} = \mathcal{K}(0)\,, \tag{6.6.16}$$

where $\mathcal{K}(0)$ is the initial value of the total kinetic energy in the beam.

The global form of the balance of entropy in $(6.2.1)_3$ and the restrictions (6.3.1), (6.3.6), (6.3.7) and (6.6.1) require

$$\frac{d}{dt}\int_P \rho\eta dv = -\int_P (\frac{\mathbf{p}\cdot\mathbf{g}}{\theta})dv \geq 0\,. \tag{6.6.17}$$

Moreover, the restriction (6.3.7) causes the total entropy to increase until the temperature becomes uniform with $\mathbf{g} = 0$. In particular, a non-uniform rate of deformation causes local temperature changes with a nonzero temperature gradient $\mathbf{g}$. However, due to assumption (6.6.14) and the result (6.6.16), the entropy cannot continue to increase without bound. This means that eventually the temperature must become uniform with $\mathbf{g} = 0$, the velocity field must go to zero and the elastic deformation metric m_{ij} must become independent of time, but it can be nonuniform due the clamped boundary. Since energy is preserved, the final values of $\mathcal{E}$ and $\mathcal{K}$ are constants given by

$$\mathcal{E}(\infty) = \mathcal{K}(0)\,, \quad \mathcal{K}(\infty) = 0\,. \tag{6.6.18}$$

This example shows the importance of the entropy in a thermomechanical process. In particular, since the entropy flux must satisfy the restriction (6.3.7), it follows that the process is thermodynamically irreversible even though the thermoelastic material is non-dissipative.

A Separation of Elastic Dilatation and Distortional Deformations

To introduce separate control over the response of the material to dilatation and distortional rates of deformation, it is convenient to use the work of Flory [6] and use the elastic dilatation J_e defined in (3.11.7), the elastic distortional deformation vectors $\mathbf{m}'_i$ defined in (3.11.14) and the elastic distortional deformation metric m'_{ij} defined in (3.11.16), which satisfy the evolution Eqs. (3.11.13), (3.11.15) and (3.11.17)

$$\begin{aligned} &J_e = \mathbf{m}_1 \times \mathbf{m}_2 \cdot \mathbf{m}_3 > 0\,, && \dot{J}_e = J_e\mathbf{D}\cdot\mathbf{I}\,, \\ &\mathbf{m}'_i = J_e^{-1/3}\,\mathbf{m}_i\,, && \dot{\mathbf{m}}'_i = \mathbf{L}''\,\mathbf{m}'_i\,, \\ &m'_{ij} = \mathbf{m}'_i\cdot\mathbf{m}'_j = m'_{ji}\,, && \dot{m}'_{ij} = 2(\mathbf{m}'_i\otimes\mathbf{m}'_j - \tfrac{1}{3}m'_{ij}\mathbf{I})\cdot\mathbf{D}\,, \end{aligned} \tag{6.6.19}$$

where $\mathbf{L}''$ is the deviatoric part of $\mathbf{L}$. Moreover, since there is no inelastic volume change for a thermoelastic material, the elastic dilatation J_e can be expressed in the form (4.1.16)

$$J_e = \frac{\rho_z}{\rho}\,, \tag{6.6.20}$$

where ρ_z is the constant zero-stress mass density at zero-stress reference temperature θ_z.

Then, for a thermoelastic material, the response functions ψ, η and $\mathbf{T}$ are specified in the forms

$$\psi = \tilde{\psi}(J_e, m'_{ij}, \theta)\,, \quad \eta = \tilde{\eta}(J_e, m'_{ij}, \theta)\,, \quad \mathbf{T} = \tilde{\mathbf{T}}(J_e, \mathbf{m}'_i, \theta)\,, \tag{6.6.21}$$

so the condition (6.6.1) requires

$$\left[\mathbf{T} - \rho J_e \frac{\partial\tilde{\psi}}{\partial J_e}\mathbf{I} - 2\rho\frac{\partial\tilde{\psi}}{\partial m'_{ij}}(\mathbf{m}'_i\otimes\mathbf{m}'_j - \frac{1}{3}m'_{ij}\mathbf{I})\right]\cdot\mathbf{D} - \rho(\frac{\partial\tilde{\psi}}{\partial\theta} + \eta)\dot{\theta} = 0\,. \tag{6.6.22}$$

Since the coefficients of $\mathbf{D}$ and $\dot{\theta}$ are independent of these rates, and the coefficient of $\mathbf{D}$ is symmetric, $\mathbf{T}$ and η must be determined by the constitutive equations

$$\begin{aligned} &\mathbf{T} = \tilde{\mathbf{T}} = -p\,\mathbf{I} + \mathbf{T}''\,, && p = \tilde{p} = -\rho_z\frac{\partial\tilde{\psi}}{\partial J_e}\,, \\ &\mathbf{T}'' = \tilde{\mathbf{T}}'' = 2J_e^{-1}\rho_z\frac{\partial\tilde{\psi}}{\partial m'_{ij}}(\mathbf{m}'_i\otimes\mathbf{m}'_j - \tfrac{1}{3}m'_{ij}\mathbf{I})\,, && \eta = \tilde{\eta} = -\frac{\partial\tilde{\psi}}{\partial\theta}\,, \end{aligned} \tag{6.6.23}$$

where use has been made of (6.6.20). Moreover, the entropy flux $\mathbf{p}$ takes the form

$$\mathbf{p} = p^{i\prime}\mathbf{m}'_i\,, \quad p^{i\prime} = \tilde{p}^{i\prime}(J_e, m'_{ij}, \theta, g'_i)\,, \quad g'_i = \mathbf{g}\cdot\mathbf{m}'_i\,, \tag{6.6.24}$$

which must satisfy the restriction (6.3.7) due to the second law of thermodynamics,

$$-\mathbf{p}\cdot\mathbf{g} = -\tilde{p}^{i\prime}g'_i > 0 \text{ for } \mathbf{g} \neq 0\,. \tag{6.6.25}$$

An Elastically Isotropic Thermoelastic Material
If the material is elastically isotropic, then the elastic dilatation J_e and the symmetric, positive-definite, unimodular elastic distortional deformation tensor $\mathbf{B}'_e$ defined by (3.11.19) satisfy the evolution Eq. (3.11.30)

$$\dot{J}_e = J_e \mathbf{D} \cdot \mathbf{I} , \quad \dot{\mathbf{B}}'_e = \dot{\mathbf{B}}'_e = \mathbf{L}''\mathbf{B}'_e + \mathbf{B}'_e\mathbf{L}''^T , \tag{6.6.26}$$

where $\mathbf{L}''$ is the deviatoric part of the velocity gradient $\mathbf{L}$. Since $\mathbf{B}'_e$ is unimodular, it has only two independent non-trivial invariants α_1 and α_2 defined in (5.8.4)

$$\alpha_1 = \mathbf{B}'_e \cdot \mathbf{I} , \quad \alpha_2 = \mathbf{B}'_e \cdot \mathbf{B}'_e , \tag{6.6.27}$$

which satisfy the evolution Eq. (5.8.7)

$$\dot{\alpha}_1 = 2\mathbf{B}''_e \cdot \mathbf{D} , \quad \dot{\alpha}_2 = 4\left(\mathbf{B}'^2_e - \frac{1}{3}\alpha_2\mathbf{I}\right) \cdot \mathbf{D} , \tag{6.6.28}$$

where $\mathbf{B}''_e$ is the deviatoric part of $\mathbf{B}'_e$. Moreover, the evolution equation for elastic dilatation can be integrated to obtain (6.6.20).

For an elastically isotropic thermoelastic material, the response functions ψ, η and $\mathbf{T}$ are specified in the forms

$$\psi = \hat{\psi}(J_e, \alpha_1, \alpha_2, \theta) , \quad \eta = \hat{\eta}(J_e, \alpha_1, \alpha_2, \theta) , \quad \mathbf{T} = \hat{\mathbf{T}}(\mathbf{B}'_e, \theta) , \tag{6.6.29}$$

so the condition (6.6.1) requires

$$\begin{aligned} &\left[\mathbf{T} - \rho_z \frac{\partial \hat{\psi}}{\partial J_e}\mathbf{I} - 2\rho_z J_e^{-1}\frac{\partial \hat{\psi}}{\partial \alpha_1}\mathbf{B}''_e - 4\rho_z J_e^{-1}\frac{\partial \hat{\psi}}{\partial \alpha_2}\left(\mathbf{B}'^2_e - \frac{1}{3}\alpha_2\,\mathbf{I}\right)\right] \cdot \mathbf{D} \\ &\quad - \rho\left(\frac{\partial \hat{\psi}}{\partial \theta} + \eta\right)\dot{\theta} = 0 , \end{aligned} \tag{6.6.30}$$

for all thermomechanical processes. Since the coefficients of $\mathbf{D}$ and $\dot{\theta}$ are independent of these rates, and the coefficient of $\mathbf{D}$ is symmetric, $\mathbf{T}$ and η must be determined by the constitutive equations

$$\begin{aligned} &\mathbf{T} = -p\,\mathbf{I} + \mathbf{T}'' , \quad p = \hat{p} = -\rho_z \frac{\partial \hat{\psi}}{\partial J_e} , \\ &\mathbf{T}'' = \hat{\mathbf{T}}'' = 2\rho_z J_e^{-1}\frac{\partial \hat{\psi}}{\partial \alpha_1}\mathbf{B}''_e + 4\rho_z J_e^{-1}\frac{\partial \hat{\psi}}{\partial \alpha_1}\left(\mathbf{B}'^2_e - \frac{1}{3}\alpha_2\,\mathbf{I}\right) , \\ &\eta = \hat{\eta} = -\frac{\partial \hat{\psi}}{\partial \theta} . \end{aligned} \tag{6.6.31}$$

Also, for isotropic response the entropy flux vector can be specified by a generalized Fourier form

$$\mathbf{p} = -\frac{\bar{\kappa}}{\theta}\mathbf{g}\,, \quad \bar{\kappa} = \bar{\kappa}(J_e, \alpha_1, \alpha_2, \theta) \geq 0\,, \tag{6.6.32}$$

where the heat conduction coefficient $\bar{\kappa}$ should not be confused with the hardening variable κ defined for inelastic response.

6.7 Thermoelastic–Inelastic Materials

For elastically anisotropic thermoelastic–inelastic materials, the microstructural vectors $\mathbf{m}_i$ are determined by integrating the evolution Eq. (5.11.30)

$$\dot{\mathbf{m}}_i = (\mathbf{L} - \mathbf{L}_p)\,\mathbf{m}_i\,, \quad \mathbf{L}_p = \Gamma\bar{\mathbf{L}}_p\,, \quad \Gamma \geq 0\,, \tag{6.7.1}$$

where $\bar{\mathbf{L}}_p$ controls the direction of the inelastic rate $\mathbf{L}_p$ and Γ is a non-negative function that controls its magnitude. In general, $\bar{\mathbf{L}}_p$ has a symmetric part $\bar{\mathbf{D}}_p$ that controls the direction of inelastic deformation rate and a skew-symmetric part $\bar{\mathbf{W}}_p$ that controls the direction of inelastic spin defined by

$$\bar{\mathbf{L}}_p = \bar{\mathbf{D}}_p + \bar{\mathbf{W}}_p\,, \quad \bar{\mathbf{D}}_p = \frac{1}{2}(\bar{\mathbf{L}}_p + \bar{\mathbf{L}}_p^T)\,, \quad \bar{\mathbf{W}}_p = \frac{1}{2}(\bar{\mathbf{L}}_p - \bar{\mathbf{L}}_p^T)\,, \tag{6.7.2}$$

both of which require constitutive equations. Also, for isotropic hardening, the hardening κ is determined by the evolution Eq. (5.11.31)

$$\dot{\kappa} = \Gamma H\,, \tag{6.7.3}$$

where H is a function that controls the rate of hardening. More general directional hardening can be modeled by introducing directional hardening variables $\beta_{ij} = \beta_{ji}$ which satisfy the evolution Eq. (5.11.32)

$$\dot{\beta}_{ij} = \Gamma H_{ij}\,, \tag{6.7.4}$$

where $H_{ij} = H_{ji}$ are functions that control the relative magnitudes of β_{ij}. In addition, the elastic deformation metric m_{ij} defined in (5.11.34) satisfies the evolution equations

$$m_{ij} = \mathbf{m}_i \cdot \mathbf{m}_j\,, \quad \dot{m}_{ij} = 2(\mathbf{D} - \Gamma\bar{\mathbf{D}}_p) \cdot (\mathbf{m}_i \otimes \mathbf{m}_j)\,. \tag{6.7.5}$$

Now, for an anisotropic thermoelastic–inelastic material, the response functions ψ, η and $\mathbf{T}$ are specified in the forms

$$\psi = \hat{\psi}(m_{ij}, \theta, \kappa, \beta_{ij}) , \quad \eta = \hat{\eta}(m_{ij}, \theta, \kappa, \beta_{ij}) , \quad \mathbf{T} = \hat{\mathbf{T}}(\mathbf{m}_i, \theta, \kappa, \beta_{ij}) , \quad (6.7.6)$$

so the rate of material dissipation (6.3.9) requires

$$\begin{aligned} \rho\theta\xi' = &\left(\mathbf{T} - 2\rho\frac{\partial\hat{\psi}}{\partial m_{ij}}\mathbf{m}_i \otimes \mathbf{m}_j\right) \cdot \mathbf{D} - \rho\left(\frac{\partial\hat{\psi}}{\partial\theta} + \eta\right)\dot{\theta} , \\ &+ \Gamma\left[2\rho\frac{\partial\hat{\psi}}{\partial m_{ij}}\mathbf{m}_i \otimes \mathbf{m}_j \cdot \bar{\mathbf{D}}_p - \rho\frac{\partial\hat{\psi}}{\partial\kappa}H - \rho\frac{\partial\hat{\psi}}{\partial\beta_{ij}}H_{ij}\right] \geq 0 . \end{aligned} \quad (6.7.7)$$

In general, without specifying details of the functional forms for Γ, $\bar{\mathbf{D}}_p$, H and H_{ij}, it is not possible to determine necessary restrictions on the constitutive equations for $\mathbf{T}$ and η. However, motivated by necessary restrictions for a rate-independent elastic–inelastic material with a yield function, the constitutive equations for $\mathbf{T}$ and η are specified by

$$\mathbf{T} = \hat{\mathbf{T}} = 2\rho\frac{\partial\hat{\psi}}{\partial m_{ij}}\mathbf{m}_i \otimes \mathbf{m}_j , \quad \eta = \hat{\eta} = -\frac{\partial\hat{\psi}}{\partial\theta} , \quad (6.7.8)$$

so the rate of material dissipation imposes the restriction

$$\rho\theta\xi' = \Gamma\left[\mathbf{T} \cdot \bar{\mathbf{D}}_p - \rho\frac{\partial\hat{\psi}}{\partial\kappa}H - \rho\frac{\partial\hat{\psi}}{\partial\beta_{ij}}H_{ij}\right] \geq 0 . \quad (6.7.9)$$

Also, the constitutive equation for stress is assumed to be restricted so that a zero-stress material state at zero-stress reference temperature θ_z [also called a Reference Lattice State (RLS)] is characterized by

$$\mathbf{T} = 0 , \quad \frac{\partial\hat{\psi}}{\partial m_{ij}} = 0 \quad \text{for} \quad m_{ij} = \delta_{ij} , \quad \theta = \theta_z , \quad (6.7.10)$$

where δ_{ij} is the Kronecker delta. This means that the triad $\mathbf{m}_i$ has been defined so that $\mathbf{m}_i$ are orthonormal vectors in a zero-stress material state at zero-stress reference temperature $\theta = \theta_z$. In addition, the entropy flux $\mathbf{p}$ can be specified in the form

$$\mathbf{p} = p^i\mathbf{m}_i , \quad p^i = \hat{p}^i(m_{ij}, \theta, \kappa, g_i) , \quad g_i = \mathbf{g} \cdot \mathbf{m}_i , \quad (6.7.11)$$

which must satisfy the restriction (6.3.7) due to the second law of thermodynamics,

$$-\hat{\mathbf{p}} \cdot \mathbf{g} = -\hat{p}^i g_i > 0 \text{ for } \mathbf{g} \neq 0 . \quad (6.7.12)$$

The evolution Eq. (6.7.1) for $\mathbf{m}_i$, (6.7.29) for κ and (6.7.4) for β_{ij} require initial conditions

$$\mathbf{m}_i(0), \kappa(0), \beta_{ij}(0) . \quad (6.7.13)$$

A Separation of Elastic Dilatation and Distortional Deformations
To introduce separate control over the response of the material to dilatation and distortional rates of deformation, it is convenient to use the elastic dilatation J_e, the distortional deformation vectors $\mathbf{m}'_i$ and the elastic distortional deformation metric m'_{ij}, which satisfy the Eq. (5.11.45)

$$\begin{aligned} &J_e = \mathbf{m}_1 \times \mathbf{m}_2 \cdot \mathbf{m}_3 > 0\,, && \dot{J}_e = J_e(\mathbf{D} - \Gamma\bar{\mathbf{D}}_p) \cdot \mathbf{I}\,, \\ &\mathbf{m}'_i = J_e^{-1/3}\mathbf{m}_i\,, && \dot{\mathbf{m}}'_i = (\mathbf{L}'' - \Gamma\bar{\mathbf{L}}''_p)\,\mathbf{m}'_i\,, \\ &m'_{ij} = \mathbf{m}'_i \cdot \mathbf{m}'_j = m'_{ji}\,, && \dot{m}'_{ij} = 2\left(\mathbf{m}'_i \otimes \mathbf{m}'_j - \tfrac{1}{3}m'_{ij}\mathbf{I}\right) \cdot (\mathbf{D} - \Gamma\bar{\mathbf{D}}_p)\,, \end{aligned} \tag{6.7.14}$$

where $\mathbf{L}''$ is the deviatoric part of the velocity gradient $\mathbf{L}$ and $\bar{\mathbf{L}}''_p$ is the deviatoric part of $\bar{\mathbf{L}}_p$. Then, for an anisotropic thermoelastic–inelastic material, the response functions ψ, η and $\mathbf{T}$ are specified in the forms

$$\begin{aligned} &\psi = \tilde{\psi}(J_e, m'_{ij}, \theta, \kappa, \beta_{ij})\,, \quad \eta = \tilde{\eta}(J_e, m'_{ij}, \theta, \kappa, \beta_{ij})\,, \\ &\mathbf{T} = \tilde{\mathbf{T}}(J_e, \mathbf{m}'_i, \theta, \kappa, \beta_{ij})\,, \end{aligned} \tag{6.7.15}$$

so the rate of material dissipation (6.3.9) requires

$$\begin{aligned} \rho\theta\xi' = &\left[\mathbf{T} - \rho J_e \frac{\partial\tilde{\psi}}{\partial J_e}\mathbf{I} - 2\rho\frac{\partial\tilde{\psi}}{\partial m'_{ij}}\left(\mathbf{m}'_i \otimes \mathbf{m}'_j - \frac{1}{3}m'_{ij}\mathbf{I}\right)\right] \cdot \mathbf{D} - \rho\left(\frac{\partial\hat{\psi}}{\partial\theta} + \eta\right)\dot{\theta}\,, \\ &+ \Gamma\left[\rho J_e \frac{\partial\tilde{\psi}}{\partial J_e}\bar{\mathbf{D}}_p \cdot \mathbf{I} + 2\rho\frac{\partial\tilde{\psi}}{\partial m'_{ij}}\left(\mathbf{m}'_i \otimes \mathbf{m}'_j - \frac{1}{3}m'_{ij}\mathbf{I}\right) \cdot \bar{\mathbf{D}}_p \right. \\ &\left. - \rho\frac{\partial\hat{\psi}}{\partial\kappa}H - \rho\frac{\partial\hat{\psi}}{\partial\beta_{ij}}H_{ij}\right] \geq 0\,, \end{aligned} \tag{6.7.16}$$

where use has been made of (6.6.20). Again, without specifying details of the rate of inelasticity and the hardening functions Γ, $\bar{\mathbf{D}}_p$, H and H_{ij}, it is not possible to obtain necessary restrictions on the constitutive equation for stress and entropy. However, motivated by the constitutive Eq. (6.6.23) for a thermoelastic–inelastic material, the constitutive equations for stress and entropy in a thermomelastic–inelastic material are specified by

$$\begin{aligned} &\mathbf{T} = -p\,\mathbf{I} + \mathbf{T}'' = \tilde{\mathbf{T}}\,, \quad p = \tilde{p} = -\rho J_e \frac{\partial\tilde{\psi}}{\partial J_e}\,, \\ &\mathbf{T}'' = \tilde{\mathbf{T}}'' = 2\rho\frac{\partial\tilde{\psi}}{\partial m'_{ij}}\left(\mathbf{m}'_i \otimes \mathbf{m}'_j - \frac{1}{3}m'_{ij}\mathbf{I}\right)\,, \\ &\eta = \tilde{\eta} = -\frac{\partial\tilde{\psi}}{\partial\theta}\,, \end{aligned} \tag{6.7.17}$$

so the rate of material dissipation imposes the restriction

$$\rho\theta\xi' = \Gamma\left[-\tilde{p}\,(\mathbf{I}\cdot\bar{\mathbf{D}}_p) + \tilde{\mathbf{T}}''\cdot\bar{\mathbf{D}}_p - \rho\frac{\partial\tilde{\psi}}{\partial\kappa}H - \rho\frac{\partial\tilde{\psi}}{\partial\beta_{ij}}H_{ij}\right] \geq 0\,. \qquad (6.7.18)$$

Also, the constitutive equation for stress is assumed to be restricted so that a zero-stress material state at zero-stress reference temperature θ_z is characterized by

$$\begin{gathered}\mathbf{T}=0\,,\quad \frac{\partial\tilde{\psi}}{\partial J_e}=0\,,\quad \frac{\partial\tilde{\psi}}{\partial m'_{ij}}=\frac{1}{3}\frac{\partial\tilde{\psi}}{\partial m'_{nn}}\delta_{ij}\\ \text{for } J_e=1\,,\quad m'_{ij}=\delta_{ij}\,,\quad \theta=\theta_z\,,\end{gathered} \qquad (6.7.19)$$

where δ_{ij} is the Kronecker delta. This means that the triad $\mathbf{m}'_i$ has been defined so that $\mathbf{m}'_i$ are orthonormal vectors in a zero-stress material state at zero-stress reference temperature $\theta = \theta_z$. In addition, the entropy flux $\mathbf{p}$ can be specified in the form (6.6.24)

$$\mathbf{p} = p^{i\prime}\mathbf{m}'_i\,,\quad p^{i\prime} = \tilde{p}^{i\prime}(J_e, m'_{ij}, \theta, g'_i)\,,\quad g'_i = \mathbf{g}\cdot\mathbf{m}'_i\,, \qquad (6.7.20)$$

which must satisfy the restriction (6.3.7) due to the second law of thermodynamics,

$$-\mathbf{p}\cdot\mathbf{g} = -\tilde{p}^{i\prime}g'_i > 0 \text{ for } \mathbf{g}\neq 0\,. \qquad (6.7.21)$$

The evolution Eq. (6.7.14) for J_e and $\mathbf{m}'_i$, (6.7.29) for κ and (6.7.4) for β_{ij} require initial conditions

$$J_e(0),\,\mathbf{m}'_i(0),\,\kappa(0),\,\beta_{ij}(0)\,. \qquad (6.7.22)$$

Examples where this formulation has been used to model elastic and inelastic anisotropy in geological materials with joints can be found in [25, 35]. Also, notice that inelastic dilatation rate $\bar{\mathbf{D}}_p\cdot\mathbf{I}\neq 0$ in (6.7.14) prevents the elastic dilatation J_e from being written in a simple form like (6.6.20) since the zero-stress density of the material at zero-stress reference temperature need not be constant.

Elastically Isotropic Thermoelastic–Inelastic Response

For elastically isotropic thermoelastic–inelastic response, the Helmholtz free energy depends on the invariants of the metric m_{ij}. Then, following the definitions of pure dilatation and pure distortion proposed by Flory [6], the elastic dilatation J_e is defined by (3.11.7)

$$J_e = \mathbf{m}_1\times\mathbf{m}_2\cdot\mathbf{m}_3 > 0\,, \qquad (6.7.23)$$

and the symmetric, positive-definite, unimodular elastic distortional deformation tensor $\mathbf{B}'_e$ is defined by (5.8.1)

$$\mathbf{B}'_e = \mathbf{m}'_i\otimes\mathbf{m}'_i\,. \qquad (6.7.24)$$

Then, the evolution equation for J_e is given by (6.7.14)

$$\dot{J}_e = J_e(\mathbf{D} - \Gamma\bar{\mathbf{D}}_p) \cdot \mathbf{I}\,, \tag{6.7.25}$$

and $\mathbf{B}_e'$ satisfies the evolution Eq. (5.11.66) with the specification

$$\dot{\mathbf{B}}_e' = \mathbf{L}''\mathbf{B}_e' + \mathbf{B}_e'\mathbf{L}''^T - \Gamma\mathbf{A}_p\,, \quad \mathbf{A}_p = \mathbf{B}_e' - \left(\frac{3}{\mathbf{B}_e'^{-1} \cdot \mathbf{I}}\right)\mathbf{I}\,. \tag{6.7.26}$$

Also, the non-trivial invariants α_1 and α_2 of $\mathbf{B}_e'$ are given by (5.11.58)

$$\alpha_1 = \mathbf{B}_e' \cdot \mathbf{I}\,, \quad \alpha_2 = \mathbf{B}_e' \cdot \mathbf{B}_e'\,, \tag{6.7.27}$$

which satisfy the evolution Eq. (5.11.59)

$$\begin{aligned}
\dot{\alpha}_1 &= 2\mathbf{B}_e'' \cdot \mathbf{D} - \Gamma\mathbf{A}_p \cdot \mathbf{I}\,, \\
\dot{\alpha}_2 &= 4\left(\mathbf{B}_e'^2 - \frac{1}{3}\alpha_2\mathbf{I}\right) \cdot \mathbf{D} - 2\Gamma\mathbf{A}_p \cdot \mathbf{B}_e'\,,
\end{aligned} \tag{6.7.28}$$

where $\mathbf{B}_e''$ is the deviatoric part of $\mathbf{B}_e'$. In addition, attention is limited to isotropic hardening κ, which satisfies the evolution Eq. (6.7.29)

$$\dot{\kappa} = \Gamma H\,. \tag{6.7.29}$$

For an elastically isotropic thermoelastic–inelastic material, the response functions ψ, η and $\mathbf{T}$ are specified in the forms

$$\psi = \psi(J_e, \alpha_1, \alpha_2, \theta, \kappa)\,, \quad \eta = \eta(J_e, \alpha_1, \alpha_2, \theta, \kappa)\,, \quad \mathbf{T} = \mathbf{T}(J_e, \mathbf{B}_e', \theta, \kappa)\,, \tag{6.7.30}$$

so the rate of material dissipation (6.3.9) requires

$$\begin{aligned}
\rho\theta\xi' = {} & \left[\mathbf{T} - \rho J_e\frac{\partial\psi}{\partial J_e}\mathbf{I} - 2\rho\frac{\partial\psi}{\partial\alpha_1}\mathbf{B}_e'' - 4\rho\frac{\partial\psi}{\partial\alpha_2}\left(\mathbf{B}_e'^2 - \frac{1}{3}\alpha_2\mathbf{I}\right)\right] \cdot \mathbf{D} - \rho\left(\frac{\partial\psi}{\partial\theta} + \eta\right)\dot{\theta}\,, \\
& + \Gamma\left[\rho J_e\frac{\partial\psi}{\partial J_e}\bar{\mathbf{D}}_p \cdot \mathbf{I} + 2\rho\frac{\partial\psi}{\partial\alpha_1}\mathbf{A}_p \cdot \mathbf{I} + 4\rho\frac{\partial\psi}{\partial\alpha_2}\mathbf{A}_p \cdot \mathbf{B}_e' - \rho\frac{\partial\psi}{\partial\kappa}H\right] \geq 0\,.
\end{aligned} \tag{6.7.31}$$

In general, without specifying details of the functional forms for Γ, $\bar{\mathbf{D}}_p$ and H, it is not possible to determine necessary restrictions on the constitutive equations for $\mathbf{T}$ and η. However, motivated by necessary restrictions for a rate-independent elastic–inelastic material with a yield function, the constitutive equations for $\mathbf{T}$ and η are specified by

$$\mathbf{T} = -p\,\mathbf{I} + \mathbf{T}'' , \quad p = -\rho J_e \frac{\partial \psi}{\partial J_e} ,$$
$$\mathbf{T}'' = 2\rho \frac{\partial \psi}{\partial \alpha_1} \mathbf{B}_e'' + 4\rho \frac{\partial \psi}{\partial \alpha_2} \left(\mathbf{B}_e'^2 - \frac{1}{3}\alpha_2 \mathbf{I} \right) , \tag{6.7.32}$$
$$\eta = -\frac{\partial \psi}{\partial \theta} ,$$

so the rate of material dissipation imposes the restriction

$$\rho\theta\xi' = \Gamma \Big[-p\,(\bar{\mathbf{D}}_p \cdot \mathbf{I}) + 2\rho \frac{\partial \psi}{\partial \alpha_1} \mathbf{A}_p \cdot \mathbf{I} + 4\rho \frac{\partial \psi}{\partial \alpha_2} \mathbf{A}_p \cdot \mathbf{B}_e' - \rho \frac{\partial \psi}{\partial \kappa} H \Big] \geq 0 . \tag{6.7.33}$$

Also, the constitutive equation for stress is assumed to be restricted so that a zero-stress material state at zero-stress reference temperature θ_z is characterized by

$$\mathbf{T} = 0 , \quad \frac{\partial \psi}{\partial J_e} = 0 , \quad \text{for } J_e = 1 , \quad \mathbf{B}_e' = \mathbf{I} , \quad \theta = \theta_z . \tag{6.7.34}$$

In addition, for isotropic response the entropy flux can be specified by a generalized Fourier form (6.6.32)

$$\mathbf{p} = -\frac{\bar{\kappa}}{\theta} \mathbf{g} , \quad \bar{\kappa} = \bar{\kappa}(J_e, \alpha_1, \alpha_2, \theta, \kappa) \geq 0 , \tag{6.7.35}$$

where the heat conduction coefficient $\bar{\kappa}$ should not be confused with the hardening variable κ.

6.8 Orthotropic Thermoelastic–Inelastic Materials

A homogeneous, isotropic elastic material deformed from a uniform zero-stress material state to a Hydrostatic State of Stress (HSS)

$$\mathbf{T} = -p\,\mathbf{I} , \quad \mathbf{T}'' = 0 \tag{6.8.1}$$

is compressed but it experiences no distortional deformation. This means that a cube of a homogeneous, isotropic elastic material in a zero-stress material state will deform to a smaller cube when it is compressed by water causing a HSS. In contrast, a homogeneous, orthotropic elastic material deformed from a uniform zero-stress material state to a compressed HSS is distorted as well as being compressed. This means that a cube of a homogeneous, orthotropic elastic material in a zero-stress material state will deform to a rectangular parallelepiped when it is compressed by water causing a HSS. This also means that additional distortional deformation is required to produce deviatoric stress.

Rubin and Jabareen [20, 21] considered this problem for the purely mechanical theory of orthotropic elastic materials and developed physically based orthotropic invariants which characterize the added distortional deformations that cause deviatoric stress.

Recently, this approach was generalized to model an orthotropic thermoelastic–inelastic material [27] where additional details can be found. Motivated by the work in [14], the equations for this theory are slightly modified relative to those reported in [27]. For this theory use is made of the elastic dilatation J_e, the elastic distortional deformation vectors $\mathbf{m}'_i$ and the elastic distortional deformation metric m'_{ij} which satisfy the evolution Eq. (6.7.14)

$$\begin{aligned}\dot{J}_e &= J_e(\mathbf{D} - \Gamma\bar{\mathbf{D}}_p)\cdot\mathbf{I}\,, \quad \dot{\mathbf{m}}'_i = (\mathbf{L}'' - \Gamma\bar{\mathbf{L}}''_p)\,\mathbf{m}'_i\,, \\ m'_{ij} &= \mathbf{m}'_i\cdot\mathbf{m}'_j = m'_{ji}\,, \quad \dot{m}'_{ij} = 2(\mathbf{m}'_i\otimes\mathbf{m}'_j - \tfrac{1}{3}m'_{ij}\mathbf{I})\cdot(\mathbf{D} - \Gamma\bar{\mathbf{D}}_p)\,,\end{aligned} \tag{6.8.2}$$

where $\mathbf{L}''$ is the deviatoric part of the velocity gradient $\mathbf{L}$, $\bar{\mathbf{L}}''_p$ is the deviatoric part of the direction $\bar{\mathbf{L}}_p$ of inelastic rate and the non-negative function Γ controls the magnitude of the inelastic deformation rate. Using (6.7.2), the direction of inelastic deformation rate $\bar{\mathbf{D}}_p$ and the direction of inelastic spin $\bar{\mathbf{W}}_p$ are defined by

$$\bar{\mathbf{D}}_p = \frac{1}{2}\left(\bar{\mathbf{L}}_p + \bar{\mathbf{L}}_p^T\right)\,, \quad \bar{\mathbf{W}}_p = \frac{1}{2}\left(\bar{\mathbf{L}}_p - \bar{\mathbf{L}}_p^T\right)\,. \tag{6.8.3}$$

In addition, the isotropic hardening variable κ is determined by the evolution Eq. (6.7.29)

$$\dot{\kappa} = \Gamma H\,, \tag{6.8.4}$$

where H is a function that controls the rate of hardening and the directional hardening variables $\beta_{ij} = \beta_{ji}$ are determined by the evolution Eq. (6.7.4)

$$\dot{\beta}_{ij} = \Gamma H_{ij}\,, \tag{6.8.5}$$

where $H_{ij} = H_{ji}$ are functions that control the relative magnitudes of β_{ij}.

A thermoelastic orthotropic material in any HSS experiences mechanical distortion and can also experience distortion due to thermal effects. Specifically, in any HSS the elastic distortional vectors $\mathbf{m}'_i$, their reciprocal vectors $\mathbf{m}^{i\prime}$ and the metrics m'_{ij} and $m^{ij\prime}$ take the forms

$$\begin{aligned}&\frac{1}{\eta_1}\mathbf{m}'_1 = \eta_1\mathbf{m}^{1\prime}\,, && \frac{1}{\eta_2}\mathbf{m}'_2 = \eta_2\mathbf{m}^{2\prime}\,, && \frac{1}{\eta_3}\mathbf{m}'_3 = \eta_3\mathbf{m}^{3\prime}\,, \\ &m'_{11} = \eta_1^2\,, && m'_{22} = \eta_2^2\,, && m'_{33} = \eta_3^2\,, \\ &m'_{12} = 0\,, && m'_{13} = 0\,, && m'_{23} = 0\,, \\ &m^{11\prime} = \frac{1}{\eta_1^2}\,, && m^{22\prime} = \frac{1}{\eta_2^2}\,, && m^{33\prime} = \frac{1}{\eta_3^2}\,, \\ &m^{12\prime} = 0\,, && m^{13\prime} = 0\,, && m^{23\prime} = 0\,,\end{aligned} \tag{6.8.6}$$

which indicate that $\mathbf{m}_i'$ and $\mathbf{m}^{i\prime}$ are orthogonal triads of vectors parallel to the principal directions of orthotropy of the material. Moreover, η_i are dependent positive constitutive distortional deformation functions of the elastic dilatation J_e and temperature θ satisfying the restrictions

$$\eta_i = \eta_i(J_e, \theta) > 0\,, \quad \eta_1\eta_2\eta_3 = 1\,, \quad \eta_i(1, \theta_z) = 1\,, \tag{6.8.7}$$

with η_i being unity in any Reference Lattice State (RLS) at reference zero-stress temperature θ_z.

Since any HSS causes a cube of the orthotropic material to deform into a rectangular parallelepiped, in principle, it is possible to use a triaxial testing machine to measure the two independent functions η_1 and η_2 for different values of J_e and θ. Of course, experimental challenges related to friction on the sides of the specimen and difficulties with applying tension while letting the sides of the specimen slide freely, currently limit the range of J_e and θ that can be explored experimentally.

The Helmholtz free energy for an orthotropic thermoelastic–inelastic material can be expressed in the form

$$\psi = \psi(J_e, m_{ij}', \theta)\,. \tag{6.8.8}$$

Specific functional forms for ψ must be proposed which match the distortions (6.8.6) in any HSS that are characterized by the measured values of the functions η_i as well as match additional experimental data for the added distortions that cause deviatoric stress.

The main idea in [27] is to rewrite the six generalized physically based orthotropic invariants β_i developed in [21] in terms of the elastic distortional deformation metric m_{ij}' and its inverse $m^{ij\prime}$, which are influenced by inelastic deformation rate, and to generalize the characterization of any HSS to include thermoelastic response. Then, the Helmholtz free energy ψ is specified in the form

$$\psi = \psi(J_e, \beta_i, \theta)\,. \tag{6.8.9}$$

Since the invariants β_i are based on the distortional deformation functions η_i, it follows that this representation of ψ automatically reproduces the distortions η_i and the results (6.8.6) in any HSS. This simplifies the constitutive modeling of an orthotropic thermoelastic–inelastic material because a specific form ψ in (6.8.9) need only consider the pressure in any HSS and experimental data for distortional deformations that cause deviatoric stress.

Specifically, using the functional forms for the distortions η_i, the physically based orthotropic invariants β_i are defined by

$$\begin{aligned}
&\beta_1 = \frac{m_{11}'}{\eta_1^2} + \eta_1^2 m^{11\prime}\,, \quad \beta_2 = \frac{m_{22}'}{\eta_2^2} + \eta_2^2 m^{22\prime}\,,\\
&\beta_3 = \frac{m_{33}'}{\eta_3^2} + \eta_3^2 m^{33\prime}\,, \quad \beta_i \geq 2 \quad \text{for} \quad i = 1, 2, 3\,,\\
&\beta_4 = \frac{m_{12}'^2}{m_{11}'m_{22}'}\,, \qquad \beta_5 = \frac{m_{13}'^2}{m_{11}'m_{33}'}\,, \qquad \beta_6 = \frac{m_{23}'^2}{m_{22}'m_{33}'}\,,
\end{aligned} \tag{6.8.10}$$

which by definition attain the values

$$\beta_1 = \beta_2 = \beta_3 = 2\,, \quad \beta_4 = \beta_5 = \beta_6 = 0 \quad \text{for any } HSS\,. \tag{6.8.11}$$

Moreover, with the help of (6.8.7) it can be shown that

$$\begin{aligned}\dot{\eta}_i &= J_e \frac{\partial \eta_i}{\partial J_e} \mathbf{I} \cdot (\mathbf{D} - \Gamma \bar{\mathbf{D}}_p) + \frac{\partial \eta_i}{\partial \theta} \dot{\theta}\,, \\ \dot{\beta}_i &= 2(-N_i \mathbf{I} + \mathbf{B}_i'') \cdot (\mathbf{D} - \Gamma \bar{\mathbf{D}}_p) + 2A_i \dot{\theta}\,,\end{aligned} \tag{6.8.12}$$

where the functions N_i and A_i are defined by

$$\begin{aligned}
N_1 &= \tfrac{J_e}{\eta_1}\left(\tfrac{m'_{11}}{\eta_1^2} - \eta_1^2 m^{11\prime}\right)\tfrac{\partial \eta_1}{\partial J_e}\,, \quad N_2 = \tfrac{J_e}{\eta_2}\left(\tfrac{m'_{22}}{\eta_2^2} - \eta_2^2 m^{22\prime}\right)\tfrac{\partial \eta_2}{\partial J_e}\,, \\
N_3 &= \tfrac{J_e}{\eta_3}\left(\tfrac{m'_{33}}{\eta_3^2} - \eta_3^2 m^{33\prime}\right)\tfrac{\partial \eta_3}{\partial J_e}\,, \quad N_4 = N_5 = N_6 = 0\,, \\
A_1 &= -\tfrac{1}{\eta_1}\left(\tfrac{m'_{11}}{\eta_1^2} - \eta_1^2 m^{11\prime}\right)\tfrac{\partial \eta_1}{\partial \theta}\,, \quad A_2 = -\tfrac{1}{\eta_2}\left(\tfrac{m'_{22}}{\eta_2^2} - \eta_2^2 m^{22\prime}\right)\tfrac{\partial \eta_2}{\partial \theta}\,, \\
A_3 &= -\tfrac{1}{\eta_3}\left(\tfrac{m'_{33}}{\eta_3^2} - \eta_3^2 m^{33\prime}\right)\tfrac{\partial \eta_3}{\partial \theta}\,, \quad A_4 = A_5 = A_6 = 0\,.
\end{aligned} \tag{6.8.13}$$

Also, the deviatoric tensors $\mathbf{B}_i''$ are defined by

$$\begin{aligned}
\mathbf{B}_1'' &= \frac{1}{\eta_1^2}\mathbf{m}_1' \otimes \mathbf{m}_1' - \eta_1^2 \mathbf{m}^{1\prime} \otimes \mathbf{m}^{1\prime} - \frac{1}{3}\left(\frac{m'_{11}}{\eta_1^2} - \eta_1^2 m^{11\prime}\right)\mathbf{I}\,, \\
\mathbf{B}_2'' &= \frac{1}{\eta_2^2}\mathbf{m}_2' \otimes \mathbf{m}_2' - \eta_2^2 \mathbf{m}^{2\prime} \otimes \mathbf{m}^{2\prime} - \frac{1}{3}\left(\frac{m'_{22}}{\eta_2^2} - \eta_2^2 m^{22\prime}\right)\mathbf{I}\,, \\
\mathbf{B}_3'' &= \frac{1}{\eta_3^2}\mathbf{m}_3' \otimes \mathbf{m}_3' - \eta_3^2 \mathbf{m}^{3\prime} \otimes \mathbf{m}^{3\prime} - \frac{1}{3}\left(\frac{m'_{33}}{\eta_3^2} - \eta_3^2 m^{33\prime}\right)\mathbf{I}\,, \\
\mathbf{B}_4'' &= \frac{m'_{12}}{m'_{11}m'_{22}}\left[(\mathbf{m}_1' \otimes \mathbf{m}_2' + \mathbf{m}_2' \otimes \mathbf{m}_1') - \frac{m'_{12}}{m'_{11}}(\mathbf{m}_1' \otimes \mathbf{m}_1') - \frac{m'_{12}}{m'_{22}}(\mathbf{m}_2' \otimes \mathbf{m}_2')\right]\,, \\
\mathbf{B}_5'' &= \frac{m'_{13}}{m'_{11}m'_{33}}\left[(\mathbf{m}_1' \otimes \mathbf{m}_3' + \mathbf{m}_3' \otimes \mathbf{m}_1') - \frac{m'_{13}}{m'_{11}}(\mathbf{m}_1' \otimes \mathbf{m}_1') - \frac{m'_{13}}{m'_{33}}(\mathbf{m}_3' \otimes \mathbf{m}_3')\right]\,, \\
\mathbf{B}_6'' &= \frac{m'_{23}}{m'_{22}m'_{33}}\left[(\mathbf{m}_2' \otimes \mathbf{m}_3' + \mathbf{m}_3' \otimes \mathbf{m}_2') - \frac{m'_{23}}{m'_{22}}(\mathbf{m}_2' \otimes \mathbf{m}_2') - \frac{m'_{23}}{m'_{33}}(\mathbf{m}_3' \otimes \mathbf{m}_3')\right]\,.
\end{aligned} \tag{6.8.14}$$

In particular, using (6.8.6), it follows that

$$N_i = 0\,, \quad A_i = 0\,, \quad \mathbf{B}_i'' = 0 \quad \text{for any } HSS\,. \tag{6.8.15}$$

Now, with the help of (6.8.9) and (6.8.12), the rate of material dissipation (6.3.9) requires

$$\rho\theta\xi' = \left[\mathbf{T}\cdot\mathbf{D} - \left\{\rho J_e \frac{\partial\psi}{\partial J_e} - 2\sum_{i=1}^{3}\rho\frac{\partial\psi}{\partial\beta_i}N_i\right\}\mathbf{I} - 2\sum_{i=1}^{6}\rho\frac{\partial\psi}{\partial\beta_i}\mathbf{B}_i''\right]\cdot(\mathbf{D}-\Gamma\bar{\mathbf{D}}_p)$$
$$-\rho\left[\frac{\partial\psi}{\partial\theta} + 2\sum_{i=1}^{6}\left(\frac{\partial\psi}{\partial\beta_i}A_i\right) + \eta\right]\dot{\theta} \geq 0\,. \tag{6.8.16}$$

Motivated by this expression, the constitutive equations for the Cauchy stress $\mathbf{T}$ and the entropy η are specified by

$$\mathbf{T} = -p\,\mathbf{I} + \mathbf{T}''\,, \qquad \mathbf{T}'' = 2\sum_{i=1}^{6}\rho\frac{\partial\psi}{\partial\beta_i}\mathbf{B}_i''\,,$$
$$p = -\rho J_e\frac{\partial\psi}{\partial J_e} + 2\sum_{i=1}^{3}\rho\frac{\partial\psi}{\partial\beta_i}N_i\,, \quad \eta = -\frac{\partial\psi}{\partial\theta} - 2\sum_{i=1}^{6}\frac{\partial\psi}{\partial\beta_i}A_i\,, \tag{6.8.17}$$

where p is the pressure and $\mathbf{T}''$ is the deviatoric stress. It then follows that the rate of material dissipation requires

$$\rho\theta\xi' = -p\,(\Gamma\bar{\mathbf{D}}_p\cdot\mathbf{I}) + \mathbf{T}''\cdot\Gamma\bar{\mathbf{D}}_p \geq 0\,. \tag{6.8.18}$$

Also, the entropy flux vector can be specified by

$$\mathbf{p} = -\frac{\mathbf{K}}{\theta}\mathbf{g}\,, \quad \mathbf{g} = \frac{\partial\theta}{\partial\mathbf{x}}\,, \quad \mathbf{K} = \mathbf{K}^T = K^{ij}\mathbf{m}_i'\otimes\mathbf{m}_j'\,, \quad K^{ji} = K^{ij}\,, \tag{6.8.19}$$

where K^{ij} is a positive-definite symmetric matrix that characterizes anisotropic heat conduction coefficients. Since K^{ij} is a positive-definite, it follows that the restriction of the second law of thermodynamics which requires heat to flow from hot to cold (6.3.7)

$$-\,\mathbf{p}\cdot\mathbf{g} > 0 \text{ for } \mathbf{g} \neq 0\,, \tag{6.8.20}$$

is automatically satisfied.

Also, notice that inelastic dilatation rate $\bar{\mathbf{D}}_p\cdot\mathbf{I}\neq 0$ in (6.8.2) prevents the elastic dilatation J_e from being written in a simple form like (6.6.20) since the zero-stress density of the material at zero-stress reference temperature θ_z need not be constant.

Specific Constitutive Equations

To simplify the discussion, consider the case of metal plasticity for which inelastic deformation rate is isochoric so that

$$\bar{\mathbf{D}}_p\cdot\mathbf{I} = 0\,, \tag{6.8.21}$$

and the elastic dilatation J_e can be expressed in the form (4.1.16).

$$J_e = \frac{\rho_z}{\rho}\,, \tag{6.8.22}$$

where ρ_z is the constant zero-stress mass density at zero-stress reference temperature θ_z.

Next, motivated by the work in [16, 26] for a material with a constant specific heat C_v, the Helmholtz free energy is specified by

$$\begin{aligned}
&\psi = \psi_1(J_e, \theta) + \psi_2(J_e, \theta, \beta_i)\,,\\
&\rho_z \psi_1 = \rho_z C_v \left[\theta - \theta_z - \theta \ln\left(\frac{\theta}{\theta_z}\right)\right] - (\theta - \theta_z) f_1(J_e) + f_2(J_e)\,,\\
&\rho_z \psi_2 = \frac{1}{2}\sum_{i=1}^{3} K_i(\beta_i - 2) + \frac{1}{2}\sum_{i=4}^{6} K_i \beta_i\,, \quad K_i \geq 0\,,
\end{aligned} \tag{6.8.23}$$

where ψ_1 controls the isotropic thermomechanical response and ψ_2 controls the orthotropic response in any HSS as well as the influence of additional distortional deformation that causes deviatoric stress. In these expressions, f_1 controls strong thermomechanical coupling and f_2 controls the isotropic response to nonlinear compression that can be determined by plate impact experiments for shock waves as discussed in Sect. 6.9. In addition, K_i control the influences of the orthotropic invariants β_i. It then follows from (6.8.17) that the associated constitutive equations are given by

$$\begin{aligned}
&p = p_1(J_e, \theta) + p_2(J_e, \theta, \beta_i)\,, \qquad \mathbf{T}'' = J_e^{-1}\sum_{i=1}^{6} K_i \mathbf{B}_i''\,,\\
&p_1 = (\theta - \theta_z)\frac{df_1}{dJ_e} - \frac{df_2}{dJ_e}\,, \qquad p_2 = J_e^{-1}\sum_{i=1}^{3} K_i N_i\,,\\
&\eta = \hat{\eta}_1(J_e, \theta) + \hat{\eta}_2(J_e, \theta, \beta_i)\,,\\
&\rho_z \hat{\eta}_1 = \rho_z C_v \ln\left(\frac{\theta}{\theta_z}\right) + f_1(J_e)\,, \quad \rho_z \hat{\eta}_2 = -\sum_{i=1}^{3} K_i A_i\,.
\end{aligned} \tag{6.8.24}$$

In these expressions, p_1 and $\hat{\eta}_1$ control the isotropic thermomechanical response and p_2 and $\hat{\eta}_2$ control the orthotropic response in any HSS as well as the influence of additional distortional deformation that causes deviatoric stress. Although this form for ψ automatically predicts the correct distortions η_i in any HSS, the functions f_1 and f_2 need to be determined to match experimental data for the pressure in each HSS. Also, the functions $\hat{\eta}_1$ and $\hat{\eta}_2$ should not be confused with the function η_1 and η_2 in (6.8.6) which characterize distortional deformations in any HSS. Furthermore, using (6.2.11) the internal energy ε can be expressed in the form

$$\begin{aligned}
&\varepsilon = \psi + \theta\eta = \varepsilon_1(J_e, \theta) + \varepsilon_2(J_e, \theta, \beta_i)\,, \\
&\rho_z\varepsilon_1 = \rho_z C_v(\theta - \theta_z) + \theta_z f_1(J_e) + f_2(J_e)\,, \\
&\rho_z\varepsilon_2 = \frac{1}{2}\sum_{i=1}^{3} K_i(\beta_i - 2) + \frac{1}{2}\sum_{i=4}^{6} K_i\beta_i - \sum_{i=1}^{3} K_i\theta A_i\,.
\end{aligned} \tag{6.8.25}$$

Now, following the work in [27] the distortions η_i in (6.8.6) for any HSS are specified by the forms

$$\begin{aligned}
&\eta_i = J_e^{n_i/3}\left(\frac{\theta}{\theta_z}\right)^{(\alpha_i\theta_z/3)}\,, \quad n_i = n_i(J_e, \theta)\,, \\
&\alpha_i = \alpha_i(J_e, \theta)\,,
\end{aligned} \tag{6.8.26}$$

where in view of (6.8.7) the functions n_i, α_i must satisfy the restrictions

$$n_1 + n_2 + n_3 = 0\,, \quad \alpha_1 + \alpha_2 + \alpha_3 = 0\,. \tag{6.8.27}$$

Since the inelastic deformation rate is isochoric (6.8.21), the rate of material dissipation (6.8.18) requires

$$\rho\theta\xi' = \mathbf{T}'' \cdot \Gamma\bar{\mathbf{D}}_p \geq 0\,. \tag{6.8.28}$$

Then, following the work in [27] the direction of inelastic deformation rate $\bar{\mathbf{D}}_p$ is specified by

$$\begin{aligned}
&\bar{\mathbf{D}}_p = \sum_{i=1}^{6} d_i \mathrm{Sign}(\mathbf{T}'' \cdot \mathbf{B}_i'')\,, \quad d_i \geq 0\,, \\
&\mathrm{Sign}(x) = -1 \;\; \text{for} \;\; x < 0\,, \quad \mathrm{Sign}(x) = 1 \;\; \text{for} \;\; x \geq 0\,,
\end{aligned} \tag{6.8.29}$$

where d_i are non-negative constants so that the rate of material dissipation (6.8.28) is automatically satisfied. These restrictions on d_i are sufficient but not necessary conditions for (6.8.28) to be satisfied. For example, the work on metal forming with inelastic orthotropic deformation rate in [14] developed more relaxed restrictions on d_i which allow some of the d_i to be negative but which satisfy (6.8.28) for small elastic distortional deformations. Also, in [14] the direction of inelastic spin $\bar{\mathbf{W}}_p$ was specified by

$$\begin{aligned}
\bar{\mathbf{W}}_p = {} & \Omega_{12}\left[\bar{\mathbf{D}}_p \cdot (\mathbf{m}_1' \otimes \mathbf{m}_2')\right](\mathbf{m}^{1'} \otimes \mathbf{m}^{2'} - \mathbf{m}^{2'} \otimes \mathbf{m}^{1'}) \\
& + \Omega_{13}\left[\bar{\mathbf{D}}_p \cdot (\mathbf{m}_1' \otimes \mathbf{m}_3')\right](\mathbf{m}^{3'} \otimes \mathbf{m}^{1'} - \mathbf{m}^{1'} \otimes \mathbf{m}^{3'}) \\
& + \Omega_{23}\left[\bar{\mathbf{D}}_p \cdot (\mathbf{m}_2' \otimes \mathbf{m}_3')\right](\mathbf{m}^{2'} \otimes \mathbf{m}^{3'} - \mathbf{m}^{3'} \otimes \mathbf{m}^{2'})\,,
\end{aligned} \tag{6.8.30}$$

where Ω_{12}, Ω_{13} and Ω_{23} are constants that control the magnitude of the inelastic spin which influences the rate of rotation of the distortional microstructural vectors $\mathbf{m}_i'$.

Furthermore, in [14] an orthotropic yield function g for purely mechanical response was proposed in the form

$$g = g(T_{ij}, \kappa)\,, \tag{6.8.31}$$

where T_{ij} are the components of $\mathbf{T}$ relative to the elastic distortional deformation base vectors $\mathbf{m}'_i$ defined by

$$\begin{aligned} &T_{ij} = -p\,m'_{ij} + T''_{ij} = T_{ji}\,, \quad T''_{ij}\,m^{ij\prime} = 0\,, \\ &T''_{ij} = \mathbf{T}'' \cdot (\mathbf{m}'_i \otimes \mathbf{m}'_j)\,, \qquad \mathbf{T}'' = T''_{ij}\,(\mathbf{m}^{i\prime} \otimes \mathbf{m}^{j\prime})\,. \end{aligned} \tag{6.8.32}$$

In this regard, it is important to emphasize that the components T_{ij} of $\mathbf{T}$ are uninfluenced by SRBM since their values T^+_{ij} in the superposed configuration are given by

$$T^+_{ij} = T_{ij}\,. \tag{6.8.33}$$

Next, the function Γ, which controls the magnitude of inelastic deformation rate, can be determined by the loading conditions (5.11.79) for rate-independent response or by (5.11.80) which models a smooth elastic–inelastic transition for rate-independent response and rate-dependent response.

Since the orthotropic invariants β_i are valid for large deformations, this formulation generalizes more standard formulations based on a quadratic strain energy function like (5.8.24) which are only accurate for moderate strains [see the end of Sect. 5.8]. Consequently, this formulation based on β_i can be used for soft materials which experience large deformations [27].

6.9 Thermoelastic–Inelastic Materials for Shock Waves

Plate impact experiments have been used for decades to study the dynamic response of materials, especially to strong shocks. Specifically, in the simplest form of a plate impact experiment, a circular cylindrical flyer plate of a known material is accelerated in a gas gun toward a circular cylindrical target plate of the material that is being studied, which has a circular cylindrical backup plate of a known transparent material. A laser is used to measure the axial velocity of a material point at the center of the back of the target plate. Since the wave propagation speed in each material is finite, there is a finite time window in which reflections from free surfaces do not reach material points on the centerline of the target plate. Moreover, since the material is shock loaded very rapidly, it is assumed that there is no time for heat conduction so the heat flux vector $\mathbf{q}$ is taken to be zero. Within this time window the deformation of the target is uniaxial strain and the equations of mass conservation and the balances of linear momentum and energy can be solved to obtain Rankine–Hugoniot jump conditions between the density, axial stress and internal energy as functions of the particle velocity and the shock velocity. The system of equations is closed by proposing a relationship between the shock velocity and the particle velocity, which are measured experimentally.

Since inelastic effects limit the magnitude of the deviatoric stress, the axial stress in strong shocks can be approximated by the pressure. The plate impact experimental data can be used to determine the Hugoniot curve for the target material which connects the pressure to the density for the equilibrium state after the shock has passed the material point of interest. For strong shocks, this creates strong thermomechanical coupling which must be modeled accurately to analyze the influence of the shock on material response.

The developments in this section are limited to an elastically isotropic thermoelastic–inelastic material. Use is made of the work in [3, 16] to propose a specific form for the Helmholtz free energy that is consistent with a Mie–Gruneisen equation of state (constitutive equation) for the pressure in shocked states. Also, for simplicity attention is limited to materials which have no inelastic dilatational deformation rate

$$\bar{\mathbf{D}}_p \cdot \mathbf{I} = 0\,, \tag{6.9.1}$$

so the elastic dilatation J_e is determined by the evolution equation

$$\frac{\dot{J}_e}{J_e} = \mathbf{D} \cdot \mathbf{I}\,, \tag{6.9.2}$$

which can be integrated and expressed in the form (4.1.16)

$$J_e = \frac{\rho_z}{\rho}\,, \tag{6.9.3}$$

where ρ_z is the constant zero-stress mass density at zero-stress reference temperature θ_z.

Also, the unimodular elastic distortional deformation tensor $\mathbf{B}_e'$ is determined by the evolution Eq. (5.11.55) with (5.11.65)

$$\dot{\mathbf{B}}_e' = \mathbf{L}''\mathbf{B}_e' + \mathbf{B}_e'\mathbf{L}''^T - \Gamma\mathbf{A}_p\,, \quad \mathbf{A}_p = \mathbf{B}_e' - \left(\frac{3}{\mathbf{B}_e'^{-1} \cdot \mathbf{I}}\right)\mathbf{I}\,, \tag{6.9.4}$$

with the elastic distortional deformation invariant α_1 satisfying the equations

$$\alpha_1 = \mathbf{B}_e' \cdot \mathbf{I}\,, \quad \dot{\alpha}_1 = 2\left(\mathbf{B}_e' - \frac{1}{3}\alpha_1\mathbf{I}\right) \cdot \mathbf{D} - \Gamma\mathbf{A}_p \cdot \mathbf{I}\,, \tag{6.9.5}$$

and the evolution equation for the isotropic hardening variable κ is given by (6.7.29)

$$\dot{\kappa} = \Gamma H\,. \tag{6.9.6}$$

Motivated by the work in [3, 16, 26] for a material with a constant specific heat C_v, the Helmholtz free energy ψ is specified by

$$\begin{aligned}
&\psi = \psi_1(J_e, \theta) + \psi_2(J_e, \alpha_1, \theta)\,,\\
&\psi_1 = C_v[\theta - \theta_z - \theta\, \ln\left(\frac{\theta}{\theta_z}\right)] + (\theta - \theta_z) f_1(J_e) + f_2(J_e)\,,\\
&\rho_z \psi_2 = \frac{1}{2}\mu(J_e, \theta)(\alpha_1 - 3)\,,
\end{aligned} \tag{6.9.7}$$

where ψ_1 controls the thermomechanical response to dilatation and μ is the shear modulus which is allowed to be a function of J_e and θ. The function f_1 controls strong thermomechanical coupling and f_2 controls nonlinear response to dilatation. These functions f_1 and f_2 need to be determined from plate impact experimental data.

Next, assuming that η and $\mathbf{T}$ are functions of the forms

$$\eta = \eta(J_e, \alpha_1, \theta)\,, \quad \mathbf{T} = \mathbf{T}(J_e, \mathbf{B}_e', \theta)\,, \tag{6.9.8}$$

it follows from (6.7.32) that

$$\begin{aligned}
&\mathbf{T} = -p\,\mathbf{I} + \mathbf{T}''\,, && p = p_1(J_e, \theta) + p_2(J_e, \theta, \alpha_1)\,,\\
&p_1 = -\rho_z[(\theta - \theta_z)\tfrac{df_1}{dJ_e} + \tfrac{df_2}{dJe}]\,, && p_2 = -\tfrac{1}{2}\tfrac{\partial \mu}{\partial J_e}(\alpha_1 - 3)\,,\\
&\mathbf{T}'' = J_e^{-1}\mu\left(\mathbf{B}_e' - \tfrac{1}{3}\alpha_1 \mathbf{I}\right)\,, && \eta = \eta_1(J_e, \theta) + \eta_2(J_e, \theta, \alpha_1)\,,\\
&\eta_1 = C_v\, \ln\left(\tfrac{\theta}{\theta_z}\right) - f_1\,, && \rho_z \eta_2 = -\tfrac{1}{2}\tfrac{\partial \mu}{\partial \theta}(\alpha_1 - 3)\,,
\end{aligned} \tag{6.9.9}$$

and from (6.7.33) that the inelastic distortional deformation rate must satisfy the restriction

$$\rho_z \theta \xi' = \frac{1}{2}\Gamma \mu \mathbf{A}_p \cdot \mathbf{I} \geq 0\,. \tag{6.9.10}$$

However, since Γ is non-negative, the shear modulus μ is positive and in [18] it was shown that $\mathbf{A}_p \cdot \mathbf{I} \geq 0$ (5.11.67), it follows that nonzero inelastic deformation rate satisfies the condition (6.9.10) and is dissipative.

Then, with the help of (6.2.11) the specific internal energy ε takes the form

$$\begin{aligned}
&\varepsilon = \varepsilon_1(J_e, \theta) + \varepsilon_2(J_e, \alpha, \theta)\,,\\
&\varepsilon_1 = C_v(\theta - \theta_z) - \theta_z f_1 + f_2\,,\\
&\rho_z \varepsilon_2 = \frac{1}{2}\left(\mu - \theta\frac{\partial \mu}{\partial \theta}\right)(\alpha_1 - 3)\,.
\end{aligned} \tag{6.9.11}$$

In these expressions, the terms ψ_1, p_1, η_1 and ε_1 depend only on J_e and θ and represent the main thermomechanical response and the terms ψ_2, p_2, η_2 and ε_2 also depend on the elastic distortional deformation α_1.

For later reference it is noted that the expression ε_1 can be solved for θ and result can be substituted into the expression for p_1 to obtain

$$p_1 = -\rho_z \left[\left(\frac{\theta_z f_1 - f_2}{C_v} \right) \frac{df_1}{dJ_e} + \frac{df_2}{dJ_e} + \left(\frac{1}{C_v} \right) \frac{df_1}{dJ_e} \varepsilon_1 \right], \tag{6.9.12}$$

which expresses p_1 as a linear function of the energy ε_1.

Next, neglecting body force, deviatoric stress, radiation and heat conduction and restricting attention to uniaxial strain in the $\mathbf{e}_1$ direction, the local equations of the conservation of mass, balance of linear momentum and balance of energy can be written in the forms

$$\rho J_e = \rho_z , \quad \rho \dot{u} = -\frac{\partial p}{\partial x} , \quad \rho \dot{\varepsilon} = -p \frac{\partial u}{\partial x} , \tag{6.9.13}$$

where x is the current position of a material point which was located at the position X in the reference configuration and u is the current velocity of the material point. Moreover, consider a steady-wave moving at the shock velocity U relative to the reference configuration into a zero-stress material state at zero-stress reference temperature and at rest with

$$J_e = 1 , \quad \theta = \theta_z , \quad \rho = \rho_z , \quad u = 0 , \quad p = 0 , \quad \varepsilon = 0 . \tag{6.9.14}$$

For this steady-wave it is convenient to introduce the auxiliary variable χ, such that

$$\chi = X - Ut . \tag{6.9.15}$$

Then, the position x and dilatation J_e are given by

$$x = X + \delta(\chi) , \quad J_e = \frac{\partial x}{\partial X} = 1 + \frac{d\delta}{dX} , \tag{6.9.16}$$

where $\delta(\chi)$ is the displacement of the material point relative to its reference position. Also, the particle velocity u, pressure p and internal energy ε are expressed in the forms

$$u = u(\chi) , \quad p = p_H(\chi) , \quad \varepsilon = \varepsilon_H(\chi) , \tag{6.9.17}$$

where p_H is the pressure and ε_H is the internal energy on the Hugoniot curve.

Differentiating the expression for x yields an expression for u given by

$$u = u(\chi) = \dot{x} = -U \frac{d\delta}{d\chi} = U(1 - J_e) . \tag{6.9.18}$$

Moreover, the balance of linear momentum and the balance of energy are expressed in the forms

$$-U\rho\frac{du}{d\chi}=-\frac{dp_H}{d\chi}J_e^{-1}\,,\quad -U\rho\frac{d\varepsilon_H}{d\chi}=-p_H\frac{du}{d\chi}J_e^{-1}\,,\tag{6.9.19}$$

which with the help of the conservation of mass (6.9.13) and the conditions (6.9.14) can be integrated to obtain

$$u=U(1-J_e)\,,\quad p_H=\rho_z U^2(1-J_e)\,,\quad \varepsilon_H=\frac{1}{2}U^2(1-J_e)^2\,,\tag{6.9.20}$$

which are the Rankine–Hugoniot jump conditions. These equations connect the states of the material on both sides of a uniaxial strain shock wave moving into a zero-stress material at rest. They are used to plot Hugoniot curves of the state variables as functions of the dilatation J_e or density $\rho=\rho_z J_e^{-1}$ for a specified shock velocity.

For a complete constitutive equation for pressure it is necessary to propose an expression for values of p off the Hugoniot curve $p=p_H$. In shock physics it is common to use a Mie–Gruneisen equation of state to determine the pressure for states off of the Hugoniot curve. This Mie–Gruneisen equation expresses the pressure as a function of dilatation and energy. Motivated by the work in [3, 16], this Mie–Gruneisen equation is written in terms of the main thermomechanical parts p_1, ε_1 of the pressure and energy in the form

$$p_1=p_H(J_e)+\rho\gamma(J_e)[\varepsilon_1-\varepsilon_H(J_e)]\,,\tag{6.9.21}$$

where the Gruneisen gamma γ is taken in the form

$$\rho\gamma=\rho_z\gamma_z\,,\tag{6.9.22}$$

with γ_z being the unshocked zero-stress value of γ.

Then, comparison of (6.9.12) with (6.9.21) shows that the constitutive Eq. (6.9.7) for the Helmholtz free energy will be consistent with the Mie–Gruneisen equation (6.9.21) provided that f_1 and f_2 satisfy the differential equations

$$\begin{aligned}&\frac{df_1}{dJ_e}=-\gamma_z C_v\,,\\&\frac{df_2}{dJ_e}+\gamma_z f_2=-\frac{p_H}{\rho_z}+\gamma_z\varepsilon_H+\gamma_z\theta_z f_1\,,\end{aligned}\tag{6.9.23}$$

which can be integrated subject to the conditions

$$f_1(1)=0\,,\quad f_2(1)=0\,,\tag{6.9.24}$$

to deduce that

$$
\begin{aligned}
f_1(J_e) &= \gamma_z C_v(1 - J_e)\,, \\
f_2(J_e) &= C_v\theta_z\left[1 + \gamma_z(1 - J_e) - \exp\{\gamma_z(1 - J_e)\}\right] + f_3\,, \\
f_3(J_e) &= \exp(-\gamma_z J_e)\int_{J_e}^{1}\left[\frac{p_H(x)}{\rho_z} - \gamma_z\varepsilon_H(x)\right]\exp(\gamma_z x)\,dx\,.
\end{aligned}
\tag{6.9.25}
$$

Experimental data for shock waves in compression is used to determine the constant coefficients S_i in the approximation

$$
U = U_z + S_1 u + S_2(\frac{u}{U})u + S_3(\frac{u}{U})^2 u\,,
\tag{6.9.26}
$$

with U_z being the zero-stress shock velocity. Substituting (6.9.20) into this expression yields

$$
U = \frac{U_z}{1 - S_1(1 - J_e) - S_2(1 - J_e)^2 - S_3(1 - J_e)^3} \quad \text{for } J_e \le 1\,,
\tag{6.9.27}
$$

for compression. This function is extended to the expansion regime using the form

$$
U = \frac{U_z}{\sqrt{1 + \frac{\gamma_z}{2}(J_e - 1)}} \quad \text{for } J_e > 1\,.
\tag{6.9.28}
$$

Then, with the help of these expressions for the shock velocity U, it can be shown that the function f_3 in (6.9.25) can be determined in closed form for expansion with U given by (6.9.28), but it is necessary to integrate the function f_3 in (6.9.25) numerically with U given by (6.9.27) for compression. Recently, it was shown in [26] that by modifying the approximation (6.9.27) for compression to take the form

$$
U = \frac{U_z \exp[\frac{1}{2}\gamma_z(1 - J_e)]}{1 - \tilde{S}_1(1 - J_e) - \tilde{S}_2(1 - J_e)^2} \quad \text{for } J_e \le 1\,,
\tag{6.9.29}
$$

it is possible to develop a closed form expression for f_3 for compression for general values of the constants $\tilde{S}_1$ and $\tilde{S}_2$. Numerous experiments have been conducted at great expense to obtain values of S_1, S_2, S_3 and γ_z for a large number of materials. Fortunately, it was shown in [26] that these values can be used without conducting additional experiments to determine values of $\tilde{S}_1$ and $\tilde{S}_2$ which yield excellent agreement with experimental data for most materials.

To complete the constitutive equations, it is necessary to specify functional forms for Γ in (6.9.4) and H in (6.9.6) for rate-independent or rate-dependent response as discussed in Sect. 5.8. Also, the entropy flux vector $\mathbf{p}$ must be specified, which could be taken in form (6.7.35).

6.10 Thermoelastic–Inelastic Porous Materials

Porosity appears naturally in a number of materials like rocks, soils, ceramics, metals and biological tissues. The $p - \alpha$ model developed in [11] and the modification in [4] have been used to model important dissipation due to porous compaction in shock wave problems. For these problems, the shock compaction occurs so quickly that even if the pores are partially or fully saturated with fluid, there is no time for the fluid to move, so fluid diffusion through the porous material can be neglected. Porous compaction, dilation and/or "bulking" (i.e., porous dilation under positive pressure) have been modeled in Nevada Tuff [19], granular media [22], sand [10] and ceramics [3]. Also, nucleation and growth of pores in metals have been modeled in [9].

Porosity in biological tissues is necessary for fluid flow that supplies essential nutrients for cell function. In contrast with shock loading, deformation of biological tissues is typically a slow process which can be significantly influenced by diffusion of fluid. An example of constitutive equations for slow deformation of biological tissues with fluid diffusion along with references to previous work can be found in [32]. In addition, prediction of the long-term quantity of production of an oil well that has been stimulated by hydrofracturing requires proper treatment of poroelasticity and inelasticity which characterizes porous compaction and the associated reduced permeability.

This section discusses the structure of constitutive equations for a porous material subjected to loading rapid enough to neglect fluid diffusion. Following the work in [10, 19, 22] an element of volume dv of the porous material is separated into an element of volume dv_s of the solid material and an element of volume dv_p of the pores

$$dv = dv_s + dv_p \,, \tag{6.10.1}$$

and the current porosity ϕ is defined by

$$\phi = \frac{dv_p}{dv} \,. \tag{6.10.2}$$

For simplicity, the pores are assumed so be evacuated so the element of mass dm of the porous material is due solely to the element of mass dm_s of the solid

$$dm = \rho dv = dm_s = \rho_s dv_s \,, \tag{6.10.3}$$

where ρ is the current mass density of the porous material and ρ_s is the current mass density of the solid matrix. Then, using (6.10.1)–(6.10.3) the mass density ρ can be expressed in the form

$$\rho = (1 - \phi)\rho_s \,. \tag{6.10.4}$$

Now, for an elastically isotropic, thermoelastic–inelastic porous material the elastic dilatation J_e satisfies the evolution equation

$$\frac{\dot{J}_e}{J_e} = \mathbf{D} \cdot \mathbf{I} - (\frac{\dot{\phi}}{1-\phi}) , \tag{6.10.5}$$

where an evolution equation for ϕ must be specified. This equation can be rewritten in the form

$$\frac{d}{dt}\Big[\ln\Big\{ \frac{J_e}{1-\phi} \Big\}\Big] = \mathbf{D} \cdot \mathbf{I} , \tag{6.10.6}$$

which with the help of the conservation of mass (6.2.5)

$$\dot{\rho} + \rho \, \mathbf{D} \cdot \mathbf{I} = 0 , \tag{6.10.7}$$

and the expression (6.10.4) for ρ can be expressed in the form

$$\frac{d}{dt}\Big[\ln\Big\{ \frac{\rho J_e}{1-\phi} \Big\}\Big] = \frac{d}{dt}\Big[\ln(\rho_s J_e) \Big] = 0 . \tag{6.10.8}$$

This equation integrates to obtain

$$J_e = \frac{\rho_{sz}}{\rho_s} , \tag{6.10.9}$$

where ρ_{sz} is the zero-stress mass density of the solid material. This shows that J_e is the solid elastic dilatation.

Also, the elastic distortional deformation tensor is characterized by the symmetric, positive-definite, unimodular tensor $\mathbf{B}'_e$ which satisfies the evolution Eq. (5.11.66)

$$\dot{\mathbf{B}}'_e = \mathbf{L}''\mathbf{B}'_e + \mathbf{B}'_e\mathbf{L}''^T - \Gamma \mathbf{A}_p , \quad \mathbf{A}_p = \mathbf{B}'_e - \left(\frac{3}{\mathbf{B}'^{-1}_e \cdot \mathbf{I}} \right) \mathbf{I} , \tag{6.10.10}$$

where Γ is a non-negative function that controls the rate of distortional inelasticity. Also, for simplicity, attention is limited to the first invariant α_1 of elastic distortional deformation defined by (6.7.27) and the evolution Eq. (6.7.28)

$$\alpha_1 = \mathbf{B}'_e \cdot \mathbf{I} , \quad \dot{\alpha}_1 = 2\mathbf{B}''_e \cdot \mathbf{D} - \Gamma \mathbf{A}_p \cdot \mathbf{I} , \tag{6.10.11}$$

where $\mathbf{B}''_e$ is the deviatoric part of $\mathbf{B}'_e$. In addition, attention is limited to isotropic hardening κ, which satisfies the evolution Eq. (6.7.29)

$$\dot{\kappa} = \Gamma H , \tag{6.10.12}$$

where H controls the hardening rate.

For an elastically isotropic thermoelastic–inelastic porous material, the response functions ψ, η and $\mathbf{T}$ are specified in the forms

$$\psi = \psi_s(J_e, \alpha_1, \theta) , \quad \eta = \eta_s(J_e, \alpha_1, \theta) , \quad \mathbf{T} = \mathbf{T}(J_e, \mathbf{B}'_e, \theta, \phi) , \tag{6.10.13}$$

where ψ_s and η_s are the Helmholtz free energy and entropy of the solid material, both per unit mass. Then, with the help of (6.10.5)–(6.10.13), the rate of material dissipation (6.3.9) requires

$$\rho\theta\xi' = \left[\mathbf{T} - \rho J_e \frac{\partial \psi_s}{\partial J_e}\mathbf{I} - 2\rho\frac{\partial \psi_s}{\partial \alpha_1}\mathbf{B}_e''\right]\cdot\mathbf{D} - \rho\left(\frac{\partial \psi_s}{\partial \theta} + \eta_s\right)\dot{\theta}\,, \\ + \left[\rho J_e \frac{\partial \psi_s}{\partial J_e}\left(\frac{\dot{\phi}}{1-\phi}\right) + \Gamma\rho\frac{\partial \psi_s}{\partial \alpha_1}\mathbf{A}_p\cdot\mathbf{I}\right] \geq 0\,. \tag{6.10.14}$$

In general, without specifying details of the functional forms for Γ, $\dot{\phi}$ and H it is not possible to determine necessary restrictions on the constitutive equations for $\mathbf{T}$ and η_s. Specifically, in [19] the added compressibility of porosity was modeled with both elastic and inelastic rates of porosity. Here, for simplicity, changes in ϕ are assumed to be inelastic only. Moreover, motivated by necessary restrictions for a rate-independent elastic–inelastic material with a yield function, the constitutive equations for $\mathbf{T}$ and η are specified by

$$\mathbf{T} = -p\,\mathbf{I} + \mathbf{T}''\,, \quad p = (1-\phi)p_s\,, \qquad p_s = -\rho_{sz}\frac{\partial \psi_s}{\partial J_e}\,, \\ \mathbf{T}'' = (1-\phi)\mathbf{T}_s''\,, \quad \mathbf{T}_s'' = 2J_e^{-1}\rho_{sz}\frac{\partial \psi_s}{\partial \alpha_1}\mathbf{B}_e''\,, \\ \eta = \eta_s = -\frac{\partial \psi_s}{\partial \theta}\,, \tag{6.10.15}$$

where p_s and $\mathbf{T}_s''$ are the pressure and deviatoric stress in the solid. It then follows that the rate of material dissipation imposes the restriction

$$\rho\theta\xi' = \rho\theta\xi_\phi' + \rho\theta\xi_d' \geq 0\,, \quad \rho\theta\xi_\phi' = -p\left(\frac{\dot{\phi}}{1-\phi}\right)\,, \quad \rho\theta\xi_d' = \Gamma\rho\frac{\partial \psi}{\partial \alpha_1}\mathbf{A}_p\cdot\mathbf{I}\,, \tag{6.10.16}$$

where $\rho\theta\xi_\phi'$ is the material dissipation rate due to porosity changes and $\rho\theta\xi_d'$ is the material dissipation rate due to inelastic distortional deformations. Assuming that the effective shear modulus is positive

$$\frac{\partial \psi}{\partial \alpha_1} > 0\,, \tag{6.10.17}$$

and using the fact that $\mathbf{A}_p \cdot \mathbf{I} \geq 0$, it follows that the inelastic distortional deformation is dissipative

$$\rho\theta\xi_d' \geq 0\,. \tag{6.10.18}$$

Also, the constitutive equation for stress is assumed to be restricted so that a zero-stress material state at zero-stress reference temperature θ_z is characterized by

$$\mathbf{T} = 0\,, \quad \frac{\partial \psi}{\partial J_e} = 0\,, \quad \text{for } J_e = 1\,, \quad \mathbf{B}_e' = \mathbf{I}\,, \quad \theta = \theta_z\,. \tag{6.10.19}$$

In addition, for isotropic response the entropy flux vector is specified by a generalized Fourier form (6.6.32)

$$\mathbf{p} = -\frac{\bar{\kappa}}{\theta}\mathbf{g}\,, \quad \mathbf{g} = \frac{\partial\theta}{\partial\mathbf{x}}\,, \quad \bar{\kappa} = \bar{\kappa}(J_e, \alpha_1, \theta, \kappa) \geq 0\,, \tag{6.10.20}$$

where the heat conduction coefficient $\bar{\kappa}$ should not be confused with the isotropic hardening variable κ.

From the definition of $\rho\theta\xi'_\phi$ in (6.10.16), it follows that porous compaction ($\dot{\phi} < 0$) at positive pressure and porous dilation ($\dot{\phi} > 0$) at negative pressure are both dissipative processes, but "bulking" (i.e., porous dilation at positive pressure) is a non-dissipative process. These three processes are modeled in the nonlinear breakage model developed in [22]. Also, the work in [10] discusses functional forms that include these effects in a thermomechanical theory.

As a special case, the Helmholtz free energy ψ_s of the solid material is specified in the form

$$\begin{aligned}
&\rho_{sz}\psi_s = \rho_{sz}C_v\left[\theta - \theta_z - \theta\ln\left(\frac{\theta}{\theta_z}\right)\right] + f_s(J_e,\theta) + \frac{1}{2}\mu_z(\alpha_1 - 3)\,,\\
&f_s = \rho_{sz}C_z^2\left[\frac{1}{S^2}\ln\left\{\frac{1}{1 - S(1 - Je)}\right\} - \left(\frac{1 - Je}{S}\right) + \alpha(1 - J_e)(\theta - \theta_z)\right] \quad \text{for} \quad S > 0\,,\\
&f_s = \rho_{sz}C_z^2\left[\frac{1}{2}(1 - J_e)^2 + \alpha(1 - J_e)(\theta - \theta_z)\right] \quad \text{for} \quad S = 0\,,
\end{aligned} \tag{6.10.21}$$

where C_v is the constant specific heat, C_z is the zero-stress shock wave speed, S is a positive constant controlling the slope of the shock velocity versus particle velocity curve, α is the constant coefficient of linear expansion and μ_z is the zero-stress shear modulus. It then follows from (6.10.15) that

$$\begin{aligned}
&p_s = \rho_{sz}C_z^2\left[\frac{(1 - J_e)}{1 - S(1 - J_e)} + \alpha(\theta - \theta_z)\right]\,, \quad \mathbf{T}''_s = J_e^{-1}\mu_z\mathbf{B}''_e\,,\\
&\eta_s = C_v\ln\left(\frac{\theta}{\theta_z}\right) - C_z^2\alpha(1 - J_e)\,.
\end{aligned} \tag{6.10.22}$$

Next, for simplicity, attention is further limited to isothermal response with $\theta = \theta_z$ for which

$$p = (1 - \phi)p_s\,, \quad p_s = p_s(J_e) = \rho_{sz}C_z^2\Big[\frac{(1 - J_e)}{1 - S(1 - J_e)}\Big]. \tag{6.10.23}$$

To motivate a form for the evolution equation for porosity consider the expression

$$\dot{p} = (1 - \phi)J_e\frac{dp_s}{dJ_e}\Big[\mathbf{D}\cdot\mathbf{I} - \frac{1}{\Gamma_\phi}\frac{\dot{\phi}}{1 - \phi}\Big], \quad \Gamma_\phi = \frac{J_e\frac{dp_s}{dJ_e}}{J_e\frac{dp_s}{dJ_e} + p_s}\,, \tag{6.10.24}$$

which with the help of (6.10.23) yields

$$J_e \frac{dp_s}{dJ_e} = -\rho_{sz} C_z^2 \left[\frac{J_e}{\{1 - S(1 - J_e)\}^2} \right] < 0\,,$$
$$\Gamma_\phi = \frac{J_e}{2J_e - 1 + S(1 - J_e)^2} > 0 \quad \text{for } J_e > \max\left(0, 1 - \frac{1}{S}\right)\,, \tag{6.10.25}$$

where J_e is restricted so that the denominators in these expressions do not vanish. Also, for later convenience it can be shown that the constitutive equation (6.10.23) for the pressure p can be solved for J_e to deduce that

$$J_e(P, \phi) = \frac{1 - \phi + (S - 1)\, P}{1 - \phi + S\, P} \quad \text{for } S \geq 0\,, \quad P = \frac{p}{\rho_{sz} C_z^2}\,, \tag{6.10.26}$$

where P is the normalized pressure.

Next, it is convenient to introduce the constant pressures $p_T < 0$, $p_c > 0$ and the maximum pressure p_{max} and the minimum porosity ϕ_{min} attained. Then, motivated by the work in [32] the evolution equation for ϕ is specified by

$$\frac{\dot{\phi}}{1 - \phi} = \Gamma_T\, \mathbf{D} \cdot \mathbf{I} \qquad \text{for} \quad p = p_T \text{ and } \mathbf{D} \cdot \mathbf{I} \geq 0\,, \tag{6.10.27a}$$

$$\dot{\phi} = 0 \qquad \text{for} \quad p_T < p < p_c\,, \tag{6.10.27b}$$

$$\frac{\dot{\phi}}{1 - \phi} = \Gamma_c\, \mathbf{D} \cdot \mathbf{I} \qquad \text{for} \quad p = p_c\,, \quad \mathbf{D} \cdot \mathbf{I} < 0 \text{ and } \phi > \phi_{min}\,, \tag{6.10.27c}$$

$$\dot{\phi} = 0 \qquad \text{for} \quad p_c < p < p_{max}\,, \tag{6.10.27d}$$

$$\frac{\dot{\phi}}{1 - \phi} = \Gamma_\phi \left(\frac{\phi}{m + \phi}\right) \mathbf{D} \cdot \mathbf{I} + \frac{\beta_d(\rho\theta_z\xi_d)}{1 + \beta_d\, p} \quad \text{for} \quad p = p_{max} \text{ and } \mathbf{D} \cdot \mathbf{I} \leq 0\,, \tag{6.10.27e}$$

where m and β_d are non-negative constants. These equations define five regions of response: porous dilatation ($\dot{\phi} \geq 0$) with $p = p_T < 0$ for (6.10.27a); elastic response for (6.10.27b); porous compaction ($\dot{\phi} \leq 0$) with $p = p_c > 0$ for (6.10.27c); elastic response with $p_c < p < p_{max}$ for (6.10.27d) and porous compaction ($\dot{\phi} \leq 0$) with $p = p_{max}$ for (6.10.27e). The non-negative functions Γ_T, Γ_c and Γ_ϕ are determined by the conditions $p = p_T$, $p = p_c$ and (6.10.24), respectively. Also, it follows from (6.10.27e) that for compaction at positive pressure, the rate of compaction, which is controlled by the constant m, competes with the rate of dilation due to bulking, which is controlled by the constant β_d. Moreover, from this evolution equation it can be seen that ϕ is automatically limited to its physical range

$$0 < \phi < 1\,. \tag{6.10.28}$$

Furthermore, for compaction at maximum pressure (6.10.27e), it can be shown that (6.10.24) reduces to

$$\dot{p} = (1-\phi)J_e \frac{dp_s}{dJ_e}\left[\left(\frac{m}{m+\phi}\right)\mathbf{D}\cdot\mathbf{I} - \frac{\beta_d(\rho\theta_z\xi_d)}{\Gamma_\phi(1+\beta_d\, p)}\right]. \tag{6.10.29}$$

In view of the restrictions (6.10.25), it follows that small values of m cause a tendency for a slow increase in pressure for $\phi >> m$, and nonzero values of β_d, with nonzero rate of dissipation due to inelastic distortional deformation $\rho\theta_z\xi_d > 0$ causing additional increase in pressure due to bulking. Also, when $\phi \to 0$ and the pressure is large, the response asymptotically approaches that of the nonporous solid matrix.

Moreover, for these evolution equations, it follows that porosity changes are dissipative

$$\rho\theta_z\xi'_\phi = -p\frac{\dot{\phi}}{1-\phi} \geq 0\,, \tag{6.10.30}$$

for the response regions without bulking (6.10.27a), (6.10.27b), (6.10.27c) and (6.10.27d). Also, for compaction with bulking (6.10.27e), it can be shown with the help of (6.10.18) that the rate of material dissipation (6.10.16) for $\theta = \theta_z$ requires

$$\rho\theta_z\xi' = -p\left(\frac{\phi}{m+\phi}\right)\Gamma_\phi\,\mathbf{D}\cdot\mathbf{I} + \frac{\rho\theta_z\xi'_d}{1+\beta_d p} \geq 0\,. \tag{6.10.31}$$

Consequently, the rate of material dissipation (6.10.16) is satisfied for all processes

$$\rho\theta_z\xi' \geq 0\,. \tag{6.10.32}$$

Numerical Integration Algorithm

Consider a time step which begins at $t = t_n$, ends at t_{n+1} with time increment $\Delta t = t_{n+1} - t_n$. A strongly objective numerical integration algorithm (5.11.89a) for $\mathbf{B}'_e$ was discussed in Sect. 5.11. Here, a strongly objective numerical integration algorithm is developed for the evolution equation for J_e. To this end, it is recalled that the relative dilatation J_r satisfies the Eq. (5.11.84)

$$\dot{J}_r = J_r\,\mathbf{D}\cdot\mathbf{I}\,, \quad J_r(t_n) = 1\,. \tag{6.10.33}$$

Then, the evolution Eq. (6.10.6) for J_e can be expressed in the form

$$\frac{d}{dt}\left[\ln\left\{\frac{J_e}{(1-\phi)J_r}\right\}\right] = 0\,, \tag{6.10.34}$$

which can be integrated to deduce that

$$J_e(t_{n+1}) = \left[\frac{1-\phi(t_{n+1})}{1-\phi(t_n)}\right] J_r(t_{n+1})\, J_e(t_n)\,. \tag{6.10.35}$$

Equating $J_e(t_{n+1})$ in (6.10.35) with J_e in (6.10.26) yields the result

$$\begin{aligned}
\phi(t_{n+1}, P) &= \frac{A(P) - \sqrt{A^2(P) - B(P)}}{2J_e(t_n)J_r(t_{n+1})}\,, \\
A(P) &= (2 + SP)J_e(t_n)J_r(t_{n+1}) + \phi(t_n) - 1\,, \\
B(P) &= 4J_e(t_n)J_r(t_{n+1})\Big[\big\{(S-1)\phi(t_n) + SJ_e(t_n)J_r(t_{n+1}) - S + 1\big\}P \\
&\quad + J_e(t_n)J_r(t_{n+1}) + \phi(t_n) - 1\Big]\,,
\end{aligned} \tag{6.10.36}$$

which can be used to determine

$$\begin{aligned}
\phi(t_{n+1}) &= \phi(t_{n+1}, P_T) \quad \text{for } P_T = \frac{p_T}{\rho_{sz}C_z^2}\,, \\
\phi(t_{n+1}) &= \phi(t_{n+1}, P_c) \quad \text{for } P_c = \frac{p_c}{\rho_{sz}C_z^2}\,.
\end{aligned} \tag{6.10.37}$$

This means that the values of $\phi(t_{n+1})$ and $J_e(t_{n+1})$ for the solutions of (6.10.27a), (6.10.27c) and (6.10.27d) at the end of the time step are determined by (6.10.35) and (6.10.37).

To obtain a solution for (6.10.27e), this equation is rewritten in the form

$$\frac{d}{dt}\left[\ln\left\{\frac{(1-\phi)^{1+m}}{\phi^m}\right\}\right] = -\Gamma_\phi \mathbf{D}\cdot\mathbf{I} - \left(\frac{m+\phi}{\phi}\right)\left[\frac{\beta_d(\rho\theta_z\xi_d)}{1+\beta_d\, p}\right] \quad \text{for } p = p_{max} \text{ and } \mathbf{D}\cdot\mathbf{I} \le 0\,. \tag{6.10.38}$$

Assuming that Γ_ϕ can be approximated as constant over the time step, this equation is rewritten in the form

$$\frac{d}{dt}\left[\ln\left\{\frac{(1-\phi)^{1+m} J_r^{\Gamma_\phi(t_{n+1})}}{\phi^m}\right\}\right] \approx -\left(\frac{m+\phi}{\phi}\right)\left[\frac{\beta_d(\rho\theta_z\xi_d)}{1+\beta_d\, p}\right]. \tag{6.10.39}$$

The solution of this equation is further approximated by the solution of the implicit expression

$$\frac{[1-\phi(t_{n+1})]^{1+m}}{\phi(t_{n+1})^m} = \frac{[1-\phi(t_n)]^{1+m}}{\phi(t_n)^m J_r(t_{n+1})^{\Gamma_\phi(t_{n+1})}} \exp\left[-\left\{\frac{m+\phi(t_{n+1})}{\phi(t_{n+1})}\right\}\frac{\Delta t\beta_d(\rho\theta_z\xi_d)(t_{n+1})}{1+\beta_d\, p(t_{n+1})}\right] \quad \text{for } J_r(t_{n+1}) \le 1\,, \tag{6.10.40}$$

where $(\rho\theta\xi_d)(t_{n+1})$ is an estimate of $\rho\theta\xi_d$ at the end of the time step, and $p(t_{n+1})$ and $\Gamma_\phi(t_{n+1})$ are determined by replacing J_e in (6.10.23) and (6.10.25), respectively, with (6.10.35) to obtain a nonlinear equation for $\phi(t_{n+1})$ to be solved numerically. Once the values $\phi(t_{n+1})$ and $J_e(t_{n+1})$ have been determined, the pressure p at the end of the time step is determined by (6.10.23).

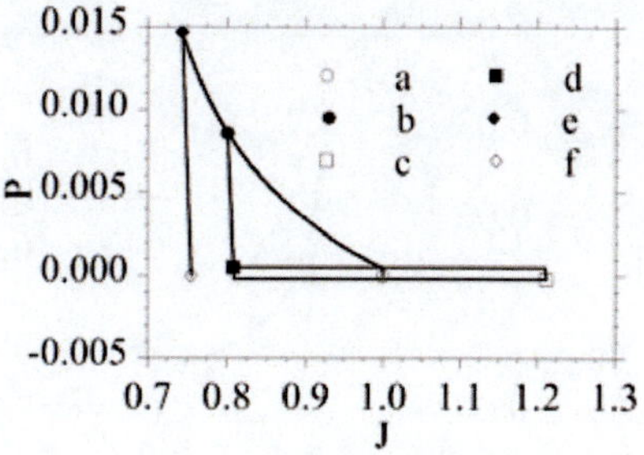

Fig. 6.1 Pure dilatation cyclic loading. Compression $a{-}b$; expansion $b{-}c$; compression $c{-}d$; compression $d{-}e$ and expansion $e{-}f$ for $S = 1.3$, $P_c = 0.005$, $P_T = -0.002$ and $m = 0.01$

To display the compaction response, it is convenient to use the total dilatation J from the initial state determined by the evolution equation and initial condition

$$\dot{J} = J\,\mathbf{D}\cdot\mathbf{I}\,, \quad J(0) = 1\,. \tag{6.10.41}$$

The exact integration of this equation over a time step is given by the expression

$$J(t_{n+1}) = J_r(t_{n+1})\,J(t_n)\,. \tag{6.10.42}$$

The following examples consider the case of no bulking with $\beta_d = 0$.

Cyclic Dilatational Loading

Next, to understand the influence of the parameter m in the evolution Eq. (6.10.27e) for ϕ, consider the case when the material is initially at zero stress with

$$J_e(0) = 1\,, \quad \mathbf{B}'_e(0) = \mathbf{I}\,, \tag{6.10.43}$$

and confine attention to pure dilatational loading for which

$$\mathbf{D} = \frac{1}{3}\,(\mathbf{D}\cdot\mathbf{I})\,\mathbf{I}\,. \tag{6.10.44}$$

For these conditions, the deviatoric stress remains zero $\mathbf{T}'' = 0$ and the pressure is determined by (6.10.23). Moreover, the initial value of porosity ϕ and the material constants S, P_T and P_c for this example are specified by

$$\phi(0) = 0.3\,, \quad S = 1.3\,, \quad P_T = -0.002\,, \quad P_c = 0.005\,. \tag{6.10.45}$$

Figure 6.1 shows the response to cyclic loading with compression $a{-}b$; expansion $b{-}c$; compression $c{-}d$; compression $d{-}e$ and expansion $e{-}f$ for $m = 0.01$. During the compression cycle $a{-}b$ the response is elastic with constant porosity ϕ until $P = P_c$ and then compaction occurs with decrease in porosity. The expansion cycle $b{-}c$ is elastic with constant porosity until $P = P_T$ and then dilation occurs at constant pressure $P = P_T$ with increase in porosity. The recompression cycle $c{-}d$ is elastic with constant porosity until $P = P_c$ and then compaction occurs at constant

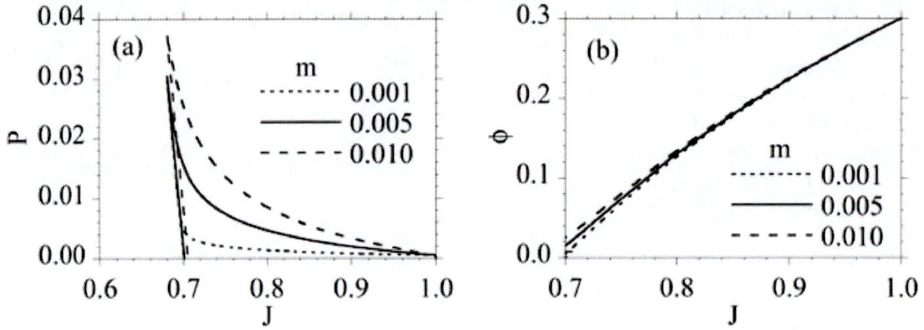

Fig. 6.2 Influence of m on pure dilatational compression followed by expansion for $S = 1.3$ and $P_c = 0.005$

pressure $P = P_c$ with decrease in porosity until $\phi = \phi_{min}$. The compression cycle $d{-}e$ is elastic with constant porosity until $P = P_{max} = p_{max}/(\rho_{sz}C_z^2)$ and continued compaction occurs with decrease in porosity. The expansion $e{-}f$ is elastic with constant porosity.

Figure 6.2 shows the influence of m on the compaction curve. Specifically, Fig. 6.2a shows the pressure response and Fig. 6.2b shows the porosity response. From these figures, it can be seen that m has a strong effect on the pressure during compaction with only small differences in the functional form of ϕ required for the range of values of m. It can also be seen that as the porosity approaches zero the response asymptotically approaches that of the nonporous solid material.

As mentioned before, more complicated evolution equations for ϕ which deal thermomechanical unloading can be found in [10, 22]. Also, to complete the constitutive equations, it is necessary to specify a functional form for Γ in (6.10.10) and H in (6.10.12) for rate-independent or rate-dependent response as discussed in Sect. 5.8.

6.11 Thermoelastic–Inelastic Theory for Growth of Biological Tissues

Biological tissues are complicated materials which are mixtures of many components that can flow relative to each other and interact mechanically, chemically and electrically (e.g., [1, 2, 12, 31]). A simplified constrained theory of mixtures with only one velocity field was developed by Humphrey and Rajagopal [12]. Also, review articles on growth and remodeling of tissues can be found in (e.g., [1, 13, 33]).

When the tissue is considered to be a homogenized solid, the standard approach to modeling growth for finite deformations is the Lagrangian formulation of growth proposed by Rodriguez et al. [15]. This formulation is based on the multiplicative form (5.11.11) by replacing the plastic deformation tensor $\mathbf{F}_p$ with a growth tensor

$\mathbf{F}_g$, such that

$$\mathbf{F}_e = \mathbf{F}\mathbf{F}_g^{-1}\,, \quad \dot{\mathbf{F}}_g = \boldsymbol{\Lambda}_g \mathbf{F}_g\,, \tag{6.11.1}$$

with the rate of growth $\boldsymbol{\Lambda}_g$ specified by a constitutive equation. However, this multiplicative formulation has the same arbitrariness as discussed previously for the Lagrangian formulation of plasticity, which can be removed by the Eulerian formulation discussed below.

Rubin et al. [23] developed an Eulerian unified theoretical structure for modeling interstitial growth and muscle activation in soft tissues. This Eulerian formulation of growth was used: in [30] to study significant differences in the mechanical modeling of confined growth predicted by the Lagrangian and Eulerian formulations; in [28] to analyze stresses in arteries and in [29] to model early cardiac morphogenesis during c-looping. This section reviews this Eulerian thermomechanical formulation of growth.

Growth requires an influx of nutrients to the tissue. Consequently, the theory developed in [23] treats the tissue as an open system with an external rate of mass supply. Specifically, the current mass density ρ of the tissue is determined by the balance of mass

$$\dot{\rho} + \rho(\mathbf{D} \cdot \mathbf{I}) = r_m \rho\,, \tag{6.11.2}$$

where r_m is the external rate of mass supply per unit mass. This simplifies the formulation by neglecting diffusion of fluid and allowing for a single velocity field to describe deformation of the tissue.

The balances of linear momentum and entropy in this theory are given by (6.2.5)

$$\rho\dot{\mathbf{v}} = \rho\mathbf{b} + \mathrm{div}\mathbf{T}\,, \quad \rho\dot{\eta} = \rho(s + \xi) - \mathrm{div}\mathbf{p}\,. \tag{6.11.3}$$

The balance of angular momentum (6.2.6) again requires the Cauchy stress to be symmetric

$$\mathbf{T}^T = \mathbf{T}\,, \tag{6.11.4}$$

but the balance of energy (6.2.7) is modified to include an external rate of energy supply b per unit mass due to mechanobiological processes

$$\rho\dot{\varepsilon} = \rho r - \mathrm{div}\mathbf{q} + \mathbf{T} \cdot \mathbf{D} + \rho b\,. \tag{6.11.5}$$

The Helmholtz free energy ψ is defined by (6.2.11)

$$\psi = \varepsilon - \theta\eta\,, \tag{6.11.6}$$

and the internal rate of entropy production $\rho\xi$ in [17] is separated into a thermal part $(-\mathbf{p} \cdot \mathbf{g})$ and a rate of material dissipation $\rho\theta\xi'$ (6.2.9)

$$\rho\theta\xi = -\mathbf{p} \cdot \mathbf{g} + \rho\theta\xi'\,, \tag{6.11.7}$$

with $\mathbf{g} = \partial\theta/\partial\mathbf{x}$ being the temperature gradient and

$$\rho\theta\xi' = \mathbf{T}\cdot\mathbf{D} - \rho(\dot{\psi} + \eta\dot{\theta}) + \rho b\,. \tag{6.11.8}$$

In these equations, the superposed $\dot{(\,)}$ denotes the standard material derivative with respect to the single velocity field.

The second law of thermodynamics for heat conduction requires the entropy flux $\mathbf{p}$ to satisfy the restriction (6.3.7) that heat flows from hot to cold regions

$$-\,\mathbf{p}\cdot\mathbf{g} > 0 \text{ for } \mathbf{g} \neq 0\,, \tag{6.11.9}$$

and the second law for the rate of material dissipation requires (6.3.9)

$$\rho\theta\xi' \geq 0\,. \tag{6.11.10}$$

This model can be used for growth of biological tissues as well as for muscle activation, both of which require an external supply of energy, which is characterized by term b in the balance of energy (6.11.5). Here, the mechanobiological processes which control growth and muscle activation are not modeled explicitly and it is assumed that b is large enough to ensure that (6.11.10) is satisfied for all thermomechanical processes with growth.

The elastic dilatation J_e for the growing tissue is determined by integrating the evolution equation

$$\frac{\dot{J}_e}{J_e} = \mathbf{D}\cdot\mathbf{I} - r_m\,, \tag{6.11.11}$$

which includes the external rate of mass supply r_m. Following the work in [23] and using the modification in [24], r_m is specified by

$$r_m = \Gamma_m \ln\left(\frac{J_e}{J_h}\right), \tag{6.11.12}$$

so the evolution equation for the elastic dilatation J_e becomes

$$\frac{\dot{J}_e}{J_e} = \mathbf{D}\cdot\mathbf{I} - \Gamma_m \ln\left(\frac{J_e}{J_h}\right), \quad \Gamma_m \geq 0\,, \quad J_h > 0\,. \tag{6.11.13}$$

Also, the symmetric, positive-definite, unimodular tensor $\mathbf{B}'_e$ that characterizes elastic distortional deformations is determined by the evolution equation

$$\begin{aligned}
&\dot{\mathbf{B}}'_e = \mathbf{L}''\mathbf{B}'_e + \mathbf{B}'_e\mathbf{L}''^T - \Gamma\mathbf{A}_g\,,\\
&\mathbf{A}_g = \mathbf{B}'_e - \left(\frac{3}{\mathbf{B}'^{-1}_e\cdot\mathbf{H}}\right)\mathbf{H}\,, \quad \Gamma \geq 0\,, \quad \mathbf{H}' = \det(\mathbf{H})^{-1/3}\mathbf{H}\,,
\end{aligned} \tag{6.11.14}$$

where $\mathbf{H}$ is a positive-definite, symmetric tensor and $\mathbf{H}'$ is its unimodular part. These evolution equations model homeostasis, which is the process that causes a tendency for J_e to approach its homeostatic value J_h and for $\mathbf{B}'_e$ to approach its homeostatic value $\mathbf{H}'$. In particular, it can be seen that in the absence of loading ($\mathbf{L} = 0$) the rates at which J_e and $\mathbf{B}'_e$ approach their homeostatic values J_h and $\mathbf{H}'$ are controlled by the functions Γ_m and Γ, respectively. Moreover, it is noted that the modified form (6.11.12) simplifies the numerical algorithm for solving the evolution Eq. (6.11.13). Also, the two nontrivial invariants of $\mathbf{B}'_e$ satisfy the equations

$$\begin{aligned} \alpha_1 &= \mathbf{B}'_e \cdot \mathbf{I}\,, \quad \dot{\alpha}_1 = 2\mathbf{B}''_e \cdot \mathbf{D} - \Gamma \mathbf{A}_g \cdot \mathbf{I}\,, \\ \alpha_2 &= \mathbf{B}'_e \cdot \mathbf{B}'_e\,, \quad \dot{\alpha}_2 = 4(\mathbf{B}'^{\,2}_e - \tfrac{1}{3}\alpha_2 \mathbf{I}) \cdot \mathbf{D} - 2\Gamma \mathbf{A}_g \cdot \mathbf{B}'_e\,, \end{aligned} \tag{6.11.15}$$

where $\mathbf{B}''_e$ is the deviatoric part of $\mathbf{B}'_e$.

In this model, the Cauchy stress $\mathbf{T}$ is a function of the elastic deformations $J_e, \mathbf{B}'_e$ and the temperature θ

$$\mathbf{T} = \mathbf{T}(J_e, \mathbf{B}'_e, \theta)\,. \tag{6.11.16}$$

This constitutive equation is restricted so that zero-stress material states occur whenever the elastic deformations are given by $J_e = 1$ and $\mathbf{B}'_e = \mathbf{I}$ and the temperature equals the zero-stress reference temperature $\theta = \theta_z$,

$$\mathbf{T} = 0 \quad \text{whenever} \quad J_e = 1\,, \quad \mathbf{B}'_e = \mathbf{I}\,, \quad \theta = \theta_z\,. \tag{6.11.17}$$

It is well known that the homeostatic state of the skin on the human body is not in a zero-stress material state. For this reason, surgeons cut the skin parallel to tension lines to minimize scarring. Within the context of this model, the stress in the homeostatic state of the tissue can be nonzero

$$\mathbf{T}(J_h, \mathbf{H}', \theta_z) \neq 0\,. \tag{6.11.18}$$

In particular, constitutive equations can be proposed for the homeostatic values J_h and $\mathbf{H}'$ of J_e and $\mathbf{B}'_e$, respectively, to ensure that the stress field in the homeostatic state of the tissue matches measured nonzero values. Additional constitutive equations need to be proposed for the homeostasis rates Γ_m and Γ.

Zero-Stress Growth:
To understand the influence of the homeostatic values J_h of J_e and $\mathbf{H}'$ of $\mathbf{B}'_e$, it is convenient to consider zero-stress growth (6.11.17) for which the evolution Eqs. (6.11.13) and (6.11.14) require

$$\begin{aligned} &\mathbf{D}_z \cdot \mathbf{I} + \Gamma_m \ln(J_h) = 0\,, \\ &2\mathbf{D}''_z - \Gamma \left[\mathbf{I} - \left(\frac{3}{\mathbf{I} \cdot \mathbf{H}} \right) \mathbf{H} \right] = 0\,, \end{aligned} \tag{6.11.19}$$

where $\mathbf{D}_z''$ denotes the deviatoric part of the zero-stress rate of deformation value $\mathbf{D}_z$ of $\mathbf{D}$ for zero-stress growth. As a special case, specify $\mathbf{H}$ in the form

$$\mathbf{H} = \mathbf{I} + \mathbf{H}'' , \quad \mathbf{H}'' \cdot \mathbf{I} = 0 , \quad \mathbf{H}'' \cdot \mathbf{H}'' < \frac{2}{3} , \tag{6.11.20}$$

where $\mathbf{H}''$ is the deviatoric part of $\mathbf{H}$ and its magnitude is bounded to ensure that $\mathbf{H}$ remains positive-definite. It then follows that (6.11.19) can be solved for the zero-stress growth rate $\mathbf{D}_z$ to obtain

$$\mathbf{D}_z = -\frac{1}{3}\Gamma_m \ln(J_h)\,\mathbf{I} + \mathbf{D}_z'' , \quad \mathbf{D}_z'' = -\frac{1}{2}\Gamma\mathbf{H}'' . \tag{6.11.21}$$

Since Γ_m is non-negative, it follows that $J_h > 1$ causes a volumetric rate of contraction and $J_h < 1$ causes a volumetric rate of expansion. Moreover, the deviatoric part $\mathbf{D}_z''$ of $\mathbf{D}_z$ is in the opposite direction to $\mathbf{H}''$. Furthermore, the limited magnitude of $\mathbf{H}''$ does not limit the magnitude of $\mathbf{D}_z''$, which is determined by the value of Γ.

In a review of growth in living systems, Kuhl [13] presented evolution equations which model volumetric, area and fiber growth. Elastic deformation measures J_e, λ_n and λ_s associated with these growth processes and used in elastically anisotropic constitutive equations were developed in [23]. In addition, expressions for the homeostatic values J_h and $\mathbf{H}'$ and the homeostasis rates Γ_m and Γ associated with these growth processes were discussed in [23].

Elastic Volumetric Growth:
The elastic dilatation J_e associated with this growth process satisfies the evolution Eq. (6.11.13).

Elastic Area Growth:
The elastic area stretch λ_n associated with growth of an area element on a material surface with unit normal vector $\mathbf{n}$ in the current configuration is motivated by expressions for the material area element da and the unit normal $\mathbf{n}$ to a material surface. To develop these expressions, it is convenient to define the second-order tensor $\mathbf{N}$ by

$$\mathbf{N} = \mathbf{n} \otimes \mathbf{n} , \tag{6.11.22}$$

which should not be confused with the unit normal vector $\mathbf{N}$ in Nanson's formula (3.3.35). Moreover, using the result (3.5.26)

$$\dot{\mathbf{n}} = -[\mathbf{L}^T - (\mathbf{D} \cdot \mathbf{n} \otimes \mathbf{n})\,\mathbf{I}]\,\mathbf{n} , \tag{6.11.23}$$

it can be shown that $\mathbf{N}$ satisfies the evolution equation

$$\dot{\mathbf{N}} = 2(\mathbf{D} \cdot \mathbf{N})\,\mathbf{N} - \mathbf{L}^T\mathbf{N} - \mathbf{N}\mathbf{L} . \tag{6.11.24}$$

Also, using (3.3.35) is can be shown that

$$\frac{da}{dA} = (J^{-4/3}\mathbf{B}' \cdot \mathbf{N})^{-1/2}, \quad \mathbf{B}' = J^{-2/3}\mathbf{F}\mathbf{F}^T. \tag{6.11.25}$$

Motivated by this expression, the elastic area stretch λ_n is defined by

$$\lambda_n = (J_e^{-4/3}\mathbf{B}'_e \cdot \mathbf{N})^{-1/2}. \tag{6.11.26}$$

Then, with the help of the evolution Eqs. (6.11.13), (6.11.14) and (6.11.24), the elastic area stretch λ_n satisfies the evolution equation

$$\frac{\dot{\lambda}_n}{\lambda_n} = (\mathbf{I} - \mathbf{N}) \cdot \mathbf{D} - \frac{2}{3}\Gamma_m \ln\left(\frac{J_e}{J_h}\right) + \frac{1}{2}\Gamma(\mathbf{B}'_e \cdot \mathbf{N})^{-1}(\mathbf{A}_g \cdot \mathbf{N}). \tag{6.11.27}$$

Elastic Fiber Growth:
The elastic fiber stretch λ_s associated with growth of a fiber in the direction of the unit vector **s** in the current configuration is motivated by expressions for the stretch λ and unit direction **s** of a material fiber. To develop these expressions, it is convenient to define the second-order tensor **S** by

$$\mathbf{S} = \mathbf{s} \otimes \mathbf{s}, \tag{6.11.28}$$

which should not be confused with the unit vector **S** in (3.3.12a) or the symmetric Piola-Kirchhoff stress in (4.6.14). Moreover, using the result (3.5.22)

$$\dot{\mathbf{s}} = [\mathbf{L} - (\mathbf{D} \cdot \mathbf{s} \otimes \mathbf{s})\,\mathbf{I}]\,\mathbf{s}, \tag{6.11.29}$$

it can be shown that **S** satisfies the evolution equation

$$\dot{\mathbf{S}} = \mathbf{L}\mathbf{S} + \mathbf{S}\mathbf{L}^T - 2(\mathbf{D} \cdot \mathbf{S})\,\mathbf{S}. \tag{6.11.30}$$

Also, using (3.3.8) and (3.3.12c) it can be shown that

$$\lambda = (J^{-2/3}\mathbf{B}'^{-1} \cdot \mathbf{S})^{-1/2}. \tag{6.11.31}$$

Motivated by this expression, the elastic fiber stretch λ_s is defined by

$$\lambda_s = (J_e^{-2/3}\mathbf{B}'^{-1}_e \cdot \mathbf{S})^{-1/2}. \tag{6.11.32}$$

Then, with the help of the evolution Eqs. (6.11.13), (6.11.14) and (6.11.30), the elastic fiber stretch λ_s satisfies the evolution equation

$$\frac{\dot{\lambda}_s}{\lambda_s} = \mathbf{S} \cdot \mathbf{D} - \frac{1}{3}\Gamma_m \ln\left(\frac{J_e}{J_h}\right) - \frac{1}{2}\Gamma(\mathbf{B}'^{-1}_e \cdot \mathbf{S})^{-1}(\mathbf{B}'^{-1}_e\mathbf{A}_g\mathbf{B}'^{-1}_e \cdot \mathbf{S}). \tag{6.11.33}$$

Constitutive Equations:
Following the work in [23], the constitutive equations for an elastically anisotropic thermoelastic material with growth are proposed in the forms

$$\begin{aligned}&\psi = \psi(J_e, \theta, \mathcal{V}), \quad \eta = \eta(J_e, \theta, \mathcal{V}), \qquad \mathbf{T} = \mathbf{T}(J_e, \theta, \mathbf{B}'_e, \mathcal{V}), \\ &\mathbf{p} = -\tfrac{\bar{\kappa}(J_e,\theta,\mathcal{V})}{\theta}\,\mathbf{g}, \quad \mathcal{V} = (\alpha_1, \alpha_2, \lambda_n, \lambda_s), \quad \mathbf{g} = \tfrac{\partial \theta}{\partial \mathbf{x}},\end{aligned} \tag{6.11.34}$$

where the non-negative function $\bar{\kappa}$ that represents the heat conduction coefficient should not be confused with an isotropic hardening variable κ. Then, using the evolution Eqs. (6.11.13), (6.11.15), (6.11.27) and (6.11.33), the stress $\mathbf{T}$ and entropy η are specified by

$$\begin{aligned}\mathbf{T} &= \rho\left[J_e\frac{\partial \psi}{\partial J_e}\mathbf{I} + 2\frac{\partial \psi}{\partial \alpha_1}\mathbf{B}''_e + 4\frac{\partial \psi}{\partial \alpha_2}\left(\mathbf{B}'^{\,2}_e - \frac{1}{3}\alpha_2\mathbf{I}\right)\right. \\ &\qquad \left. + \frac{\partial \psi}{\partial \lambda_n}\lambda_n(\mathbf{I} - \mathbf{N}) + \frac{\partial \psi}{\partial \lambda_s}\lambda_s\mathbf{S}\right], \\ \eta &= -\frac{\partial \psi}{\partial \theta}.\end{aligned} \tag{6.11.35}$$

Notice that the component of stress due to the elastic area stretch λ_n is isotropic in the plane normal to $\mathbf{n}$ and the stress due to the elastic fiber stretch λ_s is in the $\mathbf{S}$ direction. Also, the stress $\mathbf{T}$ can be written in terms of the pressure p and its deviatoric part $\mathbf{T}''$, such that

$$\begin{aligned}&\mathbf{T} = -p\,\mathbf{I} + \mathbf{T}'', \quad p = -\rho\left(J_e\frac{\partial \psi}{\partial J_e} + \frac{2}{3}\frac{\partial \psi}{\partial \lambda_n}\lambda_n + \frac{1}{3}\frac{\partial \psi}{\partial \lambda_s}\lambda_s\right), \\ &\mathbf{T}'' = \rho\left[2\frac{\partial \psi}{\partial \alpha_1}\mathbf{B}''_e + 4\frac{\partial \psi}{\partial \alpha_2}\left(\mathbf{B}'^{\,2}_e - \frac{1}{3}\alpha_2\mathbf{I}\right) + \frac{1}{3}\frac{\partial \psi}{\partial \lambda_n}\lambda_n(\mathbf{I} - 3\mathbf{N})\right. \\ &\qquad \left. + \frac{1}{3}\frac{\partial \psi}{\partial \lambda_s}\lambda_s(3\mathbf{S} - \mathbf{I})\right].\end{aligned} \tag{6.11.36}$$

Modeling Area Growth:
To model area growth, it is assumed that the growth is isotropic in a material surface that has unit normal $\mathbf{n}$ in the current configuration and $\mathbf{H}'$ is specified by

$$\mathbf{H}' = \frac{1}{h}(\mathbf{I} - \mathbf{N}) + h^2\mathbf{N}, \quad h > 0, \tag{6.11.37}$$

where $\mathbf{N}$ is defined by (6.11.22), $\mathbf{n}$ is defined by the evolution Eq. (6.11.23) and h is a positive scalar that controls the rate of area growth which needs to be specified by an evolution equation for $\dot{h}$.

To understand the implications of the constitutive form (6.11.37), consider the special case when the velocity gradient $\mathbf{L}$ is specified by

$$\mathbf{L} = \mathbf{D} = \frac{1}{2}\left(\frac{\dot{a}}{a}\right)(\mathbf{I} - \mathbf{N}) + \frac{\dot{\lambda}}{\lambda}\mathbf{N}\,, \quad \mathbf{D} \cdot \mathbf{I} = \frac{\dot{a}}{a} + \frac{\dot{\lambda}}{\lambda}\,, \tag{6.11.38}$$

with a and λ being arbitrary functions of time. For this velocity field, it follows from (6.11.23) and (6.11.24) that $\mathbf{n}$ and $\mathbf{N}$ remain constant. Using (6.11.27) and [(6.11.33) with $\mathbf{S}$ replaced by $\mathbf{N}$], it can be shown that a represents the area stretch of the material surface that is normal to $\mathbf{n}$ and λ represents the stretch of a material fiber that is normal to this material surface.

Next, $\mathbf{B}'_e$ and its inverse are specified by

$$\mathbf{B}'_e = \frac{1}{b_e}(\mathbf{I} - \mathbf{N}) + b_e^2\mathbf{N}\,, \quad \mathbf{B}'^{-1}_e = b_e(\mathbf{I} - \mathbf{N}) + \frac{1}{b_e^2}\mathbf{N}\,, \quad b_e > 0\,, \tag{6.11.39}$$

where b_e is a positive scalar to be determined. Using this expression, it follows that h is the homeostatic value of the elastic stretch b_e of the fiber normal to the material surface. Moreover, the distortional invariant α_1 in (6.11.15), the elastic area stretch λ_n in (6.11.26) and the elastic fiber stretch λ_s in (6.11.32) with $\mathbf{S}$ replaced by $\mathbf{N}$, become

$$\alpha_1 = \frac{2 + b_e^3}{b_e}\,, \quad \lambda_n = \frac{J_e^{2/3}}{b_e}\,, \quad \lambda_s = J_e^{1/3} b_e\,. \tag{6.11.40}$$

In addition, the evolution Eqs. (6.11.13) and (6.11.14) yield two scalar equations to determine J_e and b_e of the forms

$$\frac{\dot{J}_e}{J_e} = \frac{\dot{a}}{a} + \frac{\dot{\lambda}}{\lambda} - \Gamma_m \ln\left(\frac{J_e}{J_h}\right), \quad \frac{\dot{b}_e}{b_e} = -\frac{1}{3}\left(\frac{\dot{a}}{a}\right) + \frac{2}{3}\left(\frac{\dot{\lambda}}{\lambda}\right) - \Gamma\left[\frac{1 - \left(\frac{h}{b_e}\right)^3}{2 + \left(\frac{h}{b_e}\right)^3}\right]. \tag{6.11.41}$$

Therefore, steady-state solutions of these equations exist with

$$\{\, J_e\,, J_h\,, \Gamma_m\,, \frac{\dot{a}}{a}\,, \frac{\dot{\lambda}}{\lambda}\,, b_e\,, \mathbf{B}'_e\,, h\,, \Gamma \,\}\,, \tag{6.11.42}$$

being constants, such that

$$\frac{\dot{a}}{a} = \frac{2}{3}\Gamma_m \ln\left(\frac{J_e}{J_h}\right) - \Gamma\left[\frac{1 - \left(\frac{h}{b_e}\right)^3}{2 + \left(\frac{h}{b_e}\right)^3}\right], \quad \frac{\dot{\lambda}}{\lambda} = \frac{1}{3}\Gamma_m \ln\left(\frac{J_e}{J_h}\right) + \Gamma\left[\frac{1 - \left(\frac{h}{b_e}\right)^3}{2 + \left(\frac{h}{b_e}\right)^3}\right]. \tag{6.11.43}$$

In particular, area growth can occur without extension in the $\mathbf{n}$ direction with

$$\frac{\dot{a}}{a} = \Gamma_m \ln\left(\frac{J_e}{J_h}\right), \quad \frac{\dot{\lambda}}{\lambda} = 0\,, \quad \mathbf{D} \cdot \mathbf{I} = \frac{\dot{a}}{a} \quad \text{for} \quad \frac{1}{3}\Gamma_m \ln\left(\frac{J_e}{J_h}\right) = -\Gamma\left[\frac{1 - \left(\frac{h}{b_e}\right)^3}{2 + \left(\frac{h}{b_e}\right)^3}\right], \tag{6.11.44}$$

and area growth can occur without volume change with

$$\frac{\dot{a}}{a} = -\Gamma\left[\frac{1-\left(\frac{h}{b_e}\right)^3}{2+\left(\frac{h}{b_e}\right)^3}\right], \quad \frac{\dot{\lambda}}{\lambda} = -\frac{\dot{a}}{a}, \quad \mathbf{D}\cdot\mathbf{I}=0 \ \text{ for } \ \Gamma_m \ln\left(\frac{J_e}{J_h}\right)=0. \tag{6.11.45}$$

Modeling Fiber Growth:
To model fiber growth, it is assumed that the growth is isotropic in a surface normal to the unit direction **s** of the fiber in the current configuration and $\mathbf{H}'$ is specified by

$$\mathbf{H}' = \frac{1}{h}(\mathbf{I}-\mathbf{S}) + h^2\mathbf{S}, \quad h > 0, \tag{6.11.46}$$

where **S** is defined by (6.11.28), **s** is defined by the evolution Eq. (6.11.29) and h is a positive scalar that controls the rate of fiber growth which needs to be specified by an evolution equation for $\dot{h}$.

To understand the implications of the constitutive form (6.11.46), consider the special case when the velocity gradient **L** is specified by

$$\mathbf{L} = \mathbf{D} = \frac{1}{2}\left(\frac{\dot{a}}{a}\right)(\mathbf{I}-\mathbf{S}) + \frac{\dot{\lambda}}{\lambda}\mathbf{S}, \quad \mathbf{D}\cdot\mathbf{I} = \frac{\dot{a}}{a} + \frac{\dot{\lambda}}{\lambda}, \tag{6.11.47}$$

with a and λ being arbitrary functions of time. For this velocity field, it follows from (6.11.29) and (6.11.30) that **s** and **S** remain constant. Using [(6.11.27) with **N** replaced by **S**] and (6.11.33), it can be shown that a represents the area stretch of the material surface that is normal to **s** and λ represents the stretch of the material fiber that is in the direction **s**.

Next, $\mathbf{B}'_e$ and its inverse are specified in the forms (6.11.39) with **N** replaced by **S**

$$\mathbf{B}'_e = \frac{1}{b_e}(\mathbf{I}-\mathbf{S}) + b_e^2\mathbf{S}, \quad \mathbf{B}'^{-1}_e = b_e(\mathbf{I}-\mathbf{S}) + \frac{1}{b_e^2}\mathbf{S}, \quad b_e > 0, \tag{6.11.48}$$

where b_e is a positive scalar to be determined. Using this expression, it follows that h is the homeostatic value of the elastic stretch b_e of the fiber in the direction **s** normal to the material surface. Moreover, the distortional invariant α_1 in (6.11.15), the elastic area stretch λ_n in [(6.11.26) with **N** replaced by **S**] and the elastic fiber stretch λ_s in (6.11.32) are given by (6.11.40). In addition, the evolution Eqs. (6.11.13) and (6.11.14) yield two scalar Eq. (6.11.41) to determine J_e and b_e. Therefore, steady-state solutions of these equations exist with

$$J_e, J_h, \Gamma_m, \frac{\dot{a}}{a}, \frac{\dot{\lambda}}{\lambda}, b_e, \mathbf{B}'_e, h, \Gamma, \tag{6.11.49}$$

being constants, such that

$$\frac{\dot{a}}{a} = \frac{2}{3}\Gamma_m \ln\left(\frac{J_e}{J_h}\right) - \Gamma\left[\frac{1-\left(\frac{h}{b_e}\right)^3}{2+\left(\frac{h}{b_e}\right)^3}\right], \quad \frac{\dot{\lambda}}{\lambda} = \frac{1}{3}\Gamma_m \ln\left(\frac{J_e}{J_h}\right) + \Gamma\left[\frac{1-\left(\frac{h}{b_e}\right)^3}{2+\left(\frac{h}{b_e}\right)^3}\right]. \tag{6.11.50}$$

In particular, fiber growth can occur without area change normal to the fiber with

$$\frac{\dot{a}}{a} = 0, \quad \frac{\dot{\lambda}}{\lambda} = \Gamma_m \ln\left(\frac{J_e}{J_h}\right), \quad \mathbf{D}\cdot\mathbf{I} = \frac{\dot{\lambda}}{\lambda}, \quad \frac{2}{3}\Gamma_m \ln\left(\frac{J_e}{J_h}\right) = \Gamma\left[\frac{1-\left(\frac{h}{b_e}\right)^3}{2+\left(\frac{h}{b_e}\right)^3}\right], \tag{6.11.51}$$

and fiber growth can occur without volume change with

$$\frac{\dot{a}}{a} = -\frac{\dot{\lambda}}{\lambda}, \quad \frac{\dot{\lambda}}{\lambda} = \Gamma\left[\frac{1-\left(\frac{h}{b_e}\right)^3}{2+\left(\frac{h}{b_e}\right)^3}\right], \quad \mathbf{D}\cdot\mathbf{I} = 0, \quad \Gamma_m \ln\left(\frac{J_e}{J_h}\right) = 0, \tag{6.11.52}$$

which is the same as the solution (6.11.45).

Modeling Muscle Activation:
These equations have also been used to model muscle activation and details can be found in [23].

6.12 Jump Conditions for the Thermomechanical Balance Laws

The purpose of this section is to develop jump conditions for the global thermomechanical balance laws. Specifically, it is recalled from Sect. 6.2 that within the context of the thermomechanical theory proposed by Green and Naghdi [7, 8], the current mass density ρ, the current position $\mathbf{x}$ of a material point and the absolute temperature θ are determined by the global forms of the conservation of mass and the balances of linear momentum and entropy (6.2.1)

$$\frac{d}{dt}\int_P \rho dv = 0, \tag{6.12.1a}$$

$$\frac{d}{dt}\int_P \rho\mathbf{v} dv = \int_P \rho\mathbf{b} dv + \int_{\partial P} \mathbf{t} da, \tag{6.12.1b}$$

$$\frac{d}{dt}\int_P \rho\eta dv = \int_P \rho(s+\xi) dv - \int_{\partial P} \mathbf{p}\cdot\mathbf{n} da. \tag{6.12.1c}$$

Moreover, the balance of angular momentum (6.2.2) and the balance of energy (6.2.3)

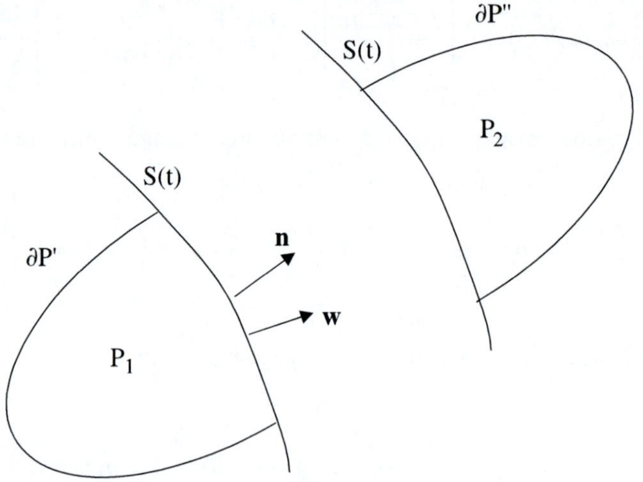

Fig. 6.3 A material region with a singular moving surface $S(t)$

$$\frac{d}{dt}\int_P (\mathbf{x}\times\rho\mathbf{v})dv = \int_P (\mathbf{x}\times\rho\mathbf{b})dv + \int_{\partial P} \mathbf{x}\times\mathbf{t}da\,, \tag{6.12.2a}$$

$$\begin{aligned}\frac{d}{dt}\int_P \left(\rho\varepsilon + \frac{1}{2}\rho\mathbf{v}\cdot\mathbf{v}\right)dv &= \int_P (\rho\mathbf{b}\cdot\mathbf{v})dv + \int_{\partial P} \mathbf{t}\cdot\mathbf{v}da\\ &+ \int_P \rho\theta s dv - \int_{\partial P} \theta\mathbf{p}\cdot\mathbf{n}da\end{aligned} \tag{6.12.2b}$$

are identically satisfied for all thermomechanical processes.

The discussion in Sect. 3.10 considered the material region P with closed material boundary ∂P to be divided into two parts P_1 and P_2 by the singular $S(t)$ that moves through the material (see Fig. 6.3). Furthermore, the intersection of ∂P_1 with ∂P was denoted by $\partial P'$ and the intersection of ∂P_2 with ∂P was denoted by $\partial P''$. Mathematically, this separation is summarized by (3.10.14)

$$\begin{aligned} &P = P_1 \cup P_2\,, && \partial P' = \partial P_1 \cap \partial P\,, && \partial P'' = \partial P_2 \cap \partial P\,,\\ &\partial P = \partial P' \cup P''\,, && \partial P_1 = \partial P' \cup S\,, && \partial P_2 = \partial P'' \cup S\,. \end{aligned} \tag{6.12.3}$$

A discussion of the motion of singular surfaces in fluid mechanics can be found in [34].

Next, the generalized transport theorem (3.10.16) is given by

$$\begin{aligned}\frac{d}{dt}\int_P \boldsymbol{\phi}(\mathbf{x},t)dv &= \int_{P_1} (\dot{\boldsymbol{\phi}} + \boldsymbol{\phi}\,\mathrm{div}\mathbf{v})dv + \int_{P_2} (\dot{\boldsymbol{\phi}} + \boldsymbol{\phi}\,\mathrm{div}\mathbf{v})dv\\ &- \int_{S(t)} [[\boldsymbol{\phi}\,\{(\mathbf{w}-\mathbf{v})\cdot\mathbf{n}\}]]da\,,\end{aligned} \tag{6.12.4}$$

where points on this singular surface move with velocity $\mathbf{w}$ and the unit normal to $S(t)$ outward from the part P_1 is denoted by $\mathbf{n}$. Also, the jump operator $[[\boldsymbol{\phi}]]$ is defined by (3.10.17)

$$[[\boldsymbol{\phi}\,\{(\mathbf{w}-\mathbf{v})\cdot\mathbf{n}\}]] = \boldsymbol{\phi}_2\{(\mathbf{w}-\mathbf{v}_2)\cdot\mathbf{n}\} - \boldsymbol{\phi}_1\{(\mathbf{w}-\mathbf{v}_1)\cdot\mathbf{n}\}\,, \tag{6.12.5}$$

where $\boldsymbol{\phi}_1$ and $\mathbf{v}_1$ are the values of $\boldsymbol{\phi}$ and $\mathbf{v}$ in part P_1 and $\boldsymbol{\phi}_2$ and $\mathbf{v}_2$ are the values of $\boldsymbol{\phi}$ and $\mathbf{v}$ in part P_2, all evaluated on the singular surface $S(t)$. In addition, $\mathbf{w}$ and $\mathbf{n}$ are the same on both sides of $S(t)$ (3.10.18)

$$\mathbf{w}_1 = \mathbf{w}_2 = \mathbf{w}\,, \quad \mathbf{n}_1 = \mathbf{n}_2 = \mathbf{n}\,. \tag{6.12.6}$$

Now, it is assumed that the local forms of the balance laws (6.12.1)

$$\dot{\rho} + \rho\text{div}\mathbf{v} = 0\,, \tag{6.12.7a}$$

$$\rho\dot{\mathbf{v}} = \rho\mathbf{b} + \text{div}\mathbf{T}\,, \tag{6.12.7b}$$

$$\rho\dot{\eta} = \rho(s+\xi) - \text{div}\mathbf{p}\,, \tag{6.12.7c}$$

and the local forms of the balance laws (6.12.2)

$$\mathbf{x}\times\rho\dot{\mathbf{v}} = \mathbf{x}\times\rho\mathbf{b} + \text{div}(\mathbf{x}\times\mathbf{T})\,, \tag{6.12.8a}$$

$$\rho\dot{\varepsilon} + \rho\dot{\mathbf{v}}\cdot\mathbf{v} = \rho\mathbf{b}\cdot\mathbf{v} + \rho\theta s + \text{div}(\mathbf{v}\cdot\mathbf{T} - \theta\mathbf{p}) \tag{6.12.8b}$$

are valid in each part P_1 and P_2 where use has been made of the expression for (4.3.24) for the traction vector $\mathbf{t}$

$$\mathbf{t} = \mathbf{T}\mathbf{n}\,. \tag{6.12.9}$$

Applying the generalized tranport theorem (6.12.4) to the global form (6.12.1a) of the conservation of mass and using the local Eq. (6.12.7a) in each of the parts P_1 and P_2 yields

$$\int_{S(t)} [[\rho\,\{(\mathbf{w}-\mathbf{v})\cdot\mathbf{n}\}]]da = 0\,. \tag{6.12.10}$$

Assuming continuity of the integrand along $S(t)$ requires the jump condition on mass

$$[[m]] = 0\,, \quad m = \rho\,[(\mathbf{w}-\mathbf{v})\cdot\mathbf{n}]\,, \tag{6.12.11}$$

to be valid for all points on $S(t)$.

Since the internal rate of entropy production ξ in the balance of entropy can be singular at the singular surface, this balance law needs special attention so it will be used as an example for the other balance laws. Specifically, due to this singularity, it follows that

$$\int_P \rho\xi dv = \int_{P_1} \rho\xi dv + \int_{P_2} \rho\xi dv + \int_{S(t)} m\,\Xi da\,, \tag{6.12.12}$$

where it is assumed that the singularity in ξ is integrable across $S(t)$ to yield the finite value Ξ. In contrast, the external rate of entropy supply s is assumed to be bounded across $S(t)$ so that

$$\int_P \rho s dv = \int_{P_1} \rho s dv + \int_{P_2} \rho s dv\,. \tag{6.12.13}$$

Now, applying the generalized transport theorem (6.12.4) to the rate of change of entropy and using (6.12.11) and the local Eqs. (6.12.7a) and (6.12.7c) yields

$$\frac{d}{dt}\int_P \rho\eta dv = \int_{P_1} [\rho(s+\xi) - \mathrm{div}\mathbf{p}]dv + \int_{P_2} [\rho(s+\xi) - \mathrm{div}\mathbf{p}]dv$$
$$- \int_{S(t)} [[m\,\eta]]da\,. \tag{6.12.14}$$

However, application of the divergence theorem yields

$$\begin{aligned} \int_{P_1} \mathrm{div}\mathbf{p} dv &= \int_{\partial P'} \mathbf{p}\cdot\mathbf{n} da + \int_{S(t)} \mathbf{p}_1\cdot\mathbf{n} da \\ \int_{P_2} \mathrm{div}\mathbf{p} dv &= \int_{\partial P''} \mathbf{p}\cdot\mathbf{n} da - \int_{S(t)} \mathbf{p}_2\cdot\mathbf{n} da\,, \end{aligned} \tag{6.12.15}$$

so that

$$\int_{P_1} \mathrm{div}\mathbf{p} dv + \int_{P_2} \mathrm{div}\mathbf{p} dv = \int_{\partial P} \mathbf{p}\cdot\mathbf{n} da - \int_{S(t)} [[\mathbf{p}\cdot\mathbf{n}]]da\,. \tag{6.12.16}$$

Thus, with the help of (6.12.12) and (6.12.13), (6.12.14) can be rewritten in the form

$$\frac{d}{dt}\int_P \rho\eta dv = \int_P \rho(s+\xi)dv - \int_{\partial P} \mathbf{p}\cdot\mathbf{n} da - \int_{S(t)} \Big(m\,\Xi + [[m\,\eta - \mathbf{p}\cdot\mathbf{n}]]\Big)da\,. \tag{6.12.17}$$

Now, using the global balance laws (6.12.1c) and assuming continuity of the integrand over $S(t)$ requires the jump condition on entropy

$$m\,\Xi + [[m\,\eta - \mathbf{p}\cdot\mathbf{n}]] = 0\,, \tag{6.12.18}$$

to be valid for all points on $S(t)$.

Following this same procedure for the other balance laws and assuming that $\rho\mathbf{b}$ and θ are bounded across $S(t)$, the jump conditions for balance laws (6.12.1) can be summarized as

$$[[m]] = 0\,, \tag{6.12.19a}$$

$$[[m\mathbf{v} + \mathbf{T}\mathbf{n}]] = 0\,, \tag{6.12.19b}$$

$$m\,\Xi + [[m\,\eta - \mathbf{p}\cdot\mathbf{n}]] = 0\,, \tag{6.12.19c}$$

and the jump conditions for the balance laws (6.12.2) can be summarized as

$$\mathbf{x} \times [[m\mathbf{v} + \mathbf{T}\mathbf{n}]] = 0\,, \tag{6.12.20a}$$

$$[[m(\varepsilon + \frac{1}{2}\mathbf{v}\cdot\mathbf{v}) + \mathbf{v}\cdot\mathbf{T}\mathbf{n} - \theta\mathbf{p}\cdot\mathbf{n}]] = 0\,, \tag{6.12.20b}$$

where m is defined by (6.12.11) and $\mathbf{x}$ is continuous across $S(t)$.

Notice that the jump in angular momentum (6.12.20a) is automatically satisfied when the jump in linear momentum (6.12.19b) is satisfied. In contrast with the local Eq. (6.12.7), the jump condition (6.12.19c) is used to determine the internal rate of entropy production Ξ due to the jump in entropy across $S(t)$ and the jump condition (6.12.20b) on energy is used to determine the jump in temperature θ.

References

1. Ambrosi D, Ateshian GA, Arruda EM, Cowin SC, Dumais J, Goriely A, Holzapfel GA, Humphrey JD, Kemkemer R, Kuhl E (2011) Perspectives on biological growth and remodeling. J Mech Phys Solids 59:863–883
2. Ateshian GA, Costa KD, Azeloglu EU, Morrison B, Hung CT (2009) Continuum modeling of biological tissue growth by cell division, and alteration of intracellular osmolytes and extracellular fixed charge density. J Biomech Eng 131:101001
3. Bar-On E, Rubin MB, Yankelevsky DZ (2003) Thermomechanical constitutive equations for the dynamic response of ceramics. Int J Solids Struct 40:4519–4548
4. Carroll M, Holt AC (1972) Suggested modification of the P-α model for porous materials. J Appl Phys 43:759–761
5. Coleman BD, Noll W (1963) The thermodynamics of elastic materials with heat conduction and viscosity. Arch Rat Mech Anal 13:167–178
6. Flory PJ (1961) Thermodynamic relations for high elastic materials. Trans Faraday Soc 57:829–838
7. Green AE, Naghdi PM (1977) On thermodynamics and the nature of the second law. Proc R Soc Lond A 357:253–270
8. Green AE, Naghdi PM (1978) The second law of thermodynamics and cyclic processes. J Appl Mech 45:487–492
9. Gurson AL (1977) Continuum theory of ductile rupture by void nucleation and growth: Part I-Yield criteria and flow rules for porous ductile media. J Eng Mater Technol 99:2–15
10. Herbold, EB, Homel MA, Rubin MB (2019) A thermomechanical breakage model for shock-loaded granular media. J Mech Phys Solids 137:103813
11. Herrmann W (1969) Constitutive equation for the dynamic compaction of ductile porous materials. J Appl Phys 40:2490–2499
12. Humphrey JD, Rajagopal KR (2002) A constrained mixture model for growth and remodeling of soft tissues. Math Model Methods Appl Sci 12:407–430
13. Kuhl E (2014) Growing matter: a review of growth in living systems. J Mech Behav Biomed Mater 29:529–543

14. Lee EH, Rubin MB (2021) Modeling anisotropic inelastic effects in sheet metal forming using microstructual vectors (Part I: Theory). Int J Plast
15. Rodriguez EK, Hoger A, McCulloch AD (1994) The constitutive equations for rate sensitive plastic materials. J Biomech 27:455–467
16. Rubin MB (1987) An elastic-viscoplastic model for metals subjected to high compression. J Appl Mech 54:532–538
17. Rubin MB (1992) Hyperbolic heat conduction and the second law. Int J Eng Sci 30:1665–1676
18. Rubin MB, Attia AV (1996) Calculation of hyperelastic response of finitely deformed elastic-viscoplastic materials. Int J Numer Methods Eng 39:309–320
19. Rubin MB, Elata D, Attia AV (1996) Modeling added compressibility of porosity and the thermomechanical response of wet porous rock with application to Mt. Helen Tuff. Int J Solids Struct 33:761–793
20. Rubin MB, Jabareen M (2008) Physically based invariants for nonlinear elastic orthotropic solids. J Elast 90:1–18
21. Rubin MB, Jabareen M (2011) Further developments of physically based invariants for nonlinear elastic orthotropic solids. J Elast 103:289–294
22. Rubin MB, Einav I (2011) A large deformation breakage model of granular materials including porosity and inelastic distortional deformation rate. Int J Eng Sci 49:1151–1169
23. Rubin MB, Safadi MM, Jabareen M (2015) A unified theoretical structure for modeling interstitial growth and muscle activation in soft tissues. Int J Eng Sci 90:1–26
24. Rubin MB (2016) A viscoplastic model for the active component in cardiac muscle. Biomech Model Mechanobiol 15:965–982
25. Rubin MB, Vorobiev O, Vitali E (2016) A thermomechanical anisotropic model for shock loading of elastic-plastic and elastic-viscoplastic materials with application to jointed rock. Comput Mech 58:107–128
26. Rubin MB, Herbold EB (2020) An analytical expression for temperature in a thermodynamically consistent model with a Mie-Gruneisen equation for pressure. Int J Impact Eng 143:103612
27. Rubin MB (2019) A new approach to modeling the thermomechanical, orthotropic, elastic-inelastic response of soft materials. Mech Soft Mater 1:3. https://doi.org/10.1007/s42558-018-0003-8
28. Safadi MM, Rubin MB (2017) A new analysis of stresses in arteries based on an Eulerian formulation of growth in tissues. Int J Eng Sci 118:40–55
29. Safadi MM, Rubin MB (2017) A new approach to modeling early cardiac morphogenesis during c-looping. Int J Eng Sci 117:1–19
30. Safadi MM, Rubin MB (2018) Significant differences in the mechanical modeling of confined growth predicted by the Lagrangian and Eulerian formulations. Int J Eng Sci 129:63–83
31. Sciumè G, Shelton S, Gray WG, Miller CT, Hussain F, Ferrari M, Decuzzi P, Schrefler BA (2013) A multiphase model for three-dimensional tumor growth. New J Phys 15:015005
32. Stracuzzi A, Rubin MB, Wahlsten A (2019) A thermomechanical theory for porous tissues with diffusion of fluid and micromechanical modeling of porosity. Mech Res Commun 97:112–122
33. Taber LA (1995) Biomechanics of growth, remodeling, and morphogenesis. Appl Mech Rev 48:487–545
34. Truesdell Clifford, Rajagopal KR (2000) An introduction to the mechanics of fluids. Birkhluser, Boston, MA
35. Vorobiev OYu, Rubin MB (2018) A thermomechanical anisotropic continuum model for geological materials with multiple joint sets. Int J Numer Anal Methods Geomech 42:1366–1388

Appendix A
Eigenvalues, Eigenvectors and Principal Invariants of a Tensor

The objective of this appendix is to briefly review some basic properties of eigenvalues and eigenvectors.

A.1 Eigenvalues and Eigenvectors

The vector $\mathbf{v}$ is said to be an eigenvector of a real second-order tensor $\mathbf{T}$ with the associated eigenvalue σ if

$$\mathbf{T}\mathbf{v} = \sigma \mathbf{v}\,, \quad T_{ij} v_j = \sigma v_i \,. \tag{A.1.1}$$

It follows that the characteristic equation for determining the three values of the eigenvalue σ is given by

$$\det(\mathbf{T} - \sigma \mathbf{I}) = -\sigma^3 + \sigma^2 I_1(\mathbf{T}) + \sigma I_2(\mathbf{T}) + I_3(\mathbf{T}) = 0\,, \tag{A.1.2}$$

where $I_1(\mathbf{T})$, $I_2(\mathbf{T})$ and $I_3(\mathbf{T})$ are the principal invariants of $\mathbf{T}$ defined by

$$\begin{aligned} I_1(\mathbf{T}) &= \mathbf{T}\cdot\mathbf{I} = \mathrm{tr}(\mathbf{T}) = T_{mm}\,, \\ I_2(\mathbf{T}) &= \frac{1}{2}[(\mathbf{T}\cdot\mathbf{I})^2 - \mathbf{T}\cdot\mathbf{T}^T] = \frac{1}{2}[T_{mm}T_{nn} - T_{mn}T_{nm}]\,, \\ I_3(\mathbf{T}) &= \det\mathbf{T} = \frac{1}{6}\varepsilon_{ijk}\varepsilon_{rst}T_{ir}T_{js}T_{kt}\,. \end{aligned} \tag{A.1.3}$$

For the remainder of this appendix, $\mathbf{T}$ is restricted to be a symmetric tensor. Then, it can be shown that the three roots of the cubic equation (A.1.2) are real. Also, it can be shown that the three independent eigenvectors $\mathbf{v}$ obtained by solving (A.1.1) can be chosen to form an orthonormal set of vectors.

M. B. Rubin, *Continuum Mechanics with Eulerian Formulations of Constitutive Equations*, Solid Mechanics and Its Applications 265,
https://doi.org/10.1007/978-3-030-57776-6

Recalling that $\mathbf{T}$ can be separated into its spherical part $T\,\mathbf{I}$ and its deviatoric part $\mathbf{T}''$ such that

$$\begin{aligned} \mathbf{T} &= T\mathbf{I} + \mathbf{T}'' , & T_{ij} &= T\delta_{ij} + T''_{ij} , \\ T &= \tfrac{1}{3}(\mathbf{T}\cdot\mathbf{I}) = \tfrac{1}{3}T_{mm} , & \mathbf{T}''\cdot\mathbf{I} &= T''_{mm} = 0 , \end{aligned} \tag{A.1.4}$$

it follows that when $\mathbf{v}$ is an eigenvector of $\mathbf{T}$, it is also an eigenvector of $\mathbf{T}''$

$$\mathbf{T}''\mathbf{v} = (\mathbf{T} - T\mathbf{I})\mathbf{v} = (\sigma - T)\mathbf{v} = \sigma''\mathbf{v} , \tag{A.1.5}$$

with the associated eigenvalue σ related to σ'' by

$$\sigma = T + \sigma'' . \tag{A.1.6}$$

A.2 Closed form Solution of the Characteristic Equation

Since the first principal invariant of the symmetric, deviatoric tensor $\mathbf{T}''$ vanishes, the characteristic equation for σ'' is given by

$$\det(\mathbf{T}'' - \sigma''\mathbf{I}) = -\sigma''^3 + \sigma''\left(\frac{\sigma_e^3}{3}\right) + J_3 = 0 , \tag{A.2.1}$$

where the alternative invariants σ_e and J_3 have been defined by

$$\sigma_e = \sqrt{\frac{3}{2}\mathbf{T}''\cdot\mathbf{T}''} = \sqrt{-3I_2(\mathbf{T}'')} , \quad J_3 = \det\mathbf{T}'' = I_3(\mathbf{T}'') . \tag{A.2.2}$$

Note that if σ_e vanishes, then $\mathbf{T}''$ vanishes so that from (A.2.1) σ'' vanishes, and from (A.1.6) it follows that all three eigenvalues are equal

$$\sigma = T \quad \text{for} \quad \sigma_e = 0 . \tag{A.2.3}$$

If σ_e does not vanish, then (A.2.1) can be divided by $(\sigma_e/3)^3$ to obtain

$$\left(\frac{3\sigma''}{\sigma_e}\right)^3 - 3\left(\frac{3\sigma''}{\sigma_e}\right) - 2\hat{J}_3 = 0 , \tag{A.2.4}$$

where the invariant $\hat{J}_3$ is defined by

$$-1 \le \hat{J}_3 = \frac{27J_3}{2\sigma_e^3} \le 1 . \tag{A.2.5}$$

Since (A.2.4) is in the standard form of a cubic equation, the solution can be obtained easily using the trigonometric form

$$\sigma_1'' = \frac{2\sigma_e}{3}\cos\left(\frac{\pi}{6}+\beta\right),$$
$$\sigma_2'' = \frac{2\sigma_e}{3}\sin(\beta), \qquad \text{(A.2.6)}$$
$$\sigma_3'' = -\frac{2\sigma_e}{3}\cos\left(\frac{\pi}{6}-\beta\right),$$

where the Lode angle [1–4] β is defined by

$$\sin(3\beta) = -\hat{J}_3, \quad -\frac{\pi}{6} \le \beta \le \frac{\pi}{6}, \qquad \text{(A.2.7)}$$

and the eigenvalues σ_1'', σ_2'' and σ_3'' are ordered so that

$$\sigma_1'' \ge \sigma_2'' \ge \sigma_3''. \qquad \text{(A.2.8)}$$

Once these values have been determined, the three solutions of (A.1.2) can be calculated using (A.1.6).

A.3 Triaxial States of Stress

Determining failure surfaces for soils, which cannot support tension, often requires triaxial tests which place an impermeable flexible membrane around the lateral surface of a circular cylindrical specimen. Then, the specimen is placed in a testing machine to apply axial contraction and extension. By applying axial contraction in conjunction with fluid pressure on the outside of the membrane, it is possible to load the specimen in a hydrostatic state of stress. Then, by maintaining the value of the pressure on the lateral surface of the cylindrical specimen and applying axial extension or contraction, it is possible to create states of Triaxial Extension (TXE) or Triaxial Compression (TXC), respectively. By applying appropriate values of axial compression and lateral pressure, it is possible to maintain the hoop stress zero, which creates a state of pure Torsion (TOR).

If **T** in the above equations represents the Cauchy stress, then these states of stress can be characterized by the Lode angle β or the invariant $\hat{J}_3$ by

$$\beta = \tfrac{\pi}{6}, \quad \hat{J}_3 = -1 \quad \text{for } TXC,$$
$$\beta = 0, \quad \hat{J}_3 = 0 \quad \text{for } TOR, \qquad \text{(A.3.1)}$$
$$\beta = -\tfrac{\pi}{6}, \quad \hat{J}_3 = 1 \quad \text{for } TXE.$$

References

1. Cristescu N (2012) Dynamic plasticity. Elsevier
2. Lode W (1926) Versuche uber den Einfluss der mittleren Hauptspannung auf das Fliessen der Metalle Eisen, Kupfer, und Nickel. Z. Physik 36:913–939
3. Rubin MB (1991) Simple, convenient isotropic failure surface. J Eng Mech 117:348–369
4. Rubin MB (2020) Erratum to: Rubin MB (1991) Simple, convenient isotropic failure surface. J Eng Mech 117:2(348). https://doi.org/10.1061/(ASCE)0733-9399

Appendix B
Consequences of Continuity

The objective of this appendix is to discuss the continuity of a function and some consequences of continuity.

B.1 Continuity of a Function

A function $\phi(\mathbf{x}, t)$ is said to be continuous with respect to the position $\mathbf{x}$ in a region R if for every position $\mathbf{y}$ in R and every $\varepsilon > 0$ there exists a $\delta > 0$, such that

$$|\phi(\mathbf{x}, t) - \phi(\mathbf{y}, t)| < \varepsilon \text{ whenever } |\mathbf{x} - \mathbf{y}| < \delta\,. \tag{B.1.1}$$

In words, this means that by reducing the radius δ of a spherical region, it is possible to reduce the maximum difference between two values of a continuous function $\phi(\mathbf{x}, t)$ in the spherical region to any positive finite value ε.

B.2 Application to the Local Form of a Global Equation

Theorem: If $\phi(\mathbf{x}, t)$ is *continuous* in R and the global equation

$$\int_P \phi dv = 0 \tag{B.2.1}$$

is valid for *every* part P in R, then the necessary and sufficient condition for the validity of (B.2.1) is that ϕ vanishes at every point in R

$$\phi(\mathbf{x}, t) = 0 \text{ on } R\,, \tag{B.2.2}$$

which is considered to be the local form of the global equation (B.2.1).

M. B. Rubin, *Continuum Mechanics with Eulerian Formulations of Constitutive Equations*, Solid Mechanics and Its Applications 265,
https://doi.org/10.1007/978-3-030-57776-6

Proof of Sufficiency:

If $\phi = 0$ on R, then (B.2.1) is trivially satisfied.
Proof of Necessity:

Necessity is proved by contradiction. Suppose that a point $\mathbf{y}$ on R exists for which $\phi(\mathbf{y}, t) > 0$. Then, by continuity of ϕ, there exists a finite region P_δ defined by the delta sphere such that

$$\begin{aligned} &|\phi(\mathbf{x}, t) - \phi(\mathbf{y}, t)| < \frac{1}{2}\phi(\mathbf{y}, t) \text{ whenever } |\mathbf{x} - \mathbf{y}| < \delta\,, \\ &-\frac{1}{2}\phi(\mathbf{y}, t) < \phi(\mathbf{x}, t) - \phi(\mathbf{y}, t) < \frac{1}{2}\phi(\mathbf{y}, t) \text{ whenever } |\mathbf{x} - \mathbf{y}| < \delta\,. \end{aligned} \tag{B.2.3}$$

This equation can be rewritten in the form

$$\frac{1}{2}\phi(\mathbf{y}, t) < \phi(\mathbf{x}, t) < \frac{3}{2}\phi(\mathbf{y}, t) \text{ whenever } |\mathbf{x} - \mathbf{y}| < \delta\,. \tag{B.2.4}$$

Since P_δ is a finite region, its volume V_δ is positive

$$V_\delta = \int_{P_\delta} dv > 0\,. \tag{B.2.5}$$

Then, taking the integral of (B.2.4) over P_δ and using (B.2.5) yields

$$\int_{P_\delta} \phi dv > \int_{P_\delta} \frac{1}{2}\phi(\mathbf{y}, t) dv = \frac{1}{2}\phi(\mathbf{y}, t) V_\delta > 0\,, \tag{B.2.6}$$

which contradicts the condition (B.2.1) so ϕ on R cannot be positive.

Similarly, suppose that a point $\mathbf{y}$ on R exits for which $\phi(\mathbf{y}, t) < 0$. Then, by continuity of ϕ, there exists a finite region P_δ defined by the delta sphere such that

$$\begin{aligned} &|\phi(\mathbf{x}, t) - \phi(\mathbf{y}, t)| < -\frac{1}{2}\phi(\mathbf{y}, t) \text{ whenever } |\mathbf{x} - \mathbf{y}| < \delta\,, \\ &\frac{1}{2}\phi(\mathbf{y}, t) < \phi(\mathbf{x}, t) - \phi(\mathbf{y}, t) < -\frac{1}{2}\phi(\mathbf{y}, t) \text{ whenever } |\mathbf{x} - \mathbf{y}| < \delta\,. \end{aligned} \tag{B.2.7}$$

This equation can be rewritten in the form

$$\frac{3}{2}\phi(\mathbf{y}, t) < \phi(\mathbf{x}, t) < \frac{1}{2}\phi(\mathbf{y}, t) \text{ whenever } |\mathbf{x} - \mathbf{y}| < \delta\,. \tag{B.2.8}$$

Since P_δ is a finite region, its volume V_δ is positive. Then, taking the integral of (B.2.8) over P_δ and using (B.2.5) yields

$$\int_{P_\delta} \phi dv < \int_{P_\delta} \frac{1}{2}\phi(\mathbf{y}, t) dv = \frac{1}{2}\phi(\mathbf{y}, t) V_\delta < 0\,, \tag{B.2.9}$$

which contradicts the condition (B.2.1) so ϕ on R cannot be negative.

Combining the results (B.2.6) and (B.2.9) yields the result that ϕ must vanish at every point on R, which proves the necessity of the local Eq. (B.2.2).

Appendix C
Lagrange Multipliers

The objective of this appendix is to discuss the use of Lagrange multipliers to find stationary values of a function subject to constraints.

C.1 Special Case

As a special case, let $f = f(x_1, x_2, x_3)$ be a real valued function of the three real valued variables x_i and assume that f is continuously differentiable. The function f has a *stationary value (extremum)* at the point $\mathbf{x} = \mathbf{x}_0$ if

$$df = \frac{\partial f}{\partial x_1} dx_1 + \frac{\partial f}{\partial x_2} dx_2 + \frac{\partial f}{\partial x_3} dx_3 = 0 \text{ at } \mathbf{x} = \mathbf{x}_0 \,. \tag{C.1.1}$$

If the variables x_i are independent, the condition (C.1.1) requires

$$\frac{\partial f}{\partial x_1} = \frac{\partial f}{\partial x_2} = \frac{\partial f}{\partial x_3} = 0 \text{ at } \mathbf{x} = \mathbf{x}_0 \,. \tag{C.1.2}$$

Next, consider the problem of finding the points $\mathbf{x} = \mathbf{x}_0$ which make f stationary and also satisfy the *kinematic constraint* condition

$$\phi(x_1, x_2, x_3) = 0 \,. \tag{C.1.3}$$

In other words, from the set of all points which satisfy the constraint (C.1.3), it is necessary to search for those points $\mathbf{x} = \mathbf{x}_0$ which also make f stationary. To this end, (C.1.3) is differentiated to obtain paths on the constraint surface

$$d\phi = \frac{\partial \phi}{\partial x_1} dx_1 + \frac{\partial \phi}{\partial x_2} dx_2 + \frac{\partial \phi}{\partial x_3} dx_3 = 0 \,. \tag{C.1.4}$$

M. B. Rubin, *Continuum Mechanics with Eulerian Formulations of Constitutive Equations*, Solid Mechanics and Its Applications 265,
https://doi.org/10.1007/978-3-030-57776-6

The condition for f to be stationary is again given by (C.1.1) but now the results (C.1.2) are not necessary conditions since x_i are dependent and must satisfy the constraint (C.1.3).

The method of Lagrange multipliers suggests that constraint (C.1.4) be multiplied by an arbitrary scalar λ. Then, the result is subtracted from the condition (C.1.1) that f be stationary to obtain

$$\left(\frac{\partial f}{\partial x_1} - \lambda\frac{\partial \phi}{\partial x_1}\right) dx_1 + \left(\frac{\partial f}{\partial x_2} - \lambda\frac{\partial \phi}{\partial x_2}\right) dx_2 + \left(\frac{\partial f}{\partial x_3} - \lambda\frac{\partial \phi}{\partial x_3}\right) dx_3 = 0 \text{ at } \mathbf{x} = \mathbf{x}_0 \,. \tag{C.1.5}$$

For the constraint (C.1.3) to be active, it is necessary that at each point at least one of the partial derivatives be nonzero $\partial\phi/\partial x_i \neq 0$. For definiteness, it is assumed that

$$\frac{\partial \phi}{\partial x_3} \neq 0 \text{ at } \mathbf{x} = \mathbf{x}_0 \,. \tag{C.1.6}$$

Next, the value of λ can be specified so that the coefficient of dx_3 in (C.1.5) vanishes

$$\frac{\partial f}{\partial x_3} = \lambda\frac{\partial \phi}{\partial x_3} \text{ at } \mathbf{x} = \mathbf{x}_0 \,, \tag{C.1.7}$$

with (C.1.5) reducing to

$$\left(\frac{\partial f}{\partial x_1} - \lambda\frac{\partial \phi}{\partial x_1}\right) dx_1 + \left(\frac{\partial f}{\partial x_2} - \lambda\frac{\partial \phi}{\partial x_2}\right) dx_2 = 0 \text{ at } \mathbf{x} = \mathbf{x}_0 \,. \tag{C.1.8}$$

Now since $\partial\phi/\partial x_3 \neq 0$, the value of dx_3 can be chosen so that the constraint Eq. (C.1.4) is satisfied for arbitrary values of dx_1 and dx_2. This means that dx_1 and dx_2 can be specified independently in (C.1.8) so it can be concluded that

$$\frac{\partial f}{\partial x_i} = \lambda\frac{\partial \phi}{\partial x_i} \text{ at } \mathbf{x} = \mathbf{x}_0 \,. \tag{C.1.9}$$

In summary, of all the points satisfying the constraint (C.1.3), the ones that correspond to stationary values of f are the ones for which $\mathbf{x}_0$ and λ are determined by the four Eqs. (C.1.3) and (C.1.9). The result (C.1.9) also means that at the stationary value of f, the gradient of f is parallel to gradient of the constraint ϕ.

An Alternative Perspective

Another way of examining the same problem is to write the function f and the constraint ϕ in the forms

$$f = f(x_\alpha, x_3) \,, \quad \phi = \phi(x_\alpha, x_3) = 0 \,, \tag{C.1.10}$$

where a Greek index is defined to take only the values $\alpha = 1, 2$. Since $\partial\phi/\partial x_3 \neq 0$, the implicit function theorem states that a function $g(x_\alpha)$ exists such that when

$x_3 = g(x_\alpha)$ the constraint (C.1.10)$_2$ is automatically satisfied

$$\phi = \phi(x_\alpha, g(x_\alpha)) = 0 \tag{C.1.11}$$

for all x_α for which $\partial\phi/\partial x_3 \neq 0$. By substituting $x_3 = g(x_\alpha)$ into the function f

$$f = f(x_\alpha, g(x_\alpha)), \tag{C.1.12}$$

it is possible to limit attention only to points which automatically satisfy the constraint (C.1.11). Since x_α are independent variables in (C.1.12), it follows that the stationary values are determined by equation

$$df = \left[\frac{\partial f}{\partial x_\alpha} + \left(\frac{\partial f}{\partial x_3}\right)\frac{\partial g}{\partial x_\alpha}\right] dx_\alpha = 0, \tag{C.1.13}$$

so for stationary points

$$\frac{\partial f}{\partial x_\alpha} = -\left(\frac{\partial f}{\partial x_3}\right)\frac{\partial g}{\partial x_\alpha} \text{ at } \mathbf{x} = \mathbf{x}_0. \tag{C.1.14}$$

However, since the constraint (C.1.11) is satisfied with x_α being independent variables, it follows that

$$d\phi = \left[\frac{\partial \phi}{\partial x_\alpha} + \left(\frac{\partial \phi}{\partial x_3}\right)\frac{\partial g}{\partial x_\alpha}\right] dx_\alpha = 0, \quad -\frac{\partial g}{\partial x_\alpha} = \frac{\frac{\partial \phi}{\partial x_\alpha}}{\frac{\partial \phi}{\partial x_3}}. \tag{C.1.15}$$

Thus, (C.1.14) can be written in the form

$$\frac{\partial f}{\partial x_\alpha} = \lambda \frac{\partial \phi}{\partial x_\alpha}, \quad \lambda = \frac{\frac{\partial f}{\partial x_3}}{\frac{\partial \phi}{\partial x_3}}. \tag{C.1.16}$$

These conditions can be simplified to obtain

$$\frac{\partial f}{\partial x_i} = \lambda \frac{\partial \phi}{\partial x_i}, \quad \phi = 0 \text{ at } \mathbf{x} = \mathbf{x}_0, \tag{C.1.17}$$

which are the same equations as (C.1.9).

Another Approach

Another approach suggests defining an auxiliary function h and a Lagrange multiplier λ, such that

$$h = h(x_1, x_2, x_3, \lambda) = f(x_1, x_2, x_3) - \lambda\phi(x_1, x_2, x_3). \tag{C.1.18}$$

Thinking of x_1, x_2, x_3 and λ as independent variables, the auxiliary function h will have a stationary value when

$$\begin{aligned} \frac{\partial h}{\partial x_1} &= \frac{\partial f}{\partial x_1} - \lambda \frac{\partial \phi}{\partial x_1} = 0\,, \\ \frac{\partial h}{\partial x_2} &= \frac{\partial f}{\partial x_2} - \lambda \frac{\partial \phi}{\partial x_2} = 0\,, \\ \frac{\partial h}{\partial x_3} &= \frac{\partial f}{\partial x_3} - \lambda \frac{\partial \phi}{\partial x_3} = 0\,, \\ \frac{\partial h}{\partial \lambda} &= -\phi = 0 \text{ at } \mathbf{x} = \mathbf{x}_0\,. \end{aligned} \tag{C.1.19}$$

These equations can be rewritten in the compact forms

$$\frac{\partial f}{\partial x_i} = \lambda \frac{\partial \phi}{\partial x_i}\,, \quad \phi = 0 \text{ at } \mathbf{x} = \mathbf{x}_0\,, \tag{C.1.20}$$

which again requires the gradient of f to be parallel to the gradient of the constraint ϕ.

C.2 A More General Case

For a more general case, let f be a real valued function of $m + n$ variables

$$f = f(x_i, y_j) \quad \text{for } i = 1, 2, \ldots, m \text{ and } j = 1, 2, \ldots, n\,, \tag{C.2.1}$$

and consider n constraint equations of the forms

$$\phi_r = \phi_r(x_i, y_j) \quad \text{for } r = 1, 2, \ldots, n\,. \tag{C.2.2}$$

Furthermore, assume that f and ϕ_r are continuously differentiable and that all the constraints ϕ_r are active so that

$$\det(\partial \phi_r / \partial y_j) \neq 0 \quad \text{for } r, j = 1, 2, \ldots, n\,. \tag{C.2.3}$$

Now form the auxiliary function h defined by

$$h = f - \sum_{r=1}^{n} \lambda_r \phi_r\,, \tag{C.2.4}$$

where λ_r are scalars called Lagrange multipliers that are independent of x_i and y_j. The method of Lagrange multipliers suggests that the points which satisfy the n

constraints (C.2.2) and which cause the function f to be stationary are determined by solving the $m + 2n$ equations

$$\begin{aligned}
\frac{\partial h}{\partial x_i} &= \frac{\partial f}{\partial x_i} - \sum_{r=1}^{n} \lambda_r \frac{\partial \phi_r}{\partial x_i} = 0 \quad \text{for } i = 1, 2, \ldots, m\,, \\
\frac{\partial h}{\partial y_j} &= \frac{\partial f}{\partial y_j} - \sum_{r=1}^{n} \lambda_r \frac{\partial \phi_r}{\partial y_j} = 0 \quad \text{for } j = 1, 2, \ldots, n\,, \\
\frac{\partial h}{\partial \lambda_r} &= -\phi_r = 0 \qquad\qquad\qquad\quad \text{for } r = 1, 2, \ldots, n\,,
\end{aligned} \tag{C.2.5}$$

for the $m + 2n$ unknowns x_i, y_j and λ_r. This method produces a *necessary* condition for f to have a stationary value. However, each stationary point must be checked individually to determine if it is a maximum, minimum or a point of inflection.

Appendix D
Stationary Values of the Normal and Shear Stresses

The objective of this appendix is to develop expressions for the stationary values of the normal and shear components of the traction vector and the associated planes.

D.1 Stationary Values of the Normal Component of Stress Vector

Letting T_{ij} be the components of the Cauchy stress $\mathbf{T}$ relative to a fixed rectangular Cartesian coordinate system with base vectors $\mathbf{e}_i$, the normal component σ of stress vector acting on the plane defined by the unit outward normal n_j is given by

$$\sigma(\mathbf{n}) = \mathbf{t} \cdot \mathbf{n} = \mathbf{T} \cdot (\mathbf{n} \otimes \mathbf{n}) = T_{ij} n_i n_j \,. \tag{D.1.1}$$

For a given value of stress $\mathbf{T}$ at a point, it is of interest to find the planes $\mathbf{n}$ for which σ is stationary. Since $\mathbf{n}$ is a unit vector, its components n_j satisfy the constraint equation

$$\phi = n_j n_j - 1 \,. \tag{D.1.2}$$

Using the method of Lagrange multipliers described in Appendix C the auxilliary function h takes the form

$$h(n_j, \lambda) = \sigma - \lambda\phi = T_{ij} n_i n_j - \lambda(n_j n_j - 1) \,, \tag{D.1.3}$$

which is used to determine the values of n_j and λ for σ to be stationary by solving equations

$$\frac{\partial h}{\partial n_k} = 2(T_{kj} - \lambda\delta_{kj}) n_j = 0 \,, \quad \frac{\partial h}{\partial \lambda} = -(n_j n_j - 1) = 0 \,. \tag{D.1.4}$$

It follows from these equations that the stationary values of σ occur when

M. B. Rubin, *Continuum Mechanics with Eulerian Formulations of Constitutive Equations*, Solid Mechanics and Its Applications 265,
https://doi.org/10.1007/978-3-030-57776-6

$$\mathbf{T}\mathbf{n} = \lambda \mathbf{n}\,, \quad \mathbf{n} \cdot \mathbf{n} = 1\,. \tag{D.1.5}$$

This means that σ attains its stationary values on the three planes that are defined by **n** parallel to the principal directions of the stress tensor **T**. The associated stationary values of σ are the principal values of the stress tensor **T**. Since **T** is a real and symmetric tensor, these principal values and directions are real so the principal values σ_i can be ordered, such that

$$\sigma_1 \geq \sigma_2 \geq \sigma_3\,. \tag{D.1.6}$$

For convenience, the base vectors $\mathbf{e}_i$ of the Cartesian coordinate system are taken to be parallel to the principal directions $\mathbf{p}_i$ of **T** so that **T** can be represented in is spectral form

$$\begin{aligned} \mathbf{T} &= \sigma_1 \mathbf{p}_1 \otimes \mathbf{p}_1 + \sigma_2 \mathbf{p}_2 \otimes \mathbf{p}_2 + \sigma_3 \mathbf{p}_3 \otimes \mathbf{p}_3\,, \\ T_{ij} &= \mathbf{T} \cdot \mathbf{p}_i \otimes \mathbf{p}_j = \begin{pmatrix} \sigma_1 & 0 & 0 \\ 0 & \sigma_2 & 0 \\ 0 & 0 & \sigma_3 \end{pmatrix}\,. \end{aligned} \tag{D.1.7}$$

Using this choice of base vectors, it follows that

$$\begin{aligned} \mathbf{t} &= \mathbf{T}\mathbf{n} = \sigma_1 n_1 \mathbf{p}_1 + \sigma_2 n_2 \mathbf{p}_2 + \sigma_3 n_3 \mathbf{p}_3\,, \\ \sigma(\mathbf{n}) &= \sigma_1 n_1^2 + \sigma_2 n_2^2 + \sigma_3 n_3^2\,. \end{aligned} \tag{D.1.8}$$

Next, with the help of the ordering (D.1.6), it follows that

$$\begin{aligned} \sigma_1 &= \sigma_1 (n_1^2 + n_2^2 + n_3^2) \geq \sigma_1 n_1^2 + \sigma_2 n_2^2 + \sigma_3 n_3^2 = \sigma(\mathbf{n})\,, \\ \sigma(\mathbf{n}) &= \sigma_1 n_1^2 + \sigma_2 n_2^2 + \sigma_3 n_3^2 \geq \sigma_3 (n_1^2 + n_2^2 + n_3^2) \geq \sigma_3\,, \end{aligned} \tag{D.1.9}$$

so $\sigma(\mathbf{n})$ is bounded above by the largest eigenvalue σ_1 and is bounded below by the smallest eigenvalue σ_3

$$\sigma_1 \geq \sigma(\mathbf{n}) \geq \sigma_3\,. \tag{D.1.10}$$

This means that the maximum value σ_1 of the normal stress σ occurs on the plane normal to the principal direction $\mathbf{p}_1$ and the minimum value σ_3 of normal stress σ occurs on the plane normal to the principal direction $\mathbf{p}_3$. The value σ_2 is called a minimax and is the value of σ on the plane normal to $\mathbf{p}_2$.

D.2 Stationary Values of the Shear Component of Stress

The shearing component $\mathbf{t}_s$ with magnitude τ of the traction vector **t** on a surface with unit normal **n** can be defined by

$$\mathbf{t}_s = \mathbf{t} - (\mathbf{t} \cdot \mathbf{n})\mathbf{n}\,, \quad \tau^2 = \mathbf{t}_s \cdot \mathbf{t}_s = \mathbf{t} \cdot \mathbf{t} - \sigma^2\,. \tag{D.2.1}$$

Thus, with the help of the spectral form of the stress **T** (D.1.7), it follows that

$$\tau^2 = \sigma_1^2 n_1^2 + \sigma_2^2 n_2^2 + \sigma_3^2 n_3^2 - (\sigma_1 n_1^2 + \sigma_2 n_2^2 + \sigma_3 n_3^2)^2\,. \tag{D.2.2}$$

Since τ and τ^2 both are stationary on the same plane, the stationary value of τ can be determined by using the method of Lagrange multiplier and taking the auxilliary function h in the form

$$h = h(n_j, \lambda) = \tau^2 - \lambda(n_j n_j - 1)\,, \tag{D.2.3}$$

which is used to determine the values of n_j and λ for τ to be stationary by solving equations

$$\begin{aligned}
\frac{\partial h}{\partial n_1} &= 2n_1\big[\sigma_1^2 - 2\sigma_1(\sigma_1 n_1^2 + \sigma_2 n_2^2 + \sigma_3 n_3^2) - \lambda\big] = 0\,,\\
\frac{\partial h}{\partial n_2} &= 2n_2\big[\sigma_2^2 - 2\sigma_2(\sigma_1 n_1^2 + \sigma_2 n_2^2 + \sigma_3 n_3^2) - \lambda\big] = 0\,,\\
\frac{\partial h}{\partial n_3} &= 2n_3\big[\sigma_3^2 - 2\sigma_3(\sigma_1 n_1^2 + \sigma_2 n_2^2 + \sigma_3 n_3^2) - \lambda\big] = 0\,,\\
\frac{\partial h}{\partial \lambda} &= -(n_j n_j - 1) = 0\,.
\end{aligned} \tag{D.2.4}$$

One solution of the Eq. (D.2.4) is given by

$$\begin{aligned}
\mathbf{n} &= \pm\mathbf{p}_1\,, \quad \tau = 0\,, \quad \sigma = \sigma_1\,,\\
\mathbf{n} &= \pm\mathbf{p}_2\,, \quad \tau = 0\,, \quad \sigma = \sigma_2\,,\\
\mathbf{n} &= \pm\mathbf{p}_3\,, \quad \tau = 0\,, \quad \sigma = \sigma_3\,.
\end{aligned} \tag{D.2.5}$$

Thus, the magnitude τ of the shear stress assumes its absolute minimum value of zero on the planes whose normals are in the principal directions of the stress tensor **T**. Furthermore, it is noted that on these same planes the normal stress σ assumes its stationary values.

A second solution of the Eq. (D.2.4) is given by

$$\begin{aligned}
\mathbf{n} &= \pm\frac{1}{\sqrt{2}}(\mathbf{p}_1 \pm \mathbf{p}_3)\,, \quad \tau = \frac{\sigma_1 - \sigma_3}{2}\,, \quad \sigma = \frac{\sigma_1 + \sigma_3}{2}\,,\\
\mathbf{n} &= \pm\frac{1}{\sqrt{2}}(\mathbf{p}_1 \pm \mathbf{p}_2)\,, \quad \tau = \frac{\sigma_1 - \sigma_2}{2}\,, \quad \sigma = \frac{\sigma_1 + \sigma_2}{2}\,,\\
\mathbf{n} &= \pm\frac{1}{\sqrt{2}}(\mathbf{p}_2 \pm \mathbf{p}_3)\,, \quad \tau = \frac{\sigma_2 - \sigma_3}{2}\,, \quad \sigma = \frac{\sigma_2 + \sigma_3}{2}\,.
\end{aligned} \tag{D.2.6}$$

Note that the maximum value of the magnitude τ of the shear stress is equal to one half of the difference between the maximum and minimum values of normal stress and it occurs on the plane whose normal bisects the angle between the normals to the planes of maximum and minimum normal stress. Also, note that the normal stress σ does not necessarily vanish on these planes.

Appendix E
Isotropic Tensors

The objective of this appendix is to develop expressions for isotropic tensors up to fourth order.

E.1 Definition of Isotropic Tensors

Let $\mathbf{e}_i$ and $\tilde{\mathbf{e}}_i$ be two sets of orthonormal base vectors that are connected by the orthogonal transformation $\mathbf{A}$

$$\begin{aligned} \mathbf{A} &= \mathbf{e}_i \otimes \tilde{\mathbf{e}}_i \,, \quad \mathbf{A}^T\mathbf{A} = \mathbf{I} \,, \\ A_{ij} &= \mathbf{A} \cdot \mathbf{e}_i \otimes \mathbf{e}_j = \tilde{\mathbf{e}}_i \cdot \mathbf{e}_j \,, \\ \tilde{A}_{ij} &= \mathbf{A} \cdot \tilde{\mathbf{e}}_i \otimes \tilde{\mathbf{e}}_j = \tilde{\mathbf{e}}_i \cdot \mathbf{e}_j \,, \end{aligned} \tag{E.1.1}$$

which shows that since $\mathbf{A}$ is a two-point tensor defined by the base vectors $\mathbf{e}_i$ and $\tilde{\mathbf{e}}_j$, its components A_{ij} relative to the basis $\mathbf{e}_i$ are the same as its components $\tilde{A}_{ij}$ relative to the basis $\tilde{\mathbf{e}}_i$.

Furthermore, let $\mathbf{T}$ be a tensor of any order with components $T_{ij...m}$ relative to the basis $\mathbf{e}_i$ and components $\tilde{T}_{ij...m}$ relative to the basis $\tilde{\mathbf{e}}_i$. Since $\mathbf{T}$ is a tensor, its components $T_{ij...m}$ and $\tilde{T}_{ij...m}$ are connected by the transformation relations

$$\tilde{T}_{ij...m} = A_{ir} A_{js} \ldots A_{mt} T_{rs...t} \,. \tag{E.1.2}$$

An Isotropic Tensor
A tensor is said to be isotropic if its components relative to any two right-handed orthonormal coordinate systems are equal. Mathematically, this means that

$$\tilde{T}_{ij...m} = T_{ij...m} \tag{E.1.3}$$

M. B. Rubin, *Continuum Mechanics with Eulerian Formulations of Constitutive Equations*, Solid Mechanics and Its Applications 265,
https://doi.org/10.1007/978-3-030-57776-6

holds for all proper orthogonal transformations $\mathbf{A}$ [$\det \mathbf{A} = +1$]. If (E.1.3) holds for all orthogonal transformations [i.e., including those with $\det \mathbf{A} = -1$], then the tensor is said to be isotropic with a center of symmetry.

E.2 Results for Specific Tensors

The Most General Zero-Order Isotropic tensors
By definition, scalar invariants satisfy the restriction (E.1.3) so they are zero-order isotropic tensors.

First-Order Isotropic Tensors
The most general first-order isotropic tensor is the zero vector

$$T_i = 0\,. \tag{E.2.1}$$

To prove this result, (E.1.2) and (E.1.3) for a first-order tensor require

$$T_i = A_{ij} T_j\,. \tag{E.2.2}$$

Taking A_{ij} to be

$$A_{ij} = \begin{pmatrix} -1 & 0 & 0 \\ 0 & -1 & 0 \\ 0 & 0 & 1 \end{pmatrix} \tag{E.2.3}$$

yields the restrictions

$$T_1 = -T_1 = 0\,, \quad T_2 = -T_2 = 0\,, \tag{E.2.4}$$

and then taking

$$A_{ij} = \begin{pmatrix} 1 & 0 & 0 \\ 0 & -1 & 0 \\ 0 & 0 & -1 \end{pmatrix} \tag{E.2.5}$$

yields the additional restriction

$$T_3 = -T_3 = 0\,, \tag{E.2.6}$$

so the first-order isotropic tensor must be the zero vector (E.2.1).

Second-Order Isotropic Tensors
The most general second-order isotropic tensor has the form

$$T_{ij} = \lambda \delta_{ij}\,, \tag{E.2.7}$$

where λ is a scalar invariant. To prove this result, (E.1.2) and (E.1.3) for a second-order tensor require

$$T_{ij} = A_{im}A_{jn}T_{mn}\,. \tag{E.2.8}$$

Taking A_{ij} to be

$$A_{ij} = \begin{pmatrix} 0 & 0 & 1 \\ 1 & 0 & 0 \\ 0 & 1 & 0 \end{pmatrix} \tag{E.2.9}$$

yields the restrictions

$$\begin{pmatrix} T_{11} & T_{12} & T_{13} \\ T_{21} & T_{22} & T_{23} \\ T_{31} & T_{32} & T_{33} \end{pmatrix} = \begin{pmatrix} T_{33} & T_{31} & T_{32} \\ T_{13} & T_{11} & T_{12} \\ T_{23} & T_{21} & T_{22} \end{pmatrix}. \tag{E.2.10}$$

Also, taking A_{ij} to be

$$A_{ij} = \begin{pmatrix} 0 & 0 & -1 \\ -1 & 0 & 0 \\ 0 & 1 & 0 \end{pmatrix} \tag{E.2.11}$$

yields the additional restrictions

$$\begin{pmatrix} T_{11} & T_{12} & T_{13} \\ T_{21} & T_{22} & T_{23} \\ T_{31} & T_{32} & T_{33} \end{pmatrix} = \begin{pmatrix} T_{33} & T_{31} & -T_{32} \\ T_{13} & T_{11} & -T_{12} \\ T_{23} & -T_{21} & T_{22} \end{pmatrix}. \tag{E.2.12}$$

Thus, from (E.2.10) and (E.2.12), it follows that

$$T_{11} = T_{22} = T_{33} = \lambda \quad \text{with all other } T_{ij} = 0\,, \tag{E.2.13}$$

so the second-order isotropic tensor must have the form (E.2.7).

Third-Order Isotropic Tensors

The most general third-order isotropic tensor has the form

$$T_{ijk} = \lambda\varepsilon_{ijk}\,, \tag{E.2.14}$$

where λ is a scalar invariant. To prove this result, (E.1.2) and (E.1.3) for a third-order tensor require

$$T_{ijk} = A_{im}A_{jn}A_{kr}T_{mnr}\,. \tag{E.2.15}$$

Denoting T_{ijk} by

$$T_{ijk} = \begin{pmatrix} T_{111} & T_{112} & T_{113} & T_{121} & T_{122} & T_{123} & T_{131} & T_{132} & T_{133} \\ T_{211} & T_{212} & T_{213} & T_{221} & T_{222} & T_{223} & T_{231} & T_{232} & T_{233} \\ T_{311} & T_{312} & T_{313} & T_{321} & T_{322} & T_{323} & T_{331} & T_{332} & T_{333} \end{pmatrix}, \tag{E.2.16}$$

and specifying A_{ij} by (E.2.9) yields the restrictions

$$T_{ijk} = \begin{pmatrix} T_{333} & T_{331} & T_{332} & T_{313} & T_{311} & T_{312} & T_{323} & T_{321} & T_{322} \\ T_{133} & T_{131} & T_{132} & T_{113} & T_{111} & T_{112} & T_{123} & T_{121} & T_{122} \\ T_{233} & T_{231} & T_{232} & T_{213} & T_{211} & T_{212} & T_{223} & T_{221} & T_{222} \end{pmatrix}. \tag{E.2.17}$$

Also, specifying A_{ij} by (E.2.11) yields the restrictions

$$T_{ijk} = \begin{pmatrix} -T_{333} & -T_{331} & T_{332} & -T_{313} & -T_{311} & T_{312} & T_{323} & T_{321} & -T_{322} \\ -T_{133} & -T_{131} & T_{132} & -T_{113} & -T_{111} & T_{112} & T_{123} & T_{121} & -T_{122} \\ T_{233} & T_{231} & -T_{232} & T_{213} & T_{211} & -T_{212} & -T_{223} & -T_{221} & -T_{222} \end{pmatrix}. \tag{E.2.18}$$

Then, using (E.2.16)–(E.2.18), it follows that

$$\begin{gathered} T_{ijk} = \begin{pmatrix} 0 & 0 & 0 & 0 & 0 & T_{123} & 0 & T_{132} & 0 \\ 0 & 0 & T_{213} & 0 & 0 & 0 & T_{231} & 0 & 0 \\ 0 & T_{312} & 0 & T_{321} & 0 & 0 & 0 & 0 & 0 \end{pmatrix}, \\ T_{123} = T_{312} = T_{231}, \quad T_{132} = T_{213} = T_{321}. \end{gathered} \tag{E.2.19}$$

Next, specifying A_{ij} by

$$A_{ij} = \begin{pmatrix} 0 & 0 & 1 \\ 0 & 1 & 0 \\ -1 & 0 & 0 \end{pmatrix} \tag{E.2.20}$$

yields the additional restriction

$$T_{123} = -T_{321}, \tag{E.2.21}$$

so the third-order isotropic tensor must have the form (E.2.14).

Fourth-Order Isotropic Tensors

The most general fourth-order isotropic tensor has the form

$$T_{ijkl} = \lambda\delta_{ij}\delta_{kl} + \mu\delta_{ik}\delta_{jl} + \gamma\delta_{il}\delta_{jk}, \tag{E.2.22}$$

where λ, μ and γ are scalar invariants. To prove this result, (E.1.2) and (E.1.3) for a fourth-order tensor require

$$T_{ijkl} = A_{im}A_{jn}A_{kr}A_{ls}T_{mnrs}. \tag{E.2.23}$$

By specifying A_{ij} in the forms (E.2.3) and (E.2.5), it can be shown that the 81 components of T_{ijkl} reduce to only 21 nonzero components which are denoted by $\bar{T}_{ijkl}$ with

$$\bar{T}_{ijkl} = \begin{pmatrix} T_{1111} & T_{1122} & T_{1133} & T_{1212} & T_{1221} & T_{1313} & T_{1331} \\ T_{2112} & T_{2121} & T_{2211} & T_{2222} & T_{2233} & T_{2323} & T_{2332} \\ T_{3113} & T_{3131} & T_{3223} & T_{3232} & T_{3311} & T_{3322} & T_{3333} \end{pmatrix} . \tag{E.2.24}$$

Specifying A_{ij} by (E.2.9) yields the restrictions

$$\bar{T}_{ijkl} = \begin{pmatrix} T_{3333} & T_{3311} & T_{3322} & T_{3131} & T_{3113} & T_{3232} & T_{3223} \\ T_{1331} & T_{1313} & T_{1133} & T_{1111} & T_{1122} & T_{1212} & T_{1221} \\ T_{2332} & T_{2323} & T_{2112} & T_{2121} & T_{2233} & T_{2211} & T_{2222} \end{pmatrix} . \tag{E.2.25}$$

Also, specifying A_{ij} by

$$A_{ij} = \begin{pmatrix} 0 & 0 & 1 \\ 0 & 1 & 0 \\ 1 & 0 & 0 \end{pmatrix} \tag{E.2.26}$$

yields the additional restrictions

$$\bar{T}_{ijkl} = \begin{pmatrix} T_{3333} & T_{3322} & T_{3311} & T_{3232} & T_{3223} & T_{3131} & T_{3113} \\ T_{2332} & T_{2323} & T_{2233} & T_{2222} & T_{2211} & T_{2121} & T_{2112} \\ T_{1331} & T_{1313} & T_{1221} & T_{1212} & T_{1133} & T_{1122} & T_{1111} \end{pmatrix} . \tag{E.2.27}$$

Thus, from (E.2.24), (E.2.25) and (E.2.27), it follows that

$$\begin{aligned} &T_{1111} = T_{2222} = T_{3333} , \\ &T_{1122} = T_{3311} = T_{2233} = T_{3322} = T_{2211} = T_{1133} = \lambda , \\ &T_{1212} = T_{3131} = T_{2323} = T_{3232} = T_{2121} = T_{1313} = \mu , \\ &T_{1331} = T_{3223} = T_{2112} = T_{3113} = T_{2332} = T_{1221} = \gamma . \end{aligned} \tag{E.2.28}$$

Next, specifying A_{ij} by

$$A_{ij} = \begin{pmatrix} \frac{1}{\sqrt{2}} & \frac{1}{\sqrt{2}} & 0 \\ -\frac{1}{\sqrt{2}} & \frac{1}{\sqrt{2}} & 0 \\ 0 & 0 & 1 \end{pmatrix} \tag{E.2.29}$$

yields the additional restriction

$$\begin{aligned} &T_{1111} = A_{1m} A_{1n} A_{1r} A_{1s} T_{mnrs} , \\ &T_{1111} = \frac{1}{4}(T_{1111} + T_{1122} + T_{1212} + T_{1221} + T_{2112} + T_{2121} + T_{2211} + T_{2222}) , \end{aligned} \tag{E.2.30}$$

so with the help of (E.2.28) and (E.2.30), it follows that

$$T_{1111} = T_{2222} = T_{3333} = \lambda + \mu + \gamma\,. \tag{E.2.31}$$

Thus, combining the results (E.2.28) and (E.2.31) shows that the most general fourth-order isotropic tensor must have the form (E.2.22).

Notice that T_{ijkl} in (E.2.22) automatically has the symmetries

$$T_{ijkl} = T_{klij}\,, \quad \mathbf{T}^{T(2)} = \mathbf{T}\,. \tag{E.2.32}$$

An Additional Restriction

If the fourth-order isotropic tensor is further restricted to be symmetric in its first two indices

$$T_{ijkl} = T_{jikl}\,, \quad {}^{LT}\mathbf{T} = \mathbf{T}\,, \tag{E.2.33}$$

then it can be shown that

$$\gamma = \mu\,, \tag{E.2.34}$$

so that T_{ijkl} reduces to

$$T_{ijkl} = \lambda\delta_{ij}\delta_{kl} + \mu(\delta_{ik}\delta_{jl} + \delta_{il}\delta_{jk})\,, \tag{E.2.35}$$

which is a doubly symmetric tensor satisfying the symmetries

$$T_{ijkl} = T_{klij} = T_{jikl} = T_{ijlk}\,, \quad \mathbf{T} = \mathbf{T}^{T(2)} = {}^{LT}\mathbf{T} = \mathbf{T}^{T}\,. \tag{E.2.36}$$

Appendix F
An Introduction to Tensors with Respect to Curvilinear Coordinates

The objective of this appendix is to provide a brief introduction to covariant and contravariant base vectors in general curvilinear coordinates as well as components of tensors relative to these base vectors. Also, the gradient, divergence, curl and Laplacian operators are introduced for general curvilinear coordinates.[1]

F.1 Covariant and Contravariant Base Vectors

The base vectors $\mathbf{e}_i$ in rectangular Cartesian coordinates are special in the sense that they are constant orthonormal vectors that specifically are independent of the rectangular Cartesian coordinates x_i. This means that the position vector $\mathbf{x}$ when expressed with respect to $\mathbf{e}_i$ can be represented in the form

$$\mathbf{x} = x_i \mathbf{e}_i \, . \tag{F.1.1}$$

Using this representation, it follows that the base vectors $\mathbf{e}_i$ can be obtained by differentiating the position vector with respect to the coordinates x_i, such that

$$\mathbf{e}_i = \frac{\partial \mathbf{x}}{\partial x_i} \, . \tag{F.1.2}$$

For general curvilinear coordinates, the notion of base vectors similar to (F.1.2) generalizes directly, but the position vector no longer admits the simple form (F.1.1) as a summation of coordinates times base vectors. For this case, the position vector $\mathbf{x}$ is expressed as a function of three independent coordinates θ^i and time t, such that

[1] Much of the content in this appendix has been adapted from [1] with permission.

M. B. Rubin, *Continuum Mechanics with Eulerian Formulations of Constitutive Equations*, Solid Mechanics and Its Applications 265,
https://doi.org/10.1007/978-3-030-57776-6

$$\mathbf{x} = \mathbf{x}(\theta^i, t)\,. \tag{F.1.3}$$

For example, $\mathbf{x}$ can be expressed in terms of the rectangular Cartesian base vectors $\mathbf{e}_i$

$$\mathbf{x} = x_m(\theta^i, t)\mathbf{e}_m\,, \tag{F.1.4}$$

with the rectangular Cartesian coordinates x_m being functions of θ^i and t.
Covariant Base Vectors
The covariant base vectors $\mathbf{g}_i$ associated with the position vector $\mathbf{x}$ in (F.1.3) are defined by

$$\mathbf{g}_i = \frac{\partial \mathbf{x}}{\partial \theta^i} = \mathbf{x}_{,i}\,, \tag{F.1.5}$$

where a comma is used to denote partial differentiation with respect to the coordinates θ^i. Geometrically, these base vectors $\mathbf{g}_i$ represent tangent vectors to the three curves defined by varying one of the coordinates θ^i while holding the other two coordinates constant.

It is important to note that since θ^i are general coordinates, they need not have the dimensions of length. Consequently, the base vectors $\mathbf{g}_i$ need not be dimensionless. Moreover, $\mathbf{g}_i$ are functions of θ^i and t and in general are not orthonormal vectors. However, since θ^i are independent coordinates, the vectors $\mathbf{g}_i$ are linearly independent vectors and the coordinates θ^i can be arranged so that $\mathbf{g}_i$ form a right-handed triad of vectors

$$g^{1/2} = \mathbf{g}_1 \times \mathbf{g}_2 \cdot \mathbf{g}_3 > 0\,. \tag{F.1.6}$$

Using the chain rule of differentiation, the length squared of a line element $d\mathbf{x}$ at a fixed time t can be related to the elemental changes in the coordinates $d\theta^i$ by the expression

$$d\mathbf{x} \cdot d\mathbf{x} = \mathbf{x}_{,i}\, d\theta^i \cdot \mathbf{x}_{,j}\, d\theta^j = (\mathbf{g}_i \cdot \mathbf{g}_j) d\theta^i d\theta^j = g_{ij} d\theta^i d\theta^j\,, \tag{F.1.7}$$

where g_{ij} is called the metric of the space and is defined by

$$g_{ij} = \mathbf{g}_i \cdot \mathbf{g}_j = g_{ji}\,. \tag{F.1.8}$$

Also, it can be shown that the quantity g defined in (F.1.6) is equal to the determinant of the metric g_{ij}

$$g = \det g_{ij}\,. \tag{F.1.9}$$

Notice in the above definitions that the coordinates θ^i have superscripts for indices and the base vectors $\mathbf{g}_i$ have subscripts for indices. This is because for analysis in curvilinear coordinates, it is necessary to distinguish between two types of bases: covariant bases which are formed using the covariant base vectors $\mathbf{g}_i$, and contravariant bases which are formed using reciprocal vectors $\mathbf{g}^i$ called contravariant base vectors. Specifically, the vectors $\mathbf{g}^i$ are defined by the cross-product operator such that

$$g^{1/2}\mathbf{g}^1 = \mathbf{g}_2 \times \mathbf{g}_3 \,, \quad g^{1/2}\mathbf{g}^2 = \mathbf{g}_3 \times \mathbf{g}_1 \,, \quad g^{1/2}\mathbf{g}^3 = \mathbf{g}_1 \times \mathbf{g}_2 \,. \tag{F.1.10}$$

Due to the properties of the cross product, the covariant vectors $\mathbf{g}_i$ and the contravariant vectors $\mathbf{g}^i$ are biorthogonal triads of vectors satisfying equations

$$\mathbf{g}_i \cdot \mathbf{g}^j = \delta_i^j \,, \tag{F.1.11}$$

where δ_i^j is the Kronecker delta taking the value 1 for $(i = j)$ and 0 for $(i \neq j)$. Using the definitions (F.1.10) and the expansion of the vector triple product

$$\mathbf{a} \times (\mathbf{b} \times \mathbf{c}) = (\mathbf{a} \cdot \mathbf{c})\mathbf{b} - (\mathbf{a} \cdot \mathbf{b})\mathbf{c} \,, \tag{F.1.12}$$

it can be shown that

$$g^{-1/2} = \mathbf{g}^1 \times \mathbf{g}^2 \cdot \mathbf{g}^3 > 0 \,, \tag{F.1.13}$$

so that $\mathbf{g}^i$ form a triad of right-handed linearly independent vectors. Also, the reciprocal metric g^{ij} is defined such that

$$g^{ij} = \mathbf{g}^i \cdot \mathbf{g}^j = g^{ji} \,. \tag{F.1.14}$$

Since $\mathbf{g}_i$ and $\mathbf{g}^i$ are both individually linearly independent triads of vectors, each of these triads spans the three-dimensional space so that both the triads $\mathbf{g}_i$ and $\mathbf{g}^i$ can be used as bases for vectors in three dimensions. In particular, the covariant vectors $\mathbf{g}_i$ can be represented in terms of the contravariant vectors and vice versa, such that

$$\mathbf{g}_i = g_{ij}\mathbf{g}^j \,, \quad \mathbf{g}^i = g^{ij}\mathbf{g}_j \,. \tag{F.1.15}$$

Using the definitions (F.1.10) and the expansion of the vector triple product (F.1.12), it can be shown that the covariant vectors $\mathbf{g}_i$ are related to cross products of the contravariant vectors $\mathbf{g}^i$ by the expressions

$$g^{-1/2}\mathbf{g}_1 = \mathbf{g}^2 \times \mathbf{g}^3, \quad g^{-1/2}\mathbf{g}_2 = \mathbf{g}^3 \times \mathbf{g}^1 \,, \quad g^{-1/2}\mathbf{g}_3 = \mathbf{g}^1 \times \mathbf{g}^2 \,. \tag{F.1.16}$$

F.2 Tensor Bases and Components of Tensors

The covariant components v_i and contravariant components v^i of an arbitrary vector $\mathbf{v}$ are defined in the usual way by taking the inner product of $\mathbf{v}$ with the appropriate base vectors

$$v_i = \mathbf{v} \cdot \mathbf{g}_i\,, \quad v^i = \mathbf{v} \cdot \mathbf{g}^i\,. \tag{F.2.1}$$

Moreover, in view of the biorthogonality of the covariant and contravariant base vectors, it follows that **v** can be represented in the equivalent forms

$$\mathbf{v} = v^i \mathbf{g}_i = v_i \mathbf{g}^i\,. \tag{F.2.2}$$

These equations express the fundamental property of a tensor that the tensor is independent of the basis with respect to which the components are evaluated. Also, it is emphasized that the components of a tensor depend explicitly on the choice of the basis.

Here, it is important to note that the covariant components are multiplied by the contravariant base vectors and vice versa. The summation convention applies when an index is repeated. For the special case of rectangular Cartesian coordinates and base vectors (F.1.1) and (F.1.2), the repeated indices are subscripts. However, for general curvilinear coordinates, the repeated indices are subscripts associated with covariant quantities and superscripts associated with contravariant quantities.

A general second-order tensor **T** has nine independent components which can be referred to a basis of nine tensors spanning the space of all second-order tensors. Using tensor products of covariant and contravariant vectors, it is possible to form four different sets of base tensors of the forms

$$\mathbf{g}_i \otimes \mathbf{g}_j\,, \quad \mathbf{g}^i \otimes \mathbf{g}^j\,, \quad \mathbf{g}_i \otimes \mathbf{g}^j\,, \quad \mathbf{g}^i \otimes \mathbf{g}_j\,. \tag{F.2.3}$$

It then follows that the covariant components T_{ij}, the contravariant components T^{ij} and the mixed components $T_i^{\ j}$ and $T^i_{\ j}$ of **T** are defined by

$$\begin{aligned} T_{ij} &= \mathbf{T} \cdot (\mathbf{g}_i \otimes \mathbf{g}_j)\,, \quad & T^{ij} &= \mathbf{T} \cdot (\mathbf{g}^i \otimes \mathbf{g}^j)\,, \\ T_i^{\ j} &= \mathbf{T} \cdot (\mathbf{g}_i \otimes \mathbf{g}^j)\,, \quad & T^i_{\ j} &= \mathbf{T} \cdot (\mathbf{g}^i \otimes \mathbf{g}_j)\,. \end{aligned} \tag{F.2.4}$$

Thus, the tensor **T** can be represented in the equivalent forms

$$\mathbf{T} = T_{ij}(\mathbf{g}^i \otimes \mathbf{g}^j) = T^{ij}(\mathbf{g}_i \otimes \mathbf{g}_j) = T_i^{\ j}(\mathbf{g}^i \otimes \mathbf{g}_j) = T^i_{\ j}(\mathbf{g}_i \otimes \mathbf{g}^j)\,. \tag{F.2.5}$$

Furthermore, because of the nature of the mixed components, it is necessary to distinguish between the locations of the first and second indices when writing subscripts or superscripts.

As an example, it is of interest to consider the second-order identity tensor **I** which has the properties that for an arbitrary vector **v**

$$\mathbf{I}\mathbf{v} = \mathbf{v}\,, \quad \mathbf{v}\mathbf{I} = \mathbf{v}\,. \tag{F.2.6}$$

In view of the properties of the covariant and contravariant vectors, it can be shown that **I** can be written in the equivalent forms

$$\mathbf{I} = \mathbf{g}_i \otimes \mathbf{g}^i = \mathbf{g}^i \otimes \mathbf{g}_i \,. \tag{F.2.7}$$

Thus, the components of $\mathbf{I}$ are given by

$$\begin{aligned} g_{ij} &= \mathbf{I} \cdot (\mathbf{g}_i \otimes \mathbf{g}_j)\,, \quad g^{ij} = \mathbf{I} \cdot (\mathbf{g}^i \otimes \mathbf{g}^j)\,, \\ \delta_i^j &= \mathbf{I} \cdot (\mathbf{g}_i \otimes \mathbf{g}^j)\,, \quad \delta_j^i = \mathbf{I} \cdot (\mathbf{g}^i \otimes \mathbf{g}_j)\,. \end{aligned} \tag{F.2.8}$$

Moreover, with the help of (F.2.7) and (F.2.8), it can be shown that

$$\delta_i^j = (\mathbf{g}_m \otimes \mathbf{g}^m) \cdot (\mathbf{g}_i \otimes \mathbf{g}^j) = g_{im} g^{mj} \,. \tag{F.2.9}$$

Consequently, taking the determinant of this expression and using (F.1.9) yields the result

$$g^{-1} = \det g^{ij} \,. \tag{F.2.10}$$

Obviously, it is possible to generalize the definitions of tensor bases (F.2.3), the components (F.2.4) and the representations (F.2.5) for tensors of general order M by taking a string of tensor products of M covariant or contravariant base vectors. For example, the covariant and contravariant bases for third-order tensors are given by

$$\mathbf{g}_i \otimes \mathbf{g}_j \otimes \mathbf{g}_k \,, \quad \mathbf{g}^i \otimes \mathbf{g}^j \otimes \mathbf{g}^k \,, \tag{F.2.11}$$

and the covariant components T_{ijk} and contravariant components T^{ijk} of a third-order tensor $\mathbf{T}$ are given by

$$T_{ijk} = \mathbf{T} \cdot (\mathbf{g}_i \otimes \mathbf{g}_j \otimes \mathbf{g}_k)\,, \quad T^{ijk} = \mathbf{T} \cdot (\mathbf{g}^i \otimes \mathbf{g}^j \otimes \mathbf{g}^k)\,. \tag{F.2.12}$$

Then, $\mathbf{T}$ admits the representations

$$\mathbf{T} = T_{ijk}(\mathbf{g}^i \otimes \mathbf{g}^j \otimes \mathbf{g}^k) = T^{ijk}(\mathbf{g}_i \otimes \mathbf{g}_j \otimes \mathbf{g}_k)\,. \tag{F.2.13}$$

Mixed components of $\mathbf{T}$ and representations of $\mathbf{T}$ in terms of these mixed components are determined in an obvious manner.

Using the previous definitions, it can be shown that the metrics g_{ij} and g^{ij} can be used to shift between covariant and contravariant components of a tensor. For example, with the help of the various representations of the second-order tensor $\mathbf{T}$ defined in (F.2.5), the result (F.1.11) and the definitions (F.1.8) and (F.1.14), it follows that

$$\begin{aligned} T_{ij} &= \mathbf{T} \cdot (\mathbf{g}_i \otimes \mathbf{g}_j) = T^{mn}(\mathbf{g}_m \otimes \mathbf{g}_n) \cdot (\mathbf{g}_i \otimes \mathbf{g}_j) = T^{mn} g_{mi} g_{nj} \,, \\ T_{ij} &= T_m{}^n(\mathbf{g}^m \otimes \mathbf{g}_n) \cdot (\mathbf{g}_i \otimes \mathbf{g}_j) = T_m{}^n \delta_i^m g_{nj} = T_i{}^n g_{nj} \,, \\ T_{ij} &= T^m{}_n(\mathbf{g}_m \otimes \mathbf{g}^n) \cdot (\mathbf{g}_i \otimes \mathbf{g}_j) = T^m{}_n g_{mi} \delta_j^n = T^m{}_j g_{mi} \,. \end{aligned} \tag{F.2.14}$$

Since the coordinates θ^i need not have the dimensions of length, the vectors $\mathbf{g}_i$ and $\mathbf{g}^i$ are not necessarily dimensionless. Therefore, the components of an arbitrary tensor $\mathbf{T}$ relative to the base tensors associated with the vectors $\mathbf{g}_i$ or $\mathbf{g}^i$ will not necessarily have the same units as the physical tensor $\mathbf{T}$. However, it is always possible to refer $\mathbf{T}$ to an orthogonal tensor basis which is associated with a right-handed set of orthonormal base vectors $\mathbf{e}_i$. Then, the components $T_{<ij...rs>}$ of $\mathbf{T}$ relative to these base tensors

$$T_{<ij...rs>} = \mathbf{T} \cdot (\mathbf{e}_i \otimes \mathbf{e}_j \otimes \cdots \otimes \mathbf{e}_r \otimes \mathbf{e}_s) \tag{F.2.15}$$

are called the physical components because they have the same dimensions as the physical quantity $\mathbf{T}$.

F.3 Basic Tensor Operations

In continuum mechanics it sometimes is desirable to refer to a material point Y by either its location $\mathbf{x}$ in the current configuration at time t or its location $\mathbf{X}$ in an arbitrary fixed reference configuration. For general coordinates θ^i, the mapping from θ^i to $\mathbf{X}$ remains a function of time

$$\mathbf{X} = \mathbf{X}(\theta^i, t) \,. \tag{F.3.1}$$

However, for the special case when θ^i are convected Lagrangian coordinates, the mapping from θ^i to $\mathbf{X}$ is independent of time

$$\mathbf{X} = \mathbf{X}(\theta^i) \,. \tag{F.3.2}$$

For either case, it is possible to define covariant vectors $\mathbf{G}_i$ and contravariant vectors $\mathbf{G}^i$ associated with the reference position vector $\mathbf{X}$ by the expressions

$$\begin{aligned} &\mathbf{G}_i = \tfrac{\partial \mathbf{X}}{\partial \theta^i}\,, && G^{1/2} = \mathbf{G}_1 \times \mathbf{G}_2 \cdot \mathbf{G}_3\,, && \mathbf{G}_i \cdot \mathbf{G}^j = \delta_i^{\,j}\,, \\ &G^{1/2}\mathbf{G}^1 = \mathbf{G}_2 \times \mathbf{G}_3\,, && G^{1/2}\mathbf{G}^2 = \mathbf{G}_3 \times \mathbf{G}_1\,, && G^{1/2}\mathbf{G}^3 = \mathbf{G}_1 \times \mathbf{G}_2\,. \end{aligned} \tag{F.3.3}$$

In the following, a number of tensor operations will be defined in terms of derivatives with respect to the present position $\mathbf{x}$ and the reference position $\mathbf{X}$ of a material point. To this end, the gradient of a scalar function f with respect to the present position $\mathbf{x}$ is a vector denoted by $\operatorname{grad} f$, and the gradient of f with respect to the reference position $\mathbf{X}$ is a vector denoted by $\operatorname{Grad} f$. These vectors can be conveniently expressed in terms of the contravariant base vectors in the forms

$$\begin{aligned} \operatorname{grad} f &= \partial f / \partial \mathbf{x} = \frac{\partial f}{\partial \theta^i} \mathbf{g}^i = f_{,i}\, \mathbf{g}^i\,, \\ \operatorname{Grad} f &= \partial f / \partial \mathbf{X} = \frac{\partial f}{\partial \theta^i} \mathbf{G}^i = f_{,i}\, \mathbf{G}^i\,. \end{aligned} \tag{F.3.4}$$

Furthermore, the gradients grad**T** and Grad**T** of an arbitrary tensor of order M ($M \geq 1$) are tensors of order $M + 1$ that can be expressed in the forms

$$\begin{aligned} \text{grad}\mathbf{T} &= \partial \mathbf{T}/\partial \mathbf{x} = \frac{\partial \mathbf{T}}{\partial \theta^i} \otimes \mathbf{g}^i = \mathbf{T},_i \otimes \mathbf{g}^i \,, \\ \text{Grad}\mathbf{T} &= \partial \mathbf{T}/\partial \mathbf{X} = \frac{\partial \mathbf{T}}{\partial \theta^i} \otimes \mathbf{G}^i = \mathbf{T},_i \otimes \mathbf{G}^i \,. \end{aligned} \tag{F.3.5}$$

Next, using these definitions, it follows that

$$\partial \theta^i/\partial \mathbf{x} = \mathbf{g}^i \,, \quad \partial \theta^i/\partial \mathbf{X} = \mathbf{G}^i \,, \tag{F.3.6}$$

since the second-order identity tensor **I** can be expressed using the chain rule of differentiation in the forms

$$\begin{aligned} \mathbf{I} &= \partial \mathbf{x}/\partial \mathbf{x} = \frac{\partial \mathbf{x}}{\partial \theta^i} \otimes \partial \theta^i/\partial \mathbf{x} = \mathbf{g}_i \otimes \mathbf{g}^i \,, \\ \mathbf{I} &= \partial \mathbf{X}/\partial \mathbf{X} = \frac{\partial \mathbf{X}}{\partial \theta^i} \otimes \partial \theta^i/\partial \mathbf{X} = \mathbf{G}_i \otimes \mathbf{G}^i \,. \end{aligned} \tag{F.3.7}$$

Also, it is noted for clarity that the gradient of a tensor function is written as a derivative with respect of **x** or **X** on a single line instead of as a fraction. This helps indicate that the gradient operator adds a tensor product on the *right-hand side* of the tensor being differentiated.

As a special case, it is possible to write the deformation gradient **F** from the reference configuration to the current configuration in the form

$$\mathbf{F} = \partial \mathbf{x}/\partial \mathbf{X} = \frac{\partial \mathbf{x}}{\partial \theta^i} \otimes \partial \theta^i/\partial \mathbf{X} = \mathbf{g}_i \otimes \mathbf{G}^i \,. \tag{F.3.8}$$

Then, the inverse $\mathbf{F}^{-1}$, the transpose $\mathbf{F}^T$ and the inverse transpose $\mathbf{F}^{-T}$ of **F** can be expressed in the forms

$$\mathbf{F}^{-1} = \mathbf{G}_i \otimes \mathbf{g}^i \,, \quad \mathbf{F}^T = \mathbf{G}^i \otimes \mathbf{g}_i \,, \quad \mathbf{F}^{-T} = \mathbf{g}^i \otimes \mathbf{G}_i \,. \tag{F.3.9}$$

Moreover, it can be shown that the determinant of **F** is given by

$$\det \mathbf{F} = (\mathbf{g}_1 \times \mathbf{g}_2 \cdot \mathbf{g}_3)(\mathbf{G}^1 \times \mathbf{G}^2 \cdot \mathbf{G}^3) = g^{1/2} G^{-1/2} \,. \tag{F.3.10}$$

The divergence of a tensor function **T** of order M ($M \geq 1$) with respect to the present position **x** is a tensor of order $M - 1$ denoted by div**T**, and the divergence of **T** with respect to the reference position **X** is a tensor of order $M - 1$ denoted by Div**T**. These tensors can be conveniently expressed in the forms

$$\text{div}\mathbf{T} = \mathbf{T},_j \cdot \mathbf{g}^j \,, \quad \text{Div}\mathbf{T} = \mathbf{T},_j \cdot \mathbf{G}^j \,. \tag{F.3.11}$$

Notice that since $\mathbf{T}$ is a tensor of order M it is necessary to differentiate the components of $\mathbf{T}$ and the M base tensors to evaluate the expression $\mathbf{T},_j$.

To simplify this operation, it is convenient to differentiate (F.1.10) to obtain

$$(g^{1/2}\mathbf{g}^j),_j = \mathbf{g}_1 \times (\mathbf{g}_{2,3} - \mathbf{g}_{3,2}) + \mathbf{g}_2 \times (\mathbf{g}_{3,1} - \mathbf{g}_{1,3}) + \mathbf{g}_3 \times (\mathbf{g}_{1,2} - \mathbf{g}_{2,1}) . \tag{F.3.12}$$

Next, assuming that the position vector $\mathbf{x}$ is sufficiently continuous, it follows that

$$\mathbf{g}_{i,j} = \frac{\partial \mathbf{g}_i}{\partial \theta^j} = \frac{\partial^2 \mathbf{x}}{\partial \theta^i \partial \theta^j} = \frac{\partial^2 \mathbf{x}}{\partial \theta^j \partial \theta^i} = \mathbf{g}_{j,i} , \tag{F.3.13}$$

which can be used to deduce that

$$(g^{1/2}\mathbf{g}^j),_j = 0 , \tag{F.3.14}$$

with a similar result for $\mathbf{G}_i$

$$(G^{1/2}\mathbf{G}^j),_j = 0 . \tag{F.3.15}$$

Then, using these results, it follows that (F.3.11) can be written in the simplified forms

$$\text{div}\mathbf{T} = g^{-1/2}(g^{1/2}\mathbf{T}\,\mathbf{g}^j),_j , \quad \text{Div}\mathbf{T} = G^{-1/2}(G^{1/2}\mathbf{T}\,\mathbf{G}^j),_j . \tag{F.3.16}$$

For example, if $\mathbf{T}$ is a second-order tensor, then the expressions (F.3.16) require differentiation of only three vectors as opposed to differentiation of the complete second-order tensor.

The curl of a tensor $\mathbf{T}$ of order M ($M \geq 1$) with respect to the present position $\mathbf{x}$ is a tensor of order M denoted by curl$\mathbf{T}$, and the curl of $\mathbf{T}$ with respect to the reference position $\mathbf{X}$ is a tensor of order M denoted by Curl$\mathbf{T}$. These tensors can be expressed in the forms

$$\text{curl}\mathbf{T} = -\mathbf{T},_j \times \mathbf{g}^j , \quad \text{Curl}\mathbf{T} = -\mathbf{T},_j \times \mathbf{G}^j . \tag{F.3.17}$$

The Laplacian of a tensor $\mathbf{T}$ of order M is a tensor of order M defined by

$$\nabla^2\mathbf{T} = \text{div}(\text{grad}\mathbf{T}) = (\mathbf{T},_i \otimes \mathbf{g}^i),_j \cdot \mathbf{g}^j . \tag{F.3.18}$$

F.4 Covariant Differentiation and Christoffel Symbols

Since the base vectors $\mathbf{g}_i$ and $\mathbf{g}^i$ depend on the coordinates θ^i, it is necessary to differentiate these base vectors when deriving expressions for partial derivatives of tensors with respect to θ^i. In particular, using (F.2.2) the derivative of the vector $\mathbf{v}$ can be written in the forms

$$\begin{aligned} \mathbf{v},_i &= (v^m \mathbf{g}_m),_i = v^m,_i \mathbf{g}_m + v^m \mathbf{g}_m,_i \ , \\ \mathbf{v},_i &= (v_m \mathbf{g}^m),_i = v_m,_i \mathbf{g}^m + v_m \mathbf{g}^m,_i \ . \end{aligned} \tag{F.4.1}$$

Using the fact that the nine vectors $\mathbf{g}_m,_i$ can be expressed in terms of the covariant base vectors $\mathbf{g}_k$, it is convenient to define the Christoffel symbol Γ^k_{mi} of the second kind, such that

$$\Gamma^k_{mi} = \mathbf{g}_m,_i \cdot \mathbf{g}^k = \Gamma^k_{im} \ , \quad \mathbf{g}_m,_i = \Gamma^k_{mi} \, \mathbf{g}_k \ . \tag{F.4.2}$$

Also, differentiating $\mathbf{g}^m \cdot \mathbf{g}_k = \delta^m_k$ by θ^i yields the result that

$$\mathbf{g}^m,_i = -\Gamma^m_{ik} \, \mathbf{g}^k \ . \tag{F.4.3}$$

These expressions allow (F.4.1) to be rewritten in the simplified forms

$$\begin{aligned} \mathbf{v},_i &= (v^k{}_{|i}) \, \mathbf{g}_k \ , \quad v^k{}_{|i} = \mathbf{v},_i \cdot \mathbf{g}^k \ , \\ \mathbf{v},_i &= (v_{k|i}) \, \mathbf{g}_k \ , \quad v_{k|i} = \mathbf{v},_i \cdot \mathbf{g}_k \ , \end{aligned} \tag{F.4.4}$$

where the covariant derivative $v^k{}_{|i}$ of the contravariant components v^k of $\mathbf{v}$ and the covariant derivative $v_{k|i}$ of the covariant components v_k of $\mathbf{v}$ are defined by

$$v^k{}_{|i} = v^k,_i + v^m \, \Gamma^k_{mi} \ , \quad v_{k|i} = v_k,_i - v_m \, \Gamma^m_{ki} \ . \tag{F.4.5}$$

Recalling the definition (F.3.5) for the gradient of a tensor, it follows that the gradient of $\mathbf{v}$ with respect to $\mathbf{x}$ can be expressed in the forms

$$\mathbf{L} = \text{grad}\mathbf{v} = \mathbf{v},_i \otimes \mathbf{g}^i = v_{k|i} (\mathbf{g}^k \otimes \mathbf{g}^i) = v^k{}_{|i} (\mathbf{g}_k \otimes \mathbf{g}^i) \ . \tag{F.4.6}$$

In these formulas, it can be seen that $v_{k|i}$ are the covariant components and $v^k{}_{|i}$ are mixed components of the second-order tensor $\mathbf{L}$.

The covariant derivatives of components of higher order tensors can be defined in a similar manner. In particular, if $\mathbf{T}$ is a second-order tensor with components T_{ij}, T^{ij}, $T_i{}^j$ and $T^i{}_j$, then

$$\begin{aligned} \mathbf{T},_m &= T_{ij|m} (\mathbf{g}^i \otimes \mathbf{g}^j) = T^{ij}{}_{|m} (\mathbf{g}_i \otimes \mathbf{g}_j) = T_i{}^j{}_{|m} (\mathbf{g}^i \otimes \mathbf{g}_j) = T^i{}_{j|m} (\mathbf{g}_i \otimes \mathbf{g}^j) \ , \\ T_{ij|m} &= T_{ij},_m - T_{kj} \, \Gamma^k_{im} - T_{ik} \, \Gamma^k_{jm} \ , \\ T^{ij}{}_{|m} &= T^{ij},_m + T^{kj} \, \Gamma^i_{km} + T^{ik} \, \Gamma^j_{km} \ , \\ T_i{}^j{}_{|m} &= T_i{}^j,_m - T_k{}^j \, \Gamma^k_{im} + T_i{}^k \, \Gamma^j_{km} \ , \\ T^i{}_{j|m} &= T^i{}_j,_m + T^k{}_j \, \Gamma^i_{km} - T^i{}_k \, \Gamma^k_{jm} \ . \end{aligned} \tag{F.4.7}$$

Similarly, it can be shown that $T_{ij|m}$ are the covariant components and $T^{ij}{}_{|m}$, $T_i{}^j{}_{|m}$ and $T^i{}_{j|m}$ are the mixed components of the third-order tensor grad$\mathbf{T}$.

Appendix G
Summary of Tensor Operations in Specific Coordinate Systems

The objective of this appendix is to record various tensor operations in cylindrical polar and spherical polar coordinates.[2]

G.1 Cylindrical Polar Coordinates

The right-handed orthonormal base vectors $\mathbf{e}_r$, $\mathbf{e}_\theta$ and $\mathbf{e}_z$ associated with the cylindrical polar coordinates r, θ and z are defined in terms of the fixed base vectors $\mathbf{e}_i$ of a rectangular Cartesian coordinate system by equations

$$\begin{aligned}\mathbf{e}_r &= \mathbf{e}_r(\theta) = \cos\theta\mathbf{e}_1 + \sin\theta\mathbf{e}_2\,,\\ \mathbf{e}_\theta &= \mathbf{e}_\theta(\theta) = -\sin\theta\mathbf{e}_1 + \cos\theta\mathbf{e}_2\,,\\ \mathbf{e}_z &= \mathbf{e}_3\,,\end{aligned} \tag{G.1.1}$$

where $\mathbf{e}_r$ and $\mathbf{e}_\theta$ lie in the $\mathbf{e}_1$, $\mathbf{e}_2$ plane and the angle θ is measured counterclockwise from the $\mathbf{e}_1$ direction about the $\mathbf{e}_3$ axis (see Fig. G.1). Also, the position vector $\mathbf{x}$ of a point in the three-dimensional space can be represented in the form

$$\mathbf{x} = r\mathbf{e}_r(\theta) + z\mathbf{e}_z\,. \tag{G.1.2}$$

In particular, notice that there is no $\theta\mathbf{e}_\theta$ term in this expression for the position vector $\mathbf{x}$ since $\mathbf{e}_r(\theta)$ contains the dependence of $\mathbf{x}$ on θ.

To present formulas for derivatives of tensors expressed in terms of these coordinates, it is convenient to first record the derivatives of the base vectors

[2] Much of the content in this appendix has been adapted from [1] with permission.

M. B. Rubin, *Continuum Mechanics with Eulerian Formulations of Constitutive Equations*, Solid Mechanics and Its Applications 265,
https://doi.org/10.1007/978-3-030-57776-6

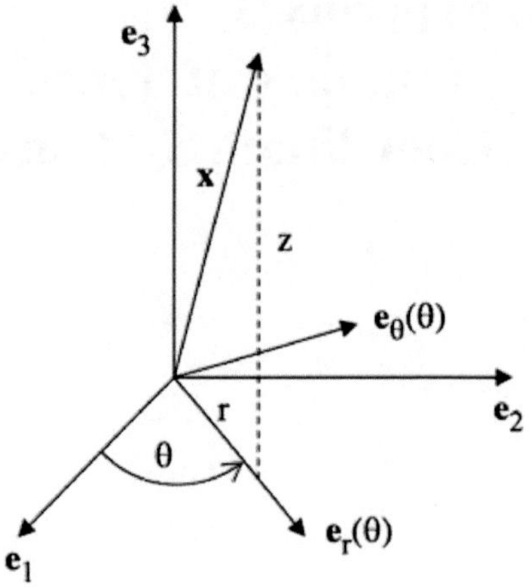

Fig. G.1 Definition of cylindrical polar coordinates

$$\frac{d\mathbf{e}_r}{d\theta} = \mathbf{e}_\theta \,, \quad \frac{d\mathbf{e}_\theta}{d\theta} = -\mathbf{e}_r \,, \quad \frac{d\mathbf{e}_z}{d\theta} = 0 \,. \tag{G.1.3}$$

Also, by taking θ^i to be general curvilinear coordinates (not necessarily convected coordinates) and setting

$$\theta^i = (r, \theta, z) \,, \tag{G.1.4}$$

it follows that the covariant base vectors $\mathbf{g}_i$, the contravariant vectors $\mathbf{g}^i$ and the scalar $g^{1/2}$ are given by

$$\begin{aligned}
&\mathbf{g}_1 = \tfrac{\partial \mathbf{x}}{\partial r} = \mathbf{e}_r \,, \quad && \mathbf{g}_2 = \tfrac{\partial \mathbf{x}}{\partial \theta} = r\mathbf{e}_\theta \,, \quad && \mathbf{g}_3 = \tfrac{\partial \mathbf{x}}{\partial z} = \mathbf{e}_z \,, \\
&\mathbf{g}^1 = \mathbf{e}_r \,, && \mathbf{g}^2 = \tfrac{1}{r}\mathbf{e}_\theta \,, && \mathbf{g}^3 = \mathbf{e}_z \,, \\
&g^{1/2} = r \,.
\end{aligned} \tag{G.1.5}$$

Next, let f, $\mathbf{v}$ and $\mathbf{T}$ be, respectively, scalar, vector and second-order tensor functions of r, θ and z. Furthermore, let $\mathbf{v}$ and $\mathbf{T}$ be expressed in terms of their physical components by

$$\begin{aligned}
\mathbf{v} &= v_r\mathbf{e}_r + v_\theta\mathbf{e}_\theta + v_z\mathbf{e}_z \,, \\
\mathbf{T} &= T_{rr}(\mathbf{e}_r \otimes \mathbf{e}_r) + T_{r\theta}(\mathbf{e}_r \otimes \mathbf{e}_\theta) + T_{rz}(\mathbf{e}_r \otimes \mathbf{e}_z) \\
&\quad + T_{\theta r}(\mathbf{e}_\theta \otimes \mathbf{e}_r) + T_{\theta\theta}(\mathbf{e}_\theta \otimes \mathbf{e}_\theta) + T_{\theta z}(\mathbf{e}_\theta \otimes \mathbf{e}_z) \\
&\quad + T_{zr}(\mathbf{e}_z \otimes \mathbf{e}_r) + T_{z\theta}(\mathbf{e}_z \otimes \mathbf{e}_\theta) + T_{zz}(\mathbf{e}_z \otimes \mathbf{e}_z) \,.
\end{aligned} \tag{G.1.6}$$

Then, the gradient operator applied to f and $\mathbf{v}$ can be expressed in the forms

$$
\begin{aligned}
\operatorname{grad} f &= \frac{\partial f}{\partial r}\,\mathbf{e}_r + \frac{1}{r}\frac{\partial f}{\partial \theta}\,\mathbf{e}_\theta + \frac{\partial f}{\partial z}\,\mathbf{e}_z\,,\\
\operatorname{grad}\mathbf{v} &= \frac{\partial v_r}{\partial r}\,(\mathbf{e}_r\otimes\mathbf{e}_r) + \frac{1}{r}(\frac{\partial v_r}{\partial \theta} - v_\theta)\,(\mathbf{e}_r\otimes\mathbf{e}_\theta) + \frac{\partial v_r}{\partial z}\,(\mathbf{e}_r\otimes\mathbf{e}_z)\\
&+ \frac{\partial v_\theta}{\partial r}\,(\mathbf{e}_\theta\otimes\mathbf{e}_r) + \frac{1}{r}\left(\frac{\partial v_\theta}{\partial \theta} + v_r)\,(\mathbf{e}_\theta\otimes\mathbf{e}_\theta\right) + \frac{\partial v_\theta}{\partial z}\,(\mathbf{e}_\theta\otimes\mathbf{e}_z)\\
&+ \frac{\partial v_z}{\partial r}\,(\mathbf{e}_z\otimes\mathbf{e}_r) + \frac{1}{r}\frac{\partial v_z}{\partial \theta}\,(\mathbf{e}_z\otimes\mathbf{e}_\theta) + \frac{\partial v_z}{\partial z}\,(\mathbf{e}_z\otimes\mathbf{e}_z)\,,
\end{aligned}
\tag{G.1.7}
$$

the divergence operator applied to **v** and **T** can be expressed in the forms

$$
\begin{aligned}
\operatorname{div}\mathbf{v} &= \frac{\partial v_r}{\partial r} + \frac{1}{r}\left(\frac{\partial v_\theta}{\partial \theta} + v_r\right) + \frac{\partial v_z}{\partial z}\,,\\
\operatorname{div}\mathbf{T} &= \left(\frac{\partial T_{rr}}{\partial r} + \frac{1}{r}\frac{\partial T_{r\theta}}{\partial \theta} + \frac{\partial T_{rz}}{\partial z} + \frac{T_{rr} - T_{\theta\theta}}{r}\right)\mathbf{e}_r\\
&+ \left(\frac{\partial T_{\theta r}}{\partial r} + \frac{1}{r}\frac{\partial T_{\theta\theta}}{\partial \theta} + \frac{\partial T_{\theta z}}{\partial z} + \frac{T_{\theta r} + T_{r\theta}}{r}\right)\mathbf{e}_\theta\\
&+ \left(\frac{\partial T_{zr}}{\partial r} + \frac{1}{r}\frac{\partial T_{z\theta}}{\partial \theta} + \frac{\partial T_{zz}}{\partial z} + \frac{T_{zr}}{r}\right)\mathbf{e}_z\,,
\end{aligned}
\tag{G.1.8}
$$

the curl operator applied to **v** can be expressed as

$$
\operatorname{curl}\mathbf{v} = \left(\frac{1}{r}\frac{\partial v_z}{\partial \theta} - \frac{\partial v_\theta}{\partial z}\right)\mathbf{e}_r + \left(\frac{\partial v_r}{\partial z} - \frac{\partial v_z}{\partial r}\right)\mathbf{e}_\theta + \left(\frac{\partial v_\theta}{\partial r} + \frac{v_\theta}{r} - \frac{1}{r}\frac{\partial v_r}{\partial \theta}\right)\mathbf{e}_z\,,
\tag{G.1.9}
$$

and the Laplacian operator applied to f can be expressed as

$$
\nabla^2 f = \operatorname{div}(\operatorname{grad} f) = \frac{\partial^2 f}{\partial r^2} + \frac{1}{r}\frac{\partial f}{\partial r} + \frac{1}{r^2}\frac{\partial^2 f}{\partial \theta^2} + \frac{\partial^2 f}{\partial z^2}\,.
\tag{G.1.10}
$$

G.2 Spherical Polar Coordinates

The right-handed orthonormal base vectors $\mathbf{e}_r$, $\mathbf{e}_\theta$ and $\mathbf{e}_\phi$ associated with the spherical polar coordinates r, θ and ϕ are defined in terms of the fixed base vectors $\mathbf{e}_i$ of a rectangular Cartesian coordinate system by the equations

$$
\begin{aligned}
\mathbf{e}_r &= \mathbf{e}_r(\theta,\phi) = \sin\theta\,(\cos\phi\mathbf{e}_1 + \sin\phi\mathbf{e}_2) + \cos\theta\mathbf{e}_3\,,\\
\mathbf{e}_\theta &= \mathbf{e}_\theta(\theta,\phi) = \cos\theta\,(\cos\phi\mathbf{e}_1 + \sin\phi\mathbf{e}_2) - \sin\theta\mathbf{e}_3\,,\\
\mathbf{e}_\phi &= \mathbf{e}_\phi(\phi) = -\sin\phi\mathbf{e}_1 + \cos\phi\mathbf{e}_2\,,
\end{aligned}
\tag{G.2.1}
$$

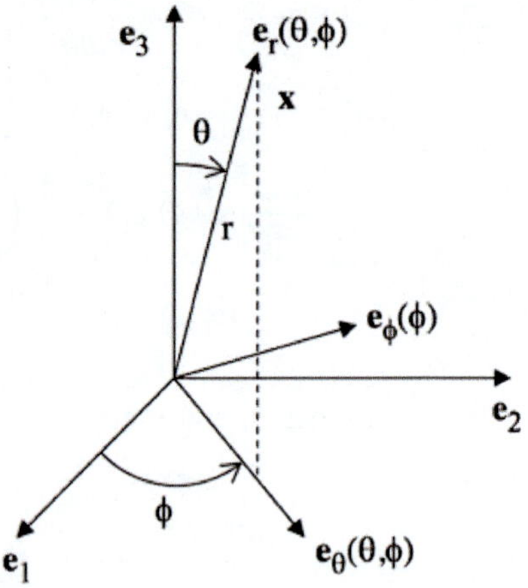

Fig. G.2 Definition of spherical polar coordinates

where the angle ϕ is measured in the horizontal plane counterclockwise from the $\mathbf{e}_1$ direction about the vertical direction $\mathbf{e}_3$ to the vertical plane which includes the position vector $\mathbf{x}$ (see Fig. G.2), and θ is the acute angle measured from the vertical direction $\mathbf{e}_3$ to the position vector. Also, the position vector $\mathbf{x}$ of a point in the three-dimensional space can be represented in the form

$$\mathbf{x} = r\mathbf{e}_r(\theta, \phi) \,. \tag{G.2.2}$$

In particular, notice that there are no terms like $\theta\mathbf{e}_\theta$ and $\phi\mathbf{e}_\phi$ in this expression for the position vector $\mathbf{x}$ since $\mathbf{e}_r(\theta, \phi)$ contains the dependence of $\mathbf{x}$ on θ and ϕ.

To present formulas for derivatives of tensors expressed in terms of these coordinates it is convenient to first record the derivatives of the base vectors

$$\begin{aligned} \frac{\partial \mathbf{e}_r}{\partial \theta} &= \mathbf{e}_\theta \,, & \frac{\partial \mathbf{e}_r}{\partial \phi} &= \sin\theta \, \mathbf{e}_\phi \,, \\ \frac{\partial \mathbf{e}_\theta}{\partial \theta} &= -\mathbf{e}_r \,, & \frac{\partial \mathbf{e}_\theta}{\partial \phi} &= \cos\theta \, \mathbf{e}_\phi \,, \\ \frac{\partial \mathbf{e}_\phi}{\partial \theta} &= 0 \,, & \frac{\partial \mathbf{e}_\phi}{\partial \phi} &= -\sin\theta \, \mathbf{e}_r - \cos\theta \, \mathbf{e}_\theta \,. \end{aligned} \tag{G.2.3}$$

Also, by taking θ^i to be general curvilinear coordinates (not necessarily convected coordinates) and setting

$$\theta^i = (r, \theta, \phi) \,, \tag{G.2.4}$$

it follows that the covariant base vectors $\mathbf{g}_i$, the contravariant vectors $\mathbf{g}^i$ and the scalar $g^{1/2}$ are given by

$$\begin{aligned} &\mathbf{g}_1 = \frac{\partial \mathbf{x}}{\partial r} = \mathbf{e}_r \,, && \mathbf{g}_2 = \frac{\partial \mathbf{x}}{\partial \theta} = r \, \mathbf{e}_\theta \,, && \mathbf{g}_3 = \frac{\partial \mathbf{x}}{\partial \phi} = r \, \sin\theta \, \mathbf{e}_\phi \,, \\ &\mathbf{g}^1 = \mathbf{e}_r \,, && \mathbf{g}^2 = \frac{1}{r} \, \mathbf{e}_\theta \,, && \mathbf{g}^3 = \frac{1}{r \, \sin\theta} \, \mathbf{e}_\phi \,, \\ &g^{1/2} = r^2 \, \sin\theta \,. \end{aligned} \tag{G.2.5}$$

Next, let f, $\mathbf{v}$ and $\mathbf{T}$ be, respectively, scalar, vector and second-order tensor functions of r, θ and ϕ. Furthermore, let $\mathbf{v}$ and $\mathbf{T}$ be expressed in terms of their physical components by

$$\begin{aligned}
\mathbf{v} &= v_r\mathbf{e}_r + v_\theta\mathbf{e}_\theta + v_\phi\mathbf{e}_\phi\,,\\
\mathbf{T} &= T_{rr}(\mathbf{e}_r\otimes\mathbf{e}_r) + T_{r\theta}(\mathbf{e}_r\otimes\mathbf{e}_\theta) + T_{r\phi}(\mathbf{e}_r\otimes\mathbf{e}_\phi)\\
&\quad + T_{\theta r}(\mathbf{e}_\theta\otimes\mathbf{e}_r) + T_{\theta\theta}(\mathbf{e}_\theta\otimes\mathbf{e}_\theta) + T_{\theta\phi}(\mathbf{e}_\theta\otimes\mathbf{e}_\phi)\\
&\quad + T_{\phi r}(\mathbf{e}_\phi\otimes\mathbf{e}_r) + T_{\phi\theta}(\mathbf{e}_\phi\otimes\mathbf{e}_\theta) + T_{\phi\phi}(\mathbf{e}_\phi\otimes\mathbf{e}_\phi)\,.
\end{aligned}\tag{G.2.6}$$

Then, the gradient operator applied to f and $\mathbf{v}$ can be expressed in the forms

$$\begin{aligned}
\operatorname{grad} f &= \frac{\partial f}{\partial r}\mathbf{e}_r + \frac{1}{r}\frac{\partial f}{\partial\theta}\mathbf{e}_\theta + \frac{1}{r\,\sin\theta}\frac{\partial f}{\partial\phi}\mathbf{e}_\phi\,,\\
\operatorname{grad}\mathbf{v} &= \frac{\partial v_r}{\partial r}(\mathbf{e}_r\otimes\mathbf{e}_r) + \frac{1}{r}\left(\frac{\partial v_r}{\partial\theta} - v_\theta\right)(\mathbf{e}_r\otimes\mathbf{e}_\theta)\\
&\quad + \frac{1}{r\,\sin\theta}\left(\frac{\partial v_r}{\partial\phi} - v_\phi\sin\theta\right)(\mathbf{e}_r\otimes\mathbf{e}_\phi)\\
&\quad + \frac{\partial v_\theta}{\partial r}(\mathbf{e}_\theta\otimes\mathbf{e}_r) + \frac{1}{r}\left(\frac{\partial v_\theta}{\partial\theta} + v_r\right)(\mathbf{e}_\theta\otimes\mathbf{e}_\theta)\\
&\quad + \frac{1}{r\,\sin\theta}\left(\frac{\partial v_\theta}{\partial\phi} - v_\phi\cos\theta\right)(\mathbf{e}_\theta\otimes\mathbf{e}_\phi)\\
&\quad + \frac{\partial v_\phi}{\partial r}(\mathbf{e}_\phi\otimes\mathbf{e}_r) + \frac{1}{r}\frac{\partial v_\phi}{\partial\theta}(\mathbf{e}_\phi\otimes\mathbf{e}_\theta)\\
&\quad + \frac{1}{r\,\sin\theta}\left(\frac{\partial v_\phi}{\partial\phi} + v_r\sin\theta + v_\theta\cos\theta\right)(\mathbf{e}_\phi\otimes\mathbf{e}_\phi)\,,
\end{aligned}\tag{G.2.7}$$

the divergence operator applied to $\mathbf{v}$ and $\mathbf{T}$ can be expressed in the forms

$$\begin{aligned}
\operatorname{div}\mathbf{v} &= \frac{\partial v_r}{\partial r} + \frac{2v_r}{r} + \frac{1}{r}\frac{\partial v_\theta}{\partial\theta} + \frac{v_\theta\cot\theta}{r} + \frac{1}{r\,\sin\theta}\frac{\partial v_\phi}{\partial\phi}\,,\\
\operatorname{div}\mathbf{T} &= \left[\frac{\partial T_{rr}}{\partial r} + \frac{1}{r}\frac{\partial T_{r\theta}}{\partial\theta} + \frac{1}{r\,\sin\theta}\frac{\partial T_{r\phi}}{\partial\phi} + \frac{2T_{rr} - T_{\theta\theta} - T_{\phi\phi} + T_{r\theta}\cot\theta}{r}\right]\mathbf{e}_r\\
&\quad + \left[\frac{\partial T_{\theta r}}{\partial r} + \frac{1}{r}\frac{\partial T_{\theta\theta}}{\partial\theta} + \frac{1}{r\,\sin\theta}\frac{\partial T_{\theta\phi}}{\partial\phi} + \frac{2T_{\theta r} + T_{r\theta} + (T_{\theta\theta} - T_{\phi\phi})\cot\theta}{r}\right]\mathbf{e}_\theta\\
&\quad + \left(\frac{\partial T_{\phi r}}{\partial r} + \frac{1}{r}\frac{\partial T_{\phi\theta}}{\partial\theta} + \frac{1}{r\,\sin\theta}\frac{\partial T_{\phi\phi}}{\partial\phi} + \frac{2T_{\phi r} + T_{r\phi} + (T_{\phi\theta} + T_{\theta\phi})\cot\theta}{r}\right]\mathbf{e}_\phi\,,
\end{aligned}\tag{G.2.8}$$

the curl operator applied to $\mathbf{v}$ can be expressed as

$$
\begin{aligned}
\operatorname{curl} \mathbf{v} = & \left(\frac{1}{r}\frac{\partial v_\phi}{\partial \theta} + \frac{v_\phi \cot\theta}{r} - \frac{1}{r\,\sin\theta}\frac{\partial v_\theta}{\partial \phi}\right)\mathbf{e}_r \\
& + \left(\frac{1}{r\,\sin\theta}\frac{\partial v_r}{\partial \phi} - \frac{\partial v_\phi}{\partial r} - \frac{v_\phi}{r}\right)\mathbf{e}_\theta \\
& + \left(\frac{\partial v_\theta}{\partial r} + \frac{v_\theta}{r} - \frac{1}{r}\frac{\partial v_r}{\partial \theta}\right)\mathbf{e}_z\,,
\end{aligned} \tag{G.2.9}
$$

and the Laplacian operator applied to f can be expressed as

$$
\nabla^2 f = \operatorname{div}(\operatorname{grad} f) = \frac{\partial^2 f}{\partial r^2} + \frac{2}{r}\frac{\partial f}{\partial r} + \frac{1}{r^2}\frac{\partial^2 f}{\partial \theta^2} + \frac{\cot\theta}{r^2}\frac{\partial f}{\partial \theta} + \frac{1}{r^2\sin^2\theta}\frac{\partial^2 f}{\partial \phi^2}\,. \tag{G.2.10}
$$

Reference

1. Rubin MB (2000) Cosserat theories: Shells, rods and points, vol 79. Springer Science & Business Media, Berlin

Index

M. B. Rubin, *Continuum Mechanics with Eulerian Formulations of Constitutive Equations*, Solid Mechanics and Its Applications 265,
https://doi.org/10.1007/978-3-030-57776-6

Q

R

S

T

U